Computer Communication Networks

ESSEX SERIES IN TELECOMMUNICATION AND INFORMATION SYSTEMS

Series Editors

Andy Downton
Ed Jones

Forthcoming Titles

Image Processing
Speech Processing
Engineering the Human–Computer Interface
Satellite and Mobile Radio Systems

COMPUTER COMMUNICATION NETWORKS

Edited by

Gill Waters

Department of Electronic Systems Engineering
University of Essex

McGRAW-HILL BOOK COMPANY

London · New York · St Louis · San Francisco · Auckland
Bogotá · Caracas · Hamburg · Lisbon · Madrid · Mexico · Milan
Montreal · New Delhi · Panama · Paris · San Juan · São Paulo
Singapore · Sydney · Tokyo · Toronto

Published by
McGRAW-HILL Book Company (UK) Limited
SHOPPENHANGERS ROAD · MAIDENHEAD · BERKSHIRE
ENGLAND
TEL: 0628 23432
FAX: 0628 770224

British Library Cataloguing in Publication Data

Computer communication networks. – (Essex series in
telecommunication and information systems)
1. Computer systems. Communication networks
I. Waters, Gill   II. Series
004.6

ISBN 0-07-707325-8

Library of Congress Cataloging-in-Publication Data

Computer communication networks / editor, Gill Waters.
    p.   cm. – (Essex series in telecommunication and information systems)
Includes bibliographical references and index.
ISBN 0-07-707325-8
1. Computer networks.   I. Waters, Gill.   II. Series.
TK5105.5.C6369   1991
004.6–dc20   90-24976 CIP

1234 CUP 94321

Typeset by Computape (Pickering) Ltd, North Yorkshire
Printed and bound in Great Britain at the University Press, Cambridge

Contents

Notes on the contributors

John Adams BSc, MSc, PhD was awarded the PhD degree for applied mathematics research by Leicester University. He works at the British Telecom Research Laboratories, Martlesham where, in 1984, he invented the Orwell protocol for integrated-services LANs. He is currently an engineering adviser in the Switched Networks Division at the laboratories where his main interests are the applications of ATM in public networks, with particular emphasis on mobility.

Mark Clark MSc, PhD is a senior lecturer in real time computer systems, software engineering and computer networks at Essex University. Before his appointment to Essex as a lecturer in 1978, he was a senior project engineer with a small company in Winchester specializing in real time control and display systems. His research interests include multimedia systems and their applications, particularly with regard to the capabilities for data traffic offered by newly emerging computer networks.

Stephen Hall BSc, PhD has studied in South Africa and Australia. He works for the Dindima Group, Australia, which designs and manufactures specialized video equipment. From 1988 to 1989 he was a lecturer in the Department of Electronic Systems Engineering, University of Essex, and from 1989 to 1990 he was a research contractor with OTC Ltd, Australia. His research interests include broadband integrated services digital networks, ATM switch design and protocols and video coding for ATM networks.

Edwin Jones BSc, MSc, PhD, CEng, MIEE, MIEEE is a senior lecturer in telecommunication systems at the University of Essex, having formerly been with GEC Research Limited. He has particular interests in digital transmission and is currently researching into efficient speech and data coding for local/wide area networks, signal processing for high capacity transmission systems and synchronization of digital multiplexes. Much of this is collaborative work with the UK telecommunications industry.

Simon Jones BSc, PhD is a research engineer in the Broadband Switching and Customer Equipment Group of the British Telecom Research Laboratories, Martlesham. Formerly, he was a lecturer in the Department of Electronic Systems Engineering at the University of Essex. His current research is in the development and design of user interfaces for broadband networks and services.

Alwyn Langsford BA, DPhil, CEng, MBCS is head of Systems and Software Engineering with AEA Technology. His team provide consultancy and technical support for reliable computer architectures, distributed systems and distributed systems management. He has been active in OSI since the inception of this standards programme, being, for a time, convener of the ISO working group concerned with application and presentation layer standards. He represents the UK interests in Open Systems Management Standards and is the international editor for the Accounting Management Standard. He is chairman of the UK Special Interest Group in Distributed Systems Management.

John Limb BEE, PhD, FIEEE is manager of Technology Analysis with Hewlett-Packard in Cupertino, California, USA. From 1971 to 1983 he was a department head in the research division of Bell Laboratories and in 1984 formed the Communications Science Research Division in Bell Communications Research. From 1986 to 1989, he was director of the newly formed Networks and Communications Laboratory at Hewlett-Packard Laboratories in Bristol, UK. His technical interests include picture coding, visual perception, networking and multimedia packet communication. He was editor-in-chief of the IEEE Journal on Selected Areas in Communications from 1984 to 1988.

Chris Smyth BA, MA, PhD is a senior lecturer in mathematics at Edinburgh University. His interests include number theory and discrete mathematics, with applications to cryptography, coding and switching. Formerly, he was a member of the Department of Electronic Systems Engineering at Essex University. He has also worked for British Telecom doing mathematical modelling and has taught in both Australia and Canada. He is a member of the American, London and Edinburgh Mathematical Societies.

Gill Waters BSc has been a lecturer in computer networks and real time systems in the Department of Electronic Systems Engineering at the University of Essex since 1984. Formerly, she worked at the Rutherford Appleton Laboratory, Oxfordshire on a wide variety of innovative computer communication projects. She has organized the short course in computer networks at Essex since its inception in 1987. Gill's research interests include multiparty protocols, broadcast and multicast applications, and the application of computer supported collaborative work to teaching. She is also interested in integrated multiservice networks, particularly in their access protocols and the scope for novel applications. She is a member of the Association for Computing Machinery and the UK Women's Engineering Society.

Series preface

This book is part of a series, the *Essex Series in Telecommunication and Information Systems*, which has developed from a set of short courses run by the Department of Electronic Systems Engineering at the University of Essex since 1987. The courses are presented as one week modules on the Department's MSc in Telecommunication and Information Systems, and are offered simultaneously as industry short courses. To date, a total of over 600 industrial personnel have attended the courses, in addition to the 70 or so postgraduate students registered each year for the MSc. The flexibility of the short course format means that the content both of individual courses, and of the courses offered from year to year have been able to develop to reflect current industrial and academic demand.

The aim of the book series is to provide readable yet authoritative coverage of key topics within the field of telecommunication and information systems. Being derived from a highly regarded university postgraduate course, the books are well suited to use in advanced taught courses at universities and polytechnics, and as a starting point and background reference for researchers. Equally, the industrial orientation of the courses ensures that both the content and presentation style are suited to the needs of the professional engineer in mid-career.

The books in the series are based largely on the course notes circulated to students, and so have been 'class-tested' several times before publication. Though primarily authored and edited by academic staff at Essex, where appropriate, each book includes chapters contributed by acknowledged experts from other universities, research establishments and industry (originally presented as seminars on the courses). Our colleagues at British Telecom Research Laboratories, Martlesham, have also provided advice and assistance in developing course syllabuses and ensuring that the material included correctly reflects industry practice as well as academic principles.

As series editors, we would like to acknowledge the tremendous support we have had in developing the concept of the series from the original idea through to the publication of the first group of books. The successful completion of this project would not have been possible without the substantial commitment shown not only by individual authors but by the Department of Electronic Systems Engineering as a whole to this project. Particular thanks go to the editors of the individual books, each of whom, in addition to authoring several chapters, was responsible for integrating the various contributors' chapters of his or her book into a coherent whole.

July 1990   Andy Downton
                                                        Ed Jones

Preface

This book provides an introduction to the techniques and applications of computer communication networks. The importance of this wide ranging subject is increasing as networking facilities become available to a large and growing number of people.

A single modern computing system typically offers a number of powerful and flexible facilities. By connecting computing systems together into networks, these facilities are enhanced and can be shared by groups of people working together. Computer communication networks enable messages and data to be transmitted rapidly around an office or around the world, and offer economical resource sharing to both small and large organizations.

The problems involved in providing computer communication networks are not trivial, but there are solutions. This book deals with those solutions—ranging from basic interconnection technology to a variety of user applications. The means of information transfer must be appropriate to the transmission medium used; computer networks are constructed from a variety of transmission media including twisted pair, coaxial cable, optical fibres or satellite channels. Network access mechanisms must be fair and must use network facilities efficiently; access strrategies often depend on the size and complexity of the network, although this has implications for the interconnection of unlike networks. Applications of computer networks range from the delivery of simple messages or files to sophisticated file management systems and distributed operating systems.

If computer communication networks are to be used widely, standards for communication must be available: computing systems which conform to the standards by following the agreed procedures can then exchange and interpret information. As the result of a substantial effort, international standards have been agreed, and are now being made available on a wide range of computing systems. Standardization has been achieved by defining a model which partitions the interconnection problems into well-defined sets that are arranged in a layered hierarchy, each layer using the facilities of existing layers. The standards model has seven such layers.

The issues of computer communication networks form a large and complex subject area, with a growing literature, which could easily have given rise to a much longer book than space permitted. Rather than being modelled directly on the layered standards approach, the book progresses from looking at the more general problems and solutions to the more specific, discussing a wide variety of issues. Consequently there are more than seven chapters: some chapters are devoted to several layers, and some layers are discussed from a variety of perspectives in different chapters. Each chapter concentrates on one topic in detail and offers plenty of reference material through which the reader can pursue the described techniques in more depth. Chapter 1 sets the scene and describes the plan of the remaining chapters in more detail.

To some extent, the book also follows the series of advances that have taken place in the field: starting with a brief historical review, dealing with wide and then local area networks, looking at topics of active expansion and refinement such as distributed systems, security and management, and concluding with the more forward-looking chapters, which include techniques for integrated service networks for multimedia applications.

This book is concerned with a very active and fascinating subject area and it is hoped that readers will be encouraged to follow up their interest, and possibly to contribute to further advances in computer communication networks.

September 1990   Gill Waters

Acknowledgements

I would like to thank everybody who has helped with the book, although space does not allow me to name them all.

Thanks are due to the authors for all their hard work. Dr Alwyn Langsford's work in the development of OSI Management standards is supported by the Information Technology Standards Unit of the UK Department of Trade and Industry.

For their helpful suggestions based on earlier drafts of material, I would like to thank Chris Cooper of the Rutherford Appleton Laboratory, Oxfordshire and Frank Coakley of Surrey University. I would also like to thank those authors who have made useful comments on my chapters. I am grateful to Stuart Johnston of Hewlett-Packard Research Laboratories, Bristol who provided valuable information on formal description techniques.

Secretarial assistance has been ably supplied by Hilda Breakspeare, Lynne Burnand, Lynne Murrell and Pat Crawford.

Earlier versions of the text have been issued as course notes and I would like to thank all those course attendees and MSc students who have made helpful comments. Particular thanks are due to Michael Hobbs and Chris Ansell.

Finally, I would like to thank my husband Ray both for his technical advice and for his moral support.

September 1990   Gill Waters

Permissions

We are grateful to the IEEE for permission to reproduce Figure 9 from 'Local area networks: a performance comparison' by Werner Bux which appeared in the *IEEE Transactions on Communications*, October 1981, Vol. COM-29 (10), 1465–1473. (Chapter 8, Fig. 8.3).

We are grateful to Addison Wesley for permission to use the following four figures from the book *Cryptography and Data Security* by Dorothy Elizabeth Robling Denning, published by Addison-Wesley in 1982 and reprinted in 1983. (The figures in this book are adapted from the original source.)

Figure 2.13 DES enciphering algorithm   (Chapter 10, Figure 10.1)
Figure 2.15 Key schedule calculation   (Chapter 10, Figure 10.2)
Figure 3.3   Synchronous stream cipher   (Chapter 10, Figure 10.3)
Figure 3.4   Linear feedback shift register   (Chapter 10, Figure 10.4)

We are grateful to the International Telecommunication Union as copyright holder for permission to use the following figures from *CCITT Recommendations*. (The figures in this book are adapted from those in the original sources.)

CCITT *Red Book*, Volume VIII – Fascicle VIII.5, (1984)
        Figure 13/X.200   (Chapter 4, Figure 4.2)
CCITT *Red Book*, Volume VIII – Fascicle VIII.3, (1984)
        Table 1/X.25   (Chapter 4, Figure 4.7)
        Table 5/X.25   (Chapter 4, Figure 4.8)
        Table 17/X.25   (Chapter 4, Figure 4.11)
        Figure 2/X.25   (Chapter 4, Figure 4.13)
        Figure 6/X.25   (Chapter 4, Figure 4.15)
        Figure B-2/X.25   (Chapter 4, Figure 4.14)
CCITT *Blue Book*, Volume VIII – Fascicle VIII.5, (1988)
        Figure in 13.3.1/X.224   (Chapter 4, Figure 4.19)
        Figure in 13.7.1/X.224   (Chapter 4, Figure 4.19)
CCITT *Blue Book*, Volume VIII – Fascicle VIII.4, (1988)
        Figure 3/X.215   (Chapter 5, Figure 5.1)
CCITT *Red Book*, Volume VIII – Fascicle VIII.7, (1984)
        Figure 1/X.400   (Chapter 5, Figure 5.7)
        Figure 5/X.400   (Chapter 5, Figure 5.8)
        Figure 6/X.400   (Chapter 5, Figure 5.8)
        Figure 7/X.400   (Chapter 5, Figure 5.8)
        Figure 12/X.400   (Chapter 5, Figure 5.9)

Trademark notices

Glossary of terms and abbreviations

(Terms in italics can also be found in the glossary.)

ACK	An acknowledgement sent from a receiver to a sender indicating the correct reception of one or more frames
ACSE	Association control service element
ALOHA	A single-channel multiple access technique based on contention
AMP	Active monitor present (token ring)
ANSA	Advanced networked systems architecture
ANSI	American National Standards Institute
ARP	Address resolution protocol
ARPA	Advanced Research Projects Agency, USA
ARPANET	Network sponsored by Advanced Research Projects Agency, USA
ASCII	American Standard Code for Information Interchange
ASE	Application service element (a set of services that are of common use to several different applications)
ASN.1	Abstract syntax notation one
ATM	Asynchronous transfer mode
atomic action	A set of activities performed by a group of systems requiring that each activity in the set must succeed or else all are deemed to have failed
AUX	Apple *UNIX*
BBP	Basic Block Protocol (Cambridge ring)
B-ISDN	Broadband *ISDN*
bit oriented	Where frames are delimited by means of bit sequences
block cipher	Cipher where text is encrypted in blocks, usually of a fixed length
boot	The process of initializing a computer system from power-on or a reset condition, usually involving the loading of some software

bridge	System interconnecting two or more Local Area Networks
broadcasting	A technique that enables a single transmitted message to reach all stations on a network
BSC	Binary synchronous communications
BSD	Berkeley System Distribution
BSI	British Standards Institute
BSP	Byte stream protocol (Cambridge ring)
byte oriented	Where frames are delimited by a special byte pattern and all characters in the frame are bytes
CATV	Community antenna television
CCA	Conceptual communication area (for virtual terminal application)
CCITT	International Consultative Committee on Telegraphy and Telephony
CCR	Commitment, concurrency and recovery (application service element)
CDCS	Cambridge distributed computing system
CEPT	Committee for European Post and Telecommunications
CFR	Cambridge fast ring
CFS	Cambridge file server
channel coding	Processing information for satisfactory transmission over a communication channel
check bits	Information added to a data signal to help in establishing the validity of received information
cipher	A secret code
ciphertext	The result of encrypting the *plaintext*
circuit switching	A network switching technique which after call set up offers a constant fixed bit rate connection until the call is cleared
CMIP	Common management information protocol
CMIS	Common management information service
congestion	A network condition in which throughput falls as the offered load is increased, due to the saturation of network resources
connectionless	Communication using single unrelated packets (*datagrams*)
connection oriented	Communication over a *virtual circuit*
CPU	Central processor unit
CRC	Cyclic redundancy check
cryptology, cryptography	Study of secret codes
CR82	Cambridge ring 1982 protocol specification
CSMA/CD	Carrier sense multiple access with collision detection

CUG	Closed user group
DAP	Data access protocol (in *DECnet*)
DARPA	Defense Advanced Research Projects Agency (USA)
datagram	A packet sent across a network that is not related to other packets and therefore contains all required addressing information
DCE	Data circuit-terminating equipment
DDCMP	Digital data communications message protocol
deadlock	An extreme form of congestion in which the throughput of a portion of a network drops to zero
DEC	Digital Equipment Corporation
decipher, decrypt	Reverse of *encipher*
DECnet	Digital Equipment Corporation's distributed computing architecture
DES	Data encryption standard
DIB	Directory information base
DNA	Digital network architecture
domain	A collection of systems attached to one or more networks treated as a single entity, for example for network management
DQDB	Distributed queuing dual bus *MAN*
DSA	Directory service agent
DTE	Data terminal equipment
DUA	Directory user agent
ECMA	European Computer Manufacturers' Association
eavesdropper	An unauthorized reader of files during transmission or storage
EFT/POS	Electronic funds transfer at point of sale
encipher, encrypt	Translate a *plaintext* message into secret code
FCS	Frame check sequence
FDDI	Fibre distributed data interface
flow control	A technique for ensuring that a sender does not transmit information faster than it can be handled by the receiver
FMD	Function management data services (in *SNA*)
frame	A unit of a message delimited by synchronization words or bits for transmission across a link
FTAM	File transfer, access and management
FTP	File transfer protocol
gateway	System connecting two or more unlike networks
HDLC	High level data link control
hybrid LAN	An *LAN* that uses different access protocols for different services

IA5	International Alphabet number 5
IBM	International Business Machines Corporation
ICMP	Internet control message protocol
IEC	International Electrotechnical Commission
IEEE	Institution of Electrical and Electronic Engineers
IEEE 802	A series of standards for *LANs*
IP	Internet protocol
ISDN	Integrated services digital network
ISO	International Standards Organization
IV	Initialization vector—a seed vector for starting various algorithms
JANET	Joint Academic NETwork (UK)
JTM	Job transfer and manipulation
key	Information known only to a cipher user; used for encryption and decryption
key space	Set of all possible keys for certain classes of ciphers
LAN	Local area network
LAPB	Link access procedure balanced (in *HDLC*)
layering	A mechanism by which a variety of communication requirements are subdivided into logical groups
LCN	Logical channel number
LLC	Logical link control
LU	Logical unit (in *SNA*)
MAC	Medium access control
MAN	Metropolitan area network
MAP	Manufacturers' Automation Protocol
message switching	A form of switching in which each complete message is stored and forwarded by the exchanges en route to its destination
modem	Modulating/demodulating equipment
modular arithmetic mod n (or mod n arithmetic)	Arithmetic where multiples of n are discarded (set to 0)
modulation	The superposition of information onto another signal to permit multiplexing and/or satisfactory transmission
MOTIS	Message oriented text interchange system
MTA	Message transfer agent (in *X.400* messaging)
MTBF	Mean time between failures
multicasting	A technique that enables a single transmitted message to reach a selected group of destinations on a network
NACK, NAK	Negative acknowledgement

nameserver	A facility providing translation between network addresses and names in the form of text strings
NAU	Network addressable unit (in *SNA*)
NBS	National Bureau of Standards (USA)
NCP	Network control program
NETBIOS	Network basic input/output system
NFS	Network file system
node	A switching exchange in a *WAN*
NOS	Network operating system
NPL	National Physical Laboratory, Teddington, UK
NPDU	Network protocol data unit
NSAP	Network service access point
NSP	Network services protocol (in *DECnet*)
ODA	Office document architecture
ODIF	Office document interchange format
ODP	Open distributed processing
one-way function	Function f whose values are easy to compute, but such that the values of f^{-1} are not easy to compute
open system	A system capable of interworking with other systems conforming to the same agreed rules
OSI	Open systems interconnection
packet	A part of a message of a suitable length to be transferred across a packet network
packet switching	A form of switching in which packets of some maximum length are stored and forwarded by the exchanges and share links in the network
PAD	Packet assembler/disassembler, enables terminal users to access a packet switched network
PBX	Private branch (telephone) exchange
PDU	Protocol data unit
PEP	Packet exchange protocol (*XNS* protocol suite)
pipelining	A technique for increasing the utilization of long links by allowing frames to be transmitted before the previous ones have been acknowledged
plaintext	A readable message or piece of data (usually before *encryption* or after *decryption*)
protocol	Agreement on the format for communication between two communicating entities
PSDN	Public switched data network
PSS	Packet switched service
PSTN	Public switched telephone network
PTT	Post, telegraph and telephone organization
PU	Physical unit (in *SNA*)

public key cipher	Cipher where encrypting *key* can be made public without revealing how to decrypt the *ciphertext*
PVC	Permanent *virtual circuit*
QPSX	Queued packet and synchronous exchange
repeater	System providing interconnection at the physical layer
ROSE	Remote operations service element
router	System connecting two or more unlike networks (also called a *gateway*)
RPC	Remote procedure call
RSA scheme	The Rivest–Shamir–Adleman *public key cipher*
RTSE	Reliable transfer service element
SAP	Service access point; the place at which an *OSI* service is made available
SCM	Session control module (in *DECnet*)
SDLC	Synchronous data link control procedure
SDU	Service data unit
Session	A formal dialogue between two *NAUs*
Slotted ALOHA	A form of *ALOHA* protocol where a single channel is divided into equal length slots that are accessed by contention
SMP	Standby monitor present (token ring)
SNA	Systems network architecture (*IBM*)
SPP	Sequenced packet protocol (*XNS* protocol suite)
SSCP	System services control point (in *SNA*)
SSP	Single shot protocol (Cambridge ring)
STE	Signalling terminal equipment
stream cipher	Cipher where message is regarded as a stream of symbols to be encrypted one at a time (compare with *block cipher*)
SVC	Switched *virtual circuit*
symbol	The basic unit of information on the transmission channel
synchronization word	A marker (usually to mark the start of a frame)
TCP	Transmission control protocol (Internet protocol suite)
TCP/IP	Frequently used to mean the Internet protocol suite (including *TCP, IP*)
TDMA	Time division multiple access
THT	Token holding time
TOP	Technical and office protocol
TPDU	Transport protocol data unit
Triple X	The three *CCITT* X-series protocols (X.3, X.28 and X.29) which are used to provide terminal access across an *X.25* network

TRT	Token rotation time
TSBSP	Transport service byte stream protocol (Cambridge ring)
UA	User agent (in *X.400* messaging)
UDP	User datagram protocol (Internet protocol suite)
UFID	Unique file identifier (in *CFS*)
ULTRIX	DEC's *UNIX* system
UNIX	A multiuser operating system
utilization	The ratio of the useful data throughput of a link to the capacity of the link
VDU	Visual display unit
virtual circuit	A logical path through a communication network
virtual filestore	A common logical view of file structure (in *FTAM*)
VLSI	Very large scale integration
VSAT	Very small aperture terminal
VTAM	Virtual telecommunications access method
VTP	Virtual terminal protocol
WAN	Wide area network
XDFS	Xerox distributed file system
XDR	External data representation
XNS	Xerox network systems
XOR	Exclusive OR (applied to bit strings)
X.25	*CCITT* recommendation for access to a packet switched network
X.400	*CCITT* recommendations for electronic messaging at the application level
X.500	*CCITT* recommendations for directory services
YP	*Yellow Pages*

1 Computer networks—an introduction

GILL WATERS

1.1. Computer networks and protocols

The ability to transfer digital information between large numbers of different computer systems and to use the information in a variety of ways has opened up a vast range of opportunities for office systems, for manufacturing and service industries and for research environments. The problems of interconnecting such diverse systems in a flexible manner are quite complex, and have required a substantial and coordinated approach to the design of communications facilities and to the ways in which computer systems make use of them. This chapter introduces some of the basic concepts that have enabled computer communication networks to be accepted and widely used.

1.1.1 What is a computer network?

A computer network is a communication system that allows computers to exchange information with each other in a meaningful way. The term *computer* may be interpreted in its widest sense as an information processing system; some techniques for computer networks can also be applied to the conveyance of voice, video and facsimile. Computer networks may link computers that are all of the same type (homogeneous networks), or they may link computers of several different types (heterogeneous networks). Networks may cover a very small area, for example they may be confined to a single room, or a large area, extending throughout a country or across international boundaries. Information is transmitted across computer networks in packets which are sequences of bits containing both signalling information and user data, each packet being no greater than a maximum length determined by the network.

There are three main categories of computer network—local, wide and metropolitan area networks.

1 A *local area network (LAN)* offers high speed communication (of the order of 10 Mbit/s) between computers situated within a limited area, typically within a building or a small site. In general, a low bit error rate can be achieved, all computers are connected to a single transmission medium and access to the medium is controlled in a distributed manner with all the computers working to agreed rules. LANs are usually owned and managed by a single organization.

2 The area covered by a *wide area network (WAN)* can vary considerably. Most WANs connect computers within a single country, but others (for example the SITA network which serves the international airline organizations) cover many countries. Wide area networks usually employ a technique called packet switching. Wide area networks may be provided centrally by post, telephone and telegraph organizations (PTTs); they may be private to a single organization or they may be operated by a group of organizations for their exclusive use (depending on an individual country's regulations). Wide area networks may also include segments connected by radio and satellite.

3 A *metropolitan area network (MAN)* covers an area of the size of a typical city. Metropolitan area networks are the newest of the three types and are still a subject for research. They offer a simple, fast way to link different organizations for the exchange of information, with a similar degree of flexibility to that provided by LANs.

1.1.2 What is a protocol?

Each type of computer system has its own method of storing information and interfacing to the outside world. A network that links different computers can only function if there is an agreement about how information is transmitted and interpreted. A *protocol* defines the rules of procedure which computers must obey when communicating with each other. At the basic level protocols define when and how packets shall be transmitted onto the network; these access protocols allow for the fundamental exchange of information. At a higher level, an agreed character code must be used and the information must be interpreted by the appropriate application software. The rules ensuring that these requirements are met are governed by various levels in a hierarchy of protocols.

1.2 Historical development of computer networks

This book is concerned with the relationship between computers and communications. This brief historical review gives a perspective on the developments and requirements that have brought us to today's position, where there is wide interconnection of computers.

The last 20 to 30 years have seen a remarkable decline in the physical size of memory for a given amount of storage, an increase in processing speeds, and a decline in computing costs. For instance, the computing power which can now be contained in an affordable desktop home microcomputer would have occupied a complete office building floor 25 years ago.

As the availability of computers has increased, so has the number of applications for which they can be used. In addition to their original use for complicated mathematical calculations, and the first commercial applications such as payroll programs, computers are now used for text processing, computer games, data bases, warehouse control, airline reservations, computer aided design using high resolution graphics, as aids to education and a wide variety of other applications. With many of these applications a need has arisen for communication between computers, and flexible communication facilities have developed along with advances in computer and communications technology.

Large projects involving many people succeed because each individual contributes his or her own skill, and there is a constant interchange of relevant information. Similarly, if a number of computer systems are available each of which can perform specific tasks, the facility to communicate allows us to organize them into more powerful systems.

Figure 1.1 illustrates how computer networks have evolved. The first computers worked in isolation on a single problem at a time. Even when multiprogramming environments were developed, any information exchange was generally accomplished by carrying the physical storage media (e.g. magnetic or paper tapes) between computers—a slow and operator intensive procedure where large amounts of data were involved. Early information transfer requirements were for scientific data collection and for the exchange of software programs. The solution was to provide direct connections between data collection computers and mainframes, for example research physicists started to use this technique in the mid to late 1960s (Bracher *et al.*, 1972). Such communication links were designed to satisfy specific needs, but they still needed to agree formats and procedures for information transfer, and these are the earliest examples of protocols.

As computer power became more accessible in the 1960s, the use of computers became more diverse, this enabled the input and output functions to be devolved away from the mainframe. A remotely located minicomputer that performed these functions could be connected to the mainframe by a telephone line (either a leased line or a dialled line). Such minicomputers were called remote job entry stations, and early examples included the IBM HASP multileaving workstation.

Following the introduction of on-line terminals and time-sharing

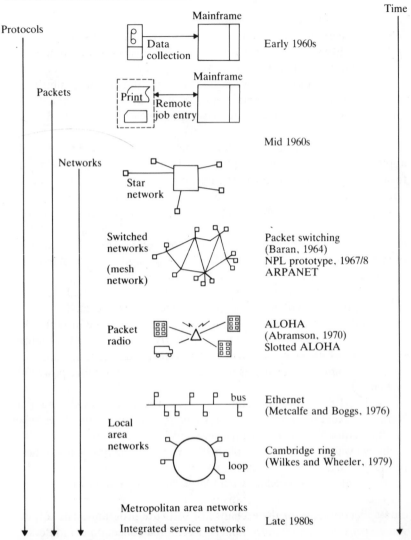

Figure 1.1 Evolution of computer networks

operating systems in the mid 1960s, remote job entry stations were adapted to include terminals enabling remote users to take advantage of these facilities. The remote link was now being shared by several very different activities at the same time—a concept that is fundamental to the way in which computer networks work. A central mainframe might have many remote links, using a variety of protocols.

As computing costs declined further, it was possible for users at remote sites to afford systems to do some of their processing, resulting in communications needs of a different kind. Communication was required with machines of the same kind for collaboration or for the use of common software, and with people at remote sites who were working on similar problems. Central computing facilities offered highly specialized programming packages, or data bases, or organized backing store and archiving. Since not all direct communication needs were with a single site, it was necessary to construct some form of network. The first networks developed from the mainframe concept and were in the form of a star, the mainframe (or a front end processor) acting as a switch.

The concept of packet switching provided the momentum behind the development of computer networks and enabled more complicated topologies to be used. It is easy for a computer to split information into packets of convenient size, and rapid switching of small amounts of data to different destinations can be achieved. The idea was first put forward in a paper by Paul Baran of the Rand Corporation (Baran, 1964); it was widely discussed in 1965. At the National Physical Laboratory (NPL) in England the first implementation of a packet subsystem with properly layered protocols was carried out (Davies *et al.*, 1967). In the United States, designers of the ARPANET (the Advanced Research Projects Agency NETwork) were pioneers of many of the techniques used in today's systems (McQuillan and Walden, 1977). ARPANET was extremely successful and is still in use today. One of the first operational systems was the SITA network built for use by the airlines (Brandt and Chretien, 1972).

Several manufacturers developed their own network architectures such as IBM's Systems Network Architecture (SNA) and Digital Equipment Corporation's DECNET. The proliferation of leased lines used in individual networks and the need for more interconnection led to the requirement for a public packet switched service, and many PTTs now offer such a service. In the United Kingdom, the Experimental Packet Switched Service (EPSS) preceded British Telecom's current Packet Switch Stream (PSS) which is capable of interworking with other packet switched networks internationally.

In the early 1970s the University of Hawaii worked on a packet radio system. The result was the ALOHA method for packet transmission using a broadcast method, applicable to radio or satellite systems (Abramson, 1970). Variants of this technique have evolved, and it also forms the basis of an important family of LAN protocols.

The development of techniques for local area networks has been carried out in several different places, with groups working on different topologies. LANs permit the sharing of expensive devices between inexpensive machines and very flexible ways of enabling several computers to work on the same problem. The idea of a common digital system for all local communications was elaborated in the early 1970s at Xerox's Palo Alto Research Centre (PARC), where the Ethernet system was developed based on broadcast transmission over a bus (Metcalfe and Boggs, 1976). At the University of Cambridge, UK, the Cambridge ring (a slotted ring) first designed in 1975 was implemented (Wilkes and

Wheeler, 1979). Developments in cable television resulted in research into broadband local area networks, which offer several different channels some with specific purposes such as television services and some that can be used for computer communication in a similar way to other LANs.

Recent developments have led to higher speeds in wide area networks and to metropolitan area networks (Klessig, 1986). The Integrated Services Digital Network (ISDN) (Ronayne, 1987) offers channels of 16 and 64 kbit/s and multiples up to 2 Mbit/s which can be used as links in a packet switched network. Interconnection between ISDN and existing packet switched services is also being provided.

1.3 Network topologies

Computers can be connected together in many different ways; the layout or topology of the network will influence how reliable the network is and how easy it is to access. A number of possible topologies can be seen in Figure 1.1.

In a *star* network, the central exchange is very important. If this fails so does the rest of the network; it can also be a bottleneck. It is easy to connect new systems, easy to police centrally, but needs a large amount of cabling.

Most LAN networks are based on a *loop* or *bus* topology, requiring a minimum of cabling. Unless special steps are taken, reliability can be a problem. Either topology can be controlled in a distributed manner.

A *mesh* network contains redundant links offering greater reliability, but a packet may have to pass through several packet switches to reach its final destination. It is reasonably easy to modify the network, but there are design problems in deciding where new packet switches or links should be placed. A mesh topology is generally used for WANs. In a *fully connected* network (not illustrated), alternative paths will generally be available even if several links fail. This is an expensive and complex alternative and is necessary only if a high degree of reliability is required.

1.4 Network protocols

Communicating computers must follow rules in order to convey useful information. The following list indicates some of the requirements which have to be resolved for a point to point link between two computers:

1 The order of progress must follow an agreed procedure.

2 Information representation (for example character code) must be agreed.

3 The recipient must be able to check that the information has arrived at its destination in the order and the format in which it was sent, and must detect missing or duplicated pieces.

4 The recipient may have to acknowledge information that has been received correctly and indicate any errors that have occurred in order for the source to retransmit information.

5 The source should be prevented from flooding the destination with information, but the link should still be used efficiently and deadlocks should be prevented.

6 Packets belonging to each of several conversations on the same link must be identified correctly.

There are further problems to be tackled when we consider not just single links but networks of computers. Some of these are indicated below:

1 Users must be able to find out where a service resides.

2 Packets belonging to a single conversation must be delivered to the correct destination and kept in sequence even if transmitted over alternative routes.

3 Access to information in transit should be confined to authenticated users.

4 The network should work efficiently and without bottlenecks.

5 Accounting and network management facilities are required.

The purpose of this book is to indicate how these requirements can be met by looking first at general techniques and then at specific examples of networks and protocols.⏋

1.4.1 Layering of protocols

A single protocol that solved all the problems mentioned above would be difficult both to comprehend and to implement. Fortunately, the problems can be separated out logically; some are related to the order of bits and the precise characters transmitted, others are related to the network as a whole and yet others are related to facilities for which the network is being used. Consequently computer protocols are designed in a layered manner. Each layer talks to its peer layer across the network using an agreed protocol, but the actual information transfer takes place across the interfaces between adjacent layers, each layer using the services offered by the next lower layer, and so on down until information is transmitted across the physical communication path (see Figure 1.2). At the other end of the network, the information is presented to each next higher layer in turn, and at each layer the relevant

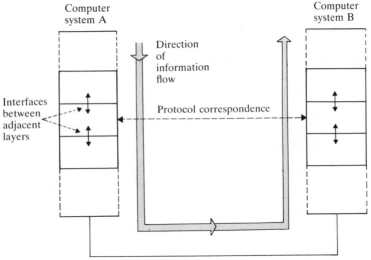

Figure 1.2   The layering concept

information is interpreted according to that layer's agreed protocol. Another advantage of this technique is that a change can be made to the protocol of an individual layer without affecting the other layers, provided that the interfaces to adjacent layers remain the same. Typically each layer will insert an extra sequence of bits (called a header) at the beginning of a block of information before passing it to the next lower layer. Some layers also insert a sequence of bits (a trailer) at the end of the block. These bit sequences are interpreted and removed by the corresponding layer at the receiving end. This is illustrated in Figure 1.3.

1.4.2 The need for standards

Protocols were initially designed on an *ad hoc* basis. Each computer manufacturer or group of users agreed its own set of protocols to ensure that its own machines could be linked, but this prevented unlike computers from being used on the same network. Under pressure from computer users and small manufacturers, the need for standards in computer networking became urgent, and standards bodies began to discuss the issue. This extensive exercise is nearing completion, and manufacturers are now providing software to the specifications laid down in the standards.

The layering principle has been embodied in a model containing *seven* layers (known as the International Standards Organization (ISO) seven–layer model for open systems interconnection (OSI). This will be discussed in detail in Chapter 4.

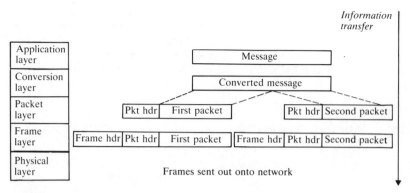

Information transfer

Figure 1.3 Implementation of layering

1.5 Packet, message and circuit switching

Packet switching is a technique for connecting a large user population together with a limited number of links, and is used on most wide area networks (see Figure 1.4). All information to be transmitted is split into packets of a maximum length. Each packet contains information that distinguishes it from other packets in the network and indicates its destination. Consecutive packets travelling on a network link may originate from and be destined for completely different hosts. Each packet is routed through the network under control of the network switches (also called exchanges or nodes) until it reaches its destination. It should be noted that as well as packets destined for different computing systems sharing the communication links of the network, packets can also be destined for different application programs within a single computing system.

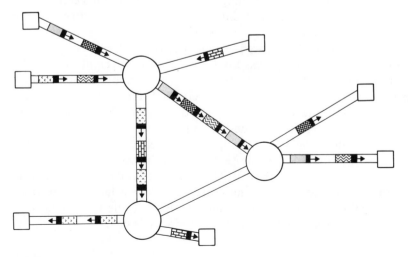

Figure 1.4 Packet switching

Table 1.1. Comparison of packet and circuit switching

Circuit switching	Packet switching
Joins two users at a common speed	Can be accessed at a speed convenient to the user
Small constant delay	Variable delay
Path must be physically available before data is sent	Can start to transmit packets immediately
No conversion undertaken by the network	Can provide conversion of represented information
Information cannot get out of sequence	Packets can arrive out of sequence (exchanges must handle this)
If more than one process is using the link, they must all talk to the same remote host	Several processes can talk to different hosts
Uses whole circuit for duration of call (charge by connect time)	Uses network only when packets are sent (charge by packets sent)
Good for constant bandwidth	Good for bursty traffic

The packet size will depend on the requirements of the traffic to be conveyed and the nature of the transmission links, for example a link with a high bit error rate would dictate a small packet size, to ensure that sufficient packets arrived uncorrupted and did not have to be retransmitted. Typical sizes in practical networks are 128 or 256 octets (8 bit sequences). Packet switching is a particularly useful technique because information transmitted in computer networks is often 'bursty', that is a variable numbers of packets may be transmitted in quick succession, frequently followed by long periods of silence.

In *message switched* systems, complete messages which include addressing information are stored at intermediate nodes before being forwarded. The extra time taken to receive and store complete messages, possibly on secondary storage such as a disk, means that messages may take some time to reach their destination.

In *circuit switching* a relationship is set up between two users across the network using a communication path which is dedicated to those users for the duration of the call. The speeds and codes used by the two endpoints of the conversation must match. If there is no information to be transmitted the capacity of the link is wasted. Table 1.1 summarizes the main differences between packet and circuit switching.

1.6 Datagrams and virtual circuits

There are two basic ways in which data can be carried across a packet switched network: either each packet can be treated separately or logically related packets can be considered together.

Separate treatment of packets is provided by *datagrams*, each of which is a packet which is totally independent of all other packets from the viewpoint of the network. Each datagram contains all the addressing information needed to reach its destination and to indicate its source. Consequently, if two parties are using datagrams to exchange information they should be aware that successive packets may take different routes and may arrive out of sequence or be lost.

In a *virtual circuit* logically related packets are considered together. A virtual circuit (or call) is set up when two parties wish to communicate across the network; it is a logical communication path set up between sender and receiver for the duration of a call. The network must maintain the sequence of packets sent across a virtual circuit even if the packets actually follow different routes through the network. This involves more overhead when setting up the call, but the protocol overhead is shorter for subsequent packets because a reduced amount of addressing information is required, and routing decisions in the switches are taken more quickly.

Datagrams are easier to transfer over internetwork boundaries. They are useful for single short messages such as periodic indications that a system is working and available. Virtual circuits are a better choice for applications such as file transfer or an interactive terminal session, which will transmit many packets over a period of time. Datagrams offer *connectionless* working; virtual circuits offer *connection-oriented* working. In general, networks in the United States have been based on datagrams and in Europe they have been based on virtual circuits.

1.7 Applications of computer networks

[Computer networks have been and will be used for an increasingly wide range of applications] Terminal or microcomputer users can benefit from access to a mainframe or large computing centre.[These may provide computing power, program libraries, data bases (e.g. flight reservations, library cataloguing), information storage systems (e.g. timetables and directories), expensive devices such as film recorders, archive stores and high speed printers.]Access may be accomplished by logging-in remotely, by submitting jobs for remote execution or by requesting transactions depending on the application.[Networks also offer the opportunity to download programs for local execution.]

[Most networks allow users to transfer files and to send messages to each other. A number of similar multiuser minis or microcomputer systems on a network can use a network to perform centralized maintenance and distribution of system software, to share resources and as an aid to collaboration. Local area networks allow sharing of expensive resources such as filestores, printers, plotters and sharing banks of processing units between users.

Networks can be used to transmit teletex, videotex images and facsimile. The speed and capacity of a network limit the feasibility of some types of application, for example uncompressed digitized colour television pictures require a bandwidth of about 140 Mbit/s. There is currently a great deal of interest in using packet networks to transmit digitized real time speech and images. Real time voice cannot easily be transmitted on a network with a low bandwidth or a variable delay. Fibre optic technology offers much higher capacities (gigabits per second) and work is proceeding on fibre optic LANs. To make good use of such capacities for MANs or WANs careful consideration of medium access, switching, topology and protocol complexity is needed. Work is also being undertaken into the use of integrated communications facilities involving voice, images and 'virtual blackboards'.

1.8 A preview of the following chapters

This book progresses from basic communication techniques through a variety of approaches to wide and local area networking and the issues of interconnection, access control and management to a description of distributed systems, and finally looks at recent advances including integrated multiservice networks. Each chapter of the book concentrates on one interesting aspect of computer communication networks. Chapters 2 and 3 look in more detail at solutions to the problems in the design of computer networks which have been introduced in this chapter. In Chapter 2, Ed Jones discusses point to point procedures, and how digital information is transmitted over individual communication links. He also covers techniques for multiple access to a broadcast channel which forms an important basis for developments in local area networks. In Chapter 3, Stephen Hall and Gill Waters look at the design solutions in the network as a whole, concentrating principally on wide area networks. This chapter includes discussions of routing, addressing, flow control and congestion control.

Chapters 4 and 5, written by Gill Waters, are devoted to describing the current standards position and giving examples and explanations of each of the seven layers in the OSI model. The lower layers described in Chapter 4 use the general solutions covered in earlier chapters to provide logical communication channels between systems. Chapter 5 covers the application oriented aspects addressed by the higher layers of the OSI model, which ensure that information sent across the network can be understood and interpreted correctly. A variety of example applications for which standards are available are introduced ranging from remote file management to electronic message delivery. For commercial and historic reasons, not all networks adhere to the standards, and in Chapter 6, Mark Clark introduces two important proprietary networking architectures, each of which is implemented in a large number of installed networks, and also shows how these architectures relate to the standards.

In Chapters 7 and 8, Simon Jones introduces a variety of local area networks which are of increasing importance as they are installed in many workplaces. In Chapter 7, he discusses the techniques available for accessing a common transmission medium shared between many users. The chapter also discusses standards for LANs; several LAN types are standardized, and these all come under the umbrella of a Committee of the Institute of Electrical and Electronics Engineers (IEEE). In Chapter 8, the performance of the various LAN protocols is compared, and the reliability aspect is discussed. The chapter continues with a look at a number of important suites of protocols which are frequently used as the higher layers above the LAN access protocols; one of these is the transmission control protocol/Internet protocol (TCP/IP) suite which is very widely available in commercial systems.

Having discussed approaches to both wide and local area network design, a number of other important aspects are covered in the following chapters.The need for communication between systems generally extends beyond the boundary of an individual network, and network interconnection is the subject of Chapter 9 by Gill Waters. It is also important that users of a network should feel that their information can be protected from unauthorized access. In Chapter 10, Chris Smyth gives a general overview of the field of encryption which can provide a high level of security. He describes a number of techniques that are particularly applicable to computer networks. In Chapter 11, Simon Jones discusses access and authentication in networks in general, including how the encryption techniques described in Chapter 10 can be used to aid authenticated access.

In a distributed system, several computer systems communicating over a computer network combine to offer a unified service. Techniques and examples of distributed systems are described by Mark Clark in Chapter 12, including distributed operating systems and network file servers. Distributed systems rely heavily on the LAN techniques discussed earlier, but their design is also related to work on individual computer operating systems. In Chapter 13, Alwyn Langsford discusses network management—an important topic that was somewhat overlooked in the early race to connect systems together. The chapter includes a review of standards for network management.

Looking at current developments and future possibilities in computer networks, John Adams, in Chapter 14, considers the ways in which packet techniques can be adapted for the integration of voice and data despite their orthogonal requirements. He describes an interesting slotted ring protocol, the Orwell protocol, which responds to these requirements; he relates his discussion to recent advances in the consideration of the broadband integrated services digital network (B-ISDN). The final chapter by Gill Waters and John Limb takes an overall look at recent advances in networking, including new architectures and applications, developments in metropolitan area networks, emerging satellite networks and formal techniques.

References and suggestions for further reading

References

Abramson, N. (1970) 'The ALOHA system—another alternative for computer communications', *Proc. AFIPS Fall Joint Computing Conf.*, Houston, Texas, 1970, 281–285.

Baran, P. (1964) 'On distributed communications networks', *IEEE Trans. Commun. Systems*, **CS-12** (1), March, 1–9.

Bracher, B. H., J. F. MacEwan and A. G. Abbott (1972) 'An on-line data collection system for film measurement', *Software Practice and Experience*, **2**, 389–396.

Brandt, G. J. and G. J. Chretien (1972) 'Methods to control and operate a message switching network', *22nd Int. Symp. on Computer Communications and Teletraffic*, New York, April 1972, 263–276.

Davies D. W., K. A. Bartlett, R. A. Scantlebury and P. T. Wilkinson (1967) 'A digital communications network for computers giving rapid response at remote terminals', *Proc ACM Symp. on Operating System Principles*, Gatlingberg, 1967.

Klessig, R. W. (1986) 'Overview of metropolitan area networks', *IEEE Commun. Mag.*, **24** (1), January, 9–15.

McQuillan, J. M. and D. C. Walden (1977) 'The ARPA network design decisions', *Computer Networks*, **1**, August, 243–289.

Metcalfe, R. M. and D. R. Boggs (1976) 'Ethernet: distributed packet switching for local computer networks', *Commun. ACM*, **19** (7), July, 395–404.

Ronayne, J. (1987) *The Integrated Services Digital Network: From Concept to Application*, Pitman, London.

Wilkes, M. V. and D. J. Wheeler (1979) 'The Cambridge digital communication ring', *Proc. Local Area Communications Network Symp.*, Boston, Mass., May 1979, 47ff.

Suggestions for further reading

Books

A very general bibliography is given here. More opportunities for further reading are given at the end of each chapter. There are currently a large number of books available on the subject of computer networks, many of which have been published or updated recently. The following three books are good general texts.

Black, U. D.: *Data Networks: Concepts, Theory and Practice*, Prentice-Hall, Englewood Cliffs, New Jersey, 1989.

Halsall, F.: *Introduction to Data Communications and Computer Networks*, 2nd edition, Addison-Wesley, UK, 1988.
Tanenbaum, A. S.: *Computer Networks*, 2nd edition, Prentice-Hall, Englewood Cliffs, New Jersey, 1988.

The following two texts are suggested for readers interested in pursuing the mathematical aspects of computer communication network performance and design.

Schwartz, M.: *Telecommunication Networks—Protocols, Modeling and Analysis*, Addison-Wesley, USA, 1987.
Hammond, J. L. and P. J. O'Reilly: *Performance Analysis of Local Computer Networks*, Addison-Wesley, USA, 1986.

Journals

The field of computer networks is a rapidly changing one, and the literature is therefore often to be found in a wide range of journals in electronics and computing including the following:

IEEE Network Magazine
IEEE Journal on Selected Areas in Communications
IEEE Transactions on Communications.
IEEE Communications Magazine
Computer Networks and ISDN Systems
Computer Communications (Butterworth Press)
ACM Computer Communication Review (this publication periodically issues bibliographies of recent work in computer communication networks)

Conference proceedings

Among the many interesting conferences, the following conferences are held regularly:

International Conference on Computer Communications (sponsored by the International Council for Computer Communication)
Distributed Computing Systems Conference (sponsored by the Computer Society of the IEEE)
ACM SIGCOMM Symposium (Association for Computing Machinery, Special Interest Group on Computer Communications)
IEEE International Conference on Computers and Communications

2 Point-to-point and broadcast communication techniques

EDWIN JONES

2.1 Introduction

Although this book is about computer networks, it is appropriate at this stage to consider briefly the communication links or channels from which such networks are usually made up. We also need to consider the formats into which the digital information is assembled prior to transmission. A number of factors will influence these issues, in particular, the bandwidth and noise characteristics of the channel. Furthermore, for reasons of economy and flexibility it may be desirable to combine or multiplex several sources of information together before coding and subsequent transmission.

In this review of transmission techniques for computer communication we deal with two types of system: *point-to-point* systems in which information is received at a single destination and *broadcast* systems in which more than one destination can receive the same information. Figure 2.1 shows a somewhat hypothetical interconnection of integrated networks in which various communication link types can be seen. These links may comprise metallic pair cable, coaxial cable, radio (including satellite) or optical fibres. They may support transmission rates from a few kilobits per second up to hundreds of megabits per second. Selection of the transmission media will depend upon the transmission rate required and also upon the distance and terrain over which the link has to be provided. Each link will be required to perform a variety of roles depending upon its position in the system.

This chapter provides a brief introduction to the concepts and techniques. It also serves to review some terms that are used in later chapters.

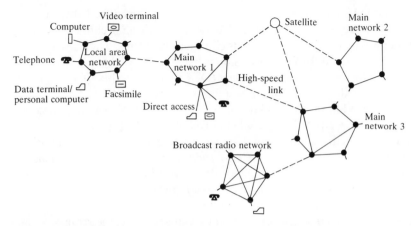

Figure 2.1   Communication links within and between networks

2.2 Signal processing requirements for a data communication link

A digital message which is to be transmitted over a communication network will usually have to undergo various processing steps; Figure 2.2 shows a common arrangement. The *data source* (known in various contexts also as the *message source* or *user*) will usually output information in a parallel format via an output *bus*. This is the preferred

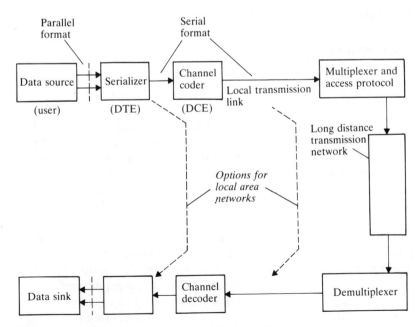

Figure 2.2   Interfaces in a data communication link

format for interconnecting within a piece of equipment or between adjacent devices but, for longer distances, the wiring becomes excessive and propagation delay variations can cause problems. Thus, a parallel-to-serial format conversion is performed by what is often termed the *data terminal equipment* (DTE). For flexibility of interconnection, serial format standards exist such as CCITT Recommendation V.24 and its US equivalent RS232C. (A simplified version known as X.21 is detailed in Chapter 4.) Such interface standards enable a variety of data sources to gain access to a network via a *channel coder* or *data circuit-terminating equipment* (DCE). This interface equipment is concerned with converting the information into a suitable form for satisfactory transmission over the *local transmission link*; in many current systems it is referred to as a *modem*. The transmission characteristics of this link will determine the requisite coding operations. In addition, several data sources may be combined (or multiplexed) together at this point before transmission.

Access to the *long distance transmission network* (if required) may also involve multiplexing and associated protocols which will enable multiple users and source types to gain an orderly access to the transmission facility. Finally, at the receiving terminal the appropriate complementary processing tasks must be performed. However, we note that for communication solely within a local area some of the interfaces may be omitted, as shown by the broken lines in Figure 2.2. For simplicity, only unidirectional transmission has been shown; for most data communication systems each device will be required to process data signals in both directions.

2.2.1 Words, frames and symbols

Figure 2.3 shows the various signal formats encountered as a message is processed prior to transmission. The parallel or bus data format is shown in Figure 2.3(a). A complete binary number (referred to as a *word*) is specified by one bit from each line. There may be from 8 to 40 or more parallel lines depending upon the particular application. The serializer then converts this parallel format to a serial form as in Figure 2.3(b). To mark the position of a word in the serial data stream a synchronizing signal is required; this can be added to each word (as in the old telegraph systems) or, more economically, in the form of a *synchronization word*. In a fully serialized arrangement this synchronization (sync) word is transmitted at known intervals, Figure 2.3(c), so that the receiver can locate the information words. A synchronization word, together with its associated data words, is known as a *frame*. In addition to synchronization and message information the frame may also contain control information. (However, we note that some 'serial formats', such as V.24 and X.21, only serialize the data and use extra wires to carry the synchronization and control signals.)

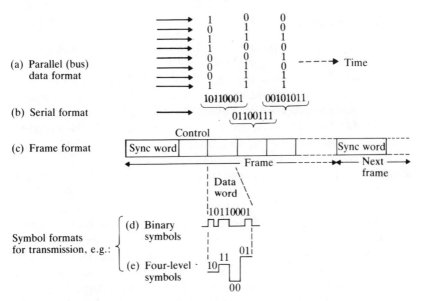

Figure 2.3 Words, frames and symbols

The resulting serial stream of binary digits can either be transmitted directly as shown in Figure 2.3(d) or encoded into another form as determined by the requirements of the channel. This further processing takes place in the channel coder. The waveform in Figure 2.3(e) shows a possible coding in which 2 bits from Figure 2.3(d) have been grouped together to form a new *symbol* for transmission. This has the effect of reducing the rate of signal transition on the line (and so saves bandwidth) but at the expense, in this case, of having to transmit four different symbols. The resulting transmission rate is expressed in symbols per second (or baud). This is a much simplified example of a channel coding function. In practice *redundancy* may be added to provide for clock extraction and for the monitoring of transmission errors at distant terminals. In addition, modulation may be required to shift the signal spectrum, if, for example, the channel has a band-pass frequency characteristic. These topics are briefly discussed later.

Frame structures that keep the bits from a given data word together (as suggested by Figure 2.3(c)) are known as *word* (or often *byte*)-*oriented protocols*. However, with modern communication systems where a frame structure may be required to accommodate a variety of data sources, possibly with different word lengths, a *bit-oriented* structure will usually be more appropriate. Here a frame accepts data bits without reference to any underlying word structure within the incoming data stream, an example is the high level data link control (*HDLC*) bit-oriented protocol described in Chapter 4.

2.2.2 Channel capacity

From the above, two important questions arise: how fast can we transmit and how many symbol levels can be reliably distinguished at the receiver? The answer to the first question will depend upon the channel bandwidth while the answer to the second will be determined by the signal-to-noise ratio at the distant receiver. Two classic results attributed to Nyquist (1928) and Shannon (1948) quantify the answers.

Figure 2.4 shows the effect on a binary signal after it has been transmitted through a bandwidth-limited low-pass channel. There will be a limit to the amount of rounding that can be tolerated; this will be a function of bandwidth. Nyquist's theorem states that for a bandwidth B, we can transmit at $2B$ symbols per second. This is a theoretical transmission rate limit which assumes an 'ideal' (flat) channel frequency characteristic over the bandwidth B. Practical channels are usually non-ideal in the Nyquist sense and so often employ frequency equalizers to make amends. However, it is usually preferable, for economic reasons, to provide an equalizer which only partially compensates for the characteristics of the channel. This will lead to an achievable transmission rate which is rather less than the Nyquist rate.

The second question relates to the decision-making process. As can be seen from Figure 2.4, once a timing reference has been established at the receiver a decision has to be made as to which symbol (0 or 1 in this case) was transmitted. The ability to do this reliably in the presence of channel noise will be determined by the signal-to-noise ratio. This decision-making process will become progressively more difficult as the number of transmitted symbol levels increases.

Shannon combined these ideas into: *for a channel of bandwidth B and signal-to-noise ratio S/N the maximum capacity is given by*

$$C_{max} = B \log_2 (1 + S/N) \qquad \text{bit/s}$$

Again, this is a maximum theoretical bound; practical systems usually achieve only a fraction of this capacity. For example, a good quality

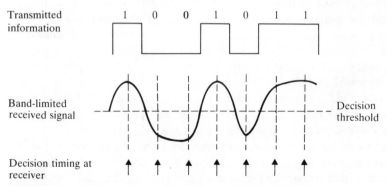

Figure 2.4 Effect of band-limiting a binary signal

telephone connection will provide a bandwidth of 3 kHz and a signal-to-noise ratio of 1000, that is a theoretical C_{max} of approximately 30 kbit/s. We contrast this with the 1.2 kbit/s data rate often provided by telephone line data modems. Higher data rates can be achieved but only if complicated modulators (involving multilevel transmitted symbols) and sophisticated equalizers are used.

2.2.3 Coding for error correction

When information is transmitted over a noisy channel, the digital decision-making process at the distant receiver will sometimes make erroneous decisions. This results in the acquisition of errors in the reconstructed information stream. If the probability of occurrence of these errors is unacceptable the user must either select a quieter transmission channel or, if this option is not available, resort to *error control coding* techniques. In most computer networks, where a very low error probability is required, such coding techniques are essential. Two possible methods are available.

Figure 2.5(a) shows a technique that is possible if a return channel is available. This method is frequently used in computer networks. Blocks of data from the source are subjected to a *check bit* (or *parity bit*) *generator* which adds redundant information before transmission. Such blocks are then passed through a *codeword store* before being transmitted. At the receiver an *error checker* uses the added check bits in each received message block to establish whether errors are present in the block. If no errors are detected the check bits are removed and the data accepted. If errors are detected a signal is sent via the return channel to the codeword store at the transmitter to request a retransmission of the erroneous block. This whole process can be repeated until the block is received without any detected errors. The required size of the data stores will depend upon the number of message blocks held in the channel during transmission. (This will be a function of the message transmission rate and the propagation delay of the channel.) Various protocols to ensure satisfactory error detection and subsequent correct reassembly of the data blocks at the receiver are in use. Some details are discussed in Chapter 3.

If a return channel is not available or if the system is such that many message blocks are held in the channel, requiring very large data stores (e.g. a satellite link), then the *forward error correcting* method of Figure 2.5(b) must be used. In block diagram form its structure closely follows that of Figure 2.5(a) except that the check bit generator at the transmitter must add a more comprehensive form of redundant information to the transmitted data stream (and so the technique is less efficient). This extra check bit information is structured so that at the error checker in the receiver there is sufficient information to detect and

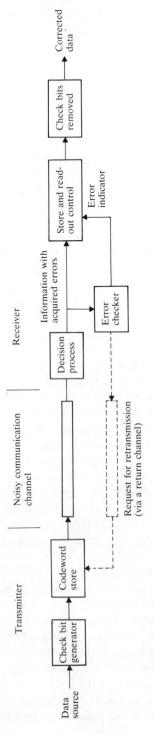

(a) Correction by retransmission

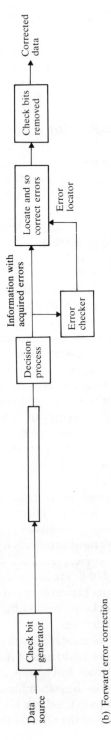

(b) Forward error correction

Figure 2.5 Error correction techniques

Decimal value	b_1	b_2	b_3	b_4	c_1		b_1	b_2	b_3	b_4	c_1	c_2	c_3
0	0	0	0	0	0		0	0	0	0	0	0	0
1	0	0	0	1	1		0	0	0	1	0	1	1
2	0	0	1	0	1		0	0	1	0	1	1	0
3	0	0	1	1	0		0	0	1	1	1	0	1
4	0	1	0	0	1		0	1	0	0	1	1	1
5	0	1	0	1	0		0	1	0	1	1	0	0
6	0	1	1	0	0		0	1	1	0	0	0	1
7	0	1	1	1	1		0	1	1	1	0	1	0
8	1	0	0	0	1		1	0	0	0	1	0	1
9	1	0	0	1	0		1	0	0	1	1	1	0
10	1	0	1	0	0		1	0	1	0	0	1	1
11	1	0	1	1	1		1	0	1	1	0	0	0
12	1	1	0	0	0		1	1	0	0	0	1	0
13	1	1	0	1	1		1	1	0	1	0	0	1
14	1	1	1	0	1		1	1	1	0	1	0	0
15	1	1	1	1	0		1	1	1	1	1	1	1

Check bit computation:

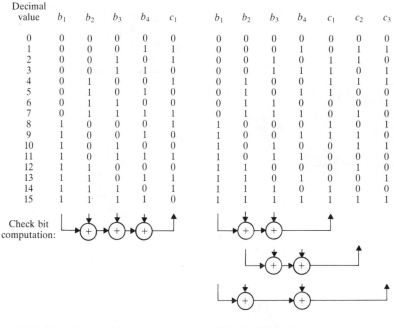

(a) A 5,4 single parity check code   (b) A 7,4 Hamming code

Figure 2.6 Codes for error detection and correction

also locate errors in each block. Once located these binary errors can be corrected by inversion.

A very simple (and rather inefficient) method of coding to permit error detection would be to send a block of source data twice. The error checker at the receiver would then be a simple comparator; if the two blocks differed a retransmission would be requested. To achieve forward error correction a data block could be sent three times—a majority decision would then select the correct block. Happily, there are much more efficient methods of coding for error detection and correction; Berlekamp *et al.* (1987) provide a good review of techniques. We consider some tutorial examples.

Figure 2.6(a) shows how a 4-bit data block (b_1 to b_4) can be protected using a *single parity check bit* (c_1). In this example, the parity check bit is computed by adding together the four data bits using modulo-2 arithmetic. This operation is specified, in diagrammatic form, at the bottom of the 16-line table. Inspection of the table shows that to move from one line of the table to another requires changes in at least two bit positions. The set of codewords is thus said to have a *minimum distance* of 2. This means that a single error in any of these 5-bit codewords during transmission can never result in another member of the codeword set being received. The receiver (which has knowledge of the valid codeword set) can thus detect all single errors in each codeword.

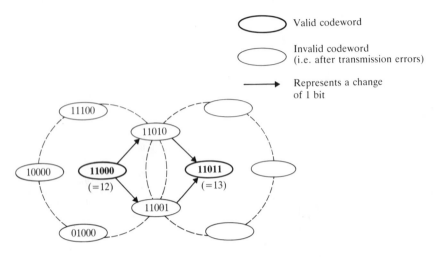

(a) Single error detection using a 5,4 code

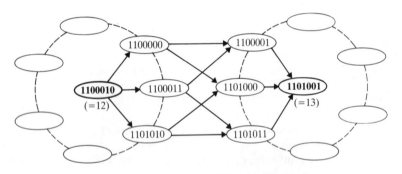

(b) Single error correction (or double error detection) using a 7,4 code

Figure 2.7 Error control mechanisms

The code is known as a 5,4 code because five bits are sent, of which four are data bits. Figure 2.7(a) summarizes what could happen if codeword 12 ($=11000$) was sent over a noisy channel. A single error could result in receiving any of the five codewords lying on the left-hand circle. All of these invalid codewords would be detectable (because none of them appear in the table of Figure 2.6(a)). However, if two errors occur in a codeword, the arrows on Figure 2.7(a) show how another valid member of the code set (data word 13 in this example) could be received. The decoder would simply see a valid codeword and know nothing about the errors; that is double errors in a codeword cannot be detected. In fact, it can be seen from the parity check mechanism that a single parity check code can detect odd numbers of errors in a codeword.

Figure 2.6(b) shows a more sophisticated example. This is the first member of the important category of codes known as *Hamming codes* (Hamming, 1950). We use the same 4-bit data block as before but now protect it with three check bits to create a 7-bit codeword for transmission. The rules generating the check bits are again given in diagrammatic form at the bottom of the coding table. Inspection of the table shows that this code requires at least three changes to a codeword to produce another member of the set; that is it has a minimum distance of 3. It follows that this code can detect up to two errors in a received word. However, Figure 2.7(b) reveals another possibility. Using the same data values (namely 12 or 13) as before, we now see that a single transmission error would move to a word on a circle surrounding the transmitted codeword. With a code of minimum distance 3, these circles can never interfere with each other and so a single error can be uniquely associated with a particular (valid) transmitted codeword, thus it can be corrected.

To complete the picture, the arrows show how a second error could move onto a circle associated with another valid codeword, and so two errors in a word would be erroneously 'corrected'. This view confirms our conclusion above that if correction were not attempted, two errors could be detected as they would always produce an invalid codeword. Finally, we see that with three errors, codeword 12 could be transformed into codeword 13; the receiver would be oblivious of the problem.

It follows that this Hamming code can either correct a single transmission error or detect two transmission errors. In general, a code with a minimum distance d can:

either *detect* $d - 1$ errors (some others may also be detected)

or   *correct* $\dfrac{d - 1}{2}$ errors (rounded down to the nearest integer)

The above coding examples are members of an important

Check bits (available after 7 clock cycles)

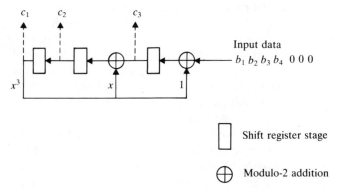

Figure 2.8   Check bit generation using a feedback shift register (with polynomial $x^3 + x + 1$)

classification known as *cyclic codes*. Such codes can be conveniently generated (and decoded) using feedback shift registers in which the rules for generating the check bits are entirely specified by the shift register feedback connections or *generator polynomial*. The Hamming code of Figure 2.6(b) requires the generator polynomial $x^3 + x + 1$ and this can be seen built into the feedback connections of Figure 2.8. After clearing the shift register, the four data bits (b_1 to b_4) are fed in, followed by three zeros. The circuit has the effect of dividing the incoming sequence by the generator polynomial and after seven clock cycles the arithmetic remainder of this division process is held in the shift register. This remainder corresponds to the required check bits c_1, c_2 and c_3.

Using the same generator polynomial, a similar division is performed at the receiver but now involving all seven received bits. After this division the contents of the 3-bit shift register are inspected. If no transmission errors have occurred the contents will be the same as the initial shift register condition at the transmitter (e.g. 000). If a single error has occurred the shift register contents or *error syndrome* will be non-zero. Now this 3-bit error syndrome will uniquely indicate the position of the error in the received 7-bit codeword. Once located, for binary coding, the error can be corrected by inverting the offending bit. The supporting mathematics is to be found in many textbooks, such as Sklar (1988).

The length of the generator polynomial can be varied to provide any number of check bits or *cyclic redundancy checks* (CRC) as they are often called. For example, the HDLC frame format discussed in Chapter 4 uses the generator polynomial $x^{16} + x^{12} + x^5 + 1$ which provides 16 check bits. This code is capable of multiple error detection which can then be used to request retransmission of an errored block. By choosing different generator polynomials, error control codes of various lengths and appropriate minimum distances can be generated. For a given distance, longer blocks will require proportionately fewer check bits but of course the probability of multiple transmission errors will increase with block length. Thus, code distance, block length and the resulting efficiency of the code will require a compromise dependent upon the error characteristics of the transmission path.

It should be noted that there can never be a guarantee that all errors in a transmission system will be detected or corrected. There will always be a finite probability that multiple transmission errors will result in a word which also belongs to the set that is acceptable to the error checker, the checker is then led to believe that it has received a correct word. However, with a parity check algorithm that is well matched to the error characteristics of the channel, the probability of missing errors can be made acceptably small. For computer data an error probability of between 10^{-8} and 10^{-12} may be required (depending on the application) whereas for speech and current video signals a higher probability of error, typically between 10^{-4} and 10^{-7}, will usually be acceptable.

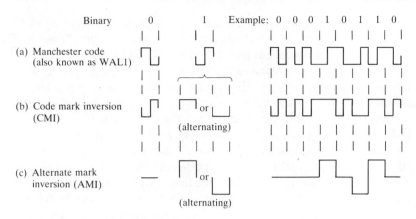

Figure 2.9   Some line codes

2.2.4 Coding for cable transmission

Metallic cables have a low-pass frequency transfer characteristic and so permit data to be transmitted directly without the need for frequency shifting. However, there is usually a need for some channel coding to ensure that the transmitted signal has the following features:

1   No d.c. component because, although metallic cables in themselves have a true low-pass characteristic, they are often coupled via transformers to their terminal equipment.

2   Embedded timing information to ensure that the distant receiver (and intermediate repeaters if used) can extract a reliable clock to time their decision making processes.

3   Sufficient added redundancy to enable the error rate of each transmission link to be monitored. This will usually be in addition to any end-to-end (data source-to-sink) error coding of the information content using the methods described in the previous section.

Coding to ensure that the transmitted signal has these properties is known as *line coding*. Figure 2.9 gives three examples to be found in computer networks.

The *Manchester code* of Figure 2.9(a) is used in the magnetic recording of digital signals and in the Ethernet local area network system. Each binary digit is converted into a 2-bit symbol for transmission. No d.c. component is generated because the first half of each transmitted symbol is complemented by the second half. Also, the resulting signal transition in the centre of each transmitted symbol ensures an adequate clock content. The absence of such a transition indicates a transmission error and so the third requirement, error monitoring, is also possible.

Code mark inversion of Figure 2.9(b) is specified by CCITT as an interface code in certain types of transmission equipment. In this code the d.c. component present in the transmitted symbol representing a mark (binary '1') is not cancelled until the next mark occurs. Thus, the code achieves a balance, but only over a run of transmitted symbols.

The above two codes achieve the desired properties, but at the cost of having to signal at twice the binary digit rate, they thus require a large channel bandwidth. Another approach is to build in the desired features by using a multilevel transmitted symbol; the *alternate mark inversion* (AMI) code of Figure 2.9(c) does this. The zero d.c. property is achieved by sending marks as + 1 and − 1 alternately. An estimate of the link error rate is achieved by counting violations of the alternating mark inversion rule. However, we note that AMI has a potential weakness; if many successive binary zeros are to be sent, then the clock content will fail. Either the number of successive zeros has to be restricted, as in early pulse code modulation (PCM) systems, or a code substitution has to be made to break up this potential hazard (as in many current 24 and 30 channel PCM systems).

Many line codes, designed to meet the requirements of a variety of transmission systems are known; the *International Journal of Electronics* (1983) provides a comprehensive survey.

2.2.5 Modulation for band-pass channels

Optical fibres, radio channels and the analogue telephone network are band-pass channels in that they cannot directly pass low frequency signals. Thus a data source that contains low frequency energy must be frequency shifted or *modulated* before transmission. This is achieved by selecting an appropriate *carrier* waveform and allowing its *amplitude*, *frequency* or *phase* to vary in sympathy with the data source, such that the resulting spectrum lies within the requisite limits for the channel in question. Figure 2.10 shows waveforms for simple digital modulation (or keying) onto sinusoidal carriers.

Current optical systems use amplitude shift keying, simply turning a laser on and off in sympathy with the data to be transmitted. Modern digital radio systems favour combinations of amplitude and phase shift keying; these hybrid modulation schemes make efficient use of *signal space* but require quite sophisticated terminal equipment. For the transmission of data over the analogue telephone network various modulator/demodulator (modem) designs are in use. Data rates up to 1.2 kbit/s are common using frequency shift keying; for higher data rates the more sophisticated multistate phase shift keying and hybrid methods must be used. The analogue telephone network is currently an important means of digital communication, a review of the methods used is to be found in Pahlavan and Holsinger (1988) with further details in their extensive list of references. However, the transmission of data by this means will decline as dedicated digital transmission facilities are gradually extended towards the customer's premises.

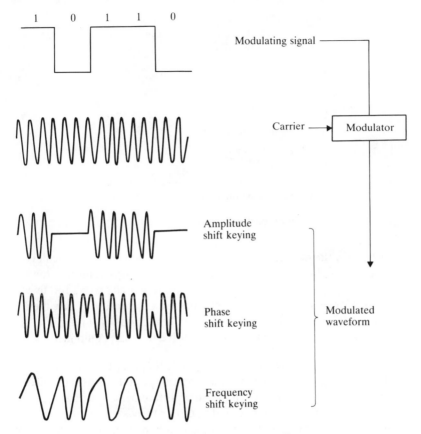

Figure 2.10 Basic methods of digital modulation

Finally, it should be noted that data will usually be encoded in accordance with the line coding ideas of the previous section before modulation for transmission. This ensures that after demodulation at the receiver a clock can be extracted to time the decision making process. The line code will also permit link error rate monitoring which is usually required for maintenance purposes.

2.3 Digital multiplexing

The concept of combining or multiplexing data traffic onto a common transmission medium has already been mentioned. There are two principal reasons for doing this:

1 Multiplexing can be used to ensure that the transmission medium is operated at its most economical information rate. Most transmission media, together with their associated terminal equipment, will possess a minimum cost transmission rate as shown in Figure 2.11.

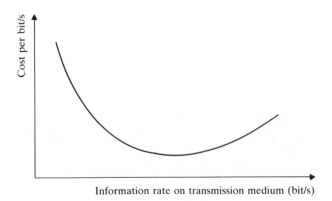

Information rate on transmission medium (bit/s)

Figure 2.11 Economics of multiplexing

Usually, data sources will have to be combined together to achieve this rate. If channels are operated at a lower information rate the cost of equipment provision will have to be shared among too few users. Conversely, if too many sources are combined together then bandwidth and/or signal-to-noise ratio limitations may require excessively sophisticated and so expensive terminal equipment.

2 In systems where users receive all signals (i.e. they are not demultiplexed first), then multiplexing can permit the user to select the signal required. For example, in entertainment broadcasting the user is able to select the required audio or television channel. A similar approach is used on most local area networks where selective demultiplexing (either sender or receiver controlled) is akin to the switching and routing functions of conventional telephone exchanges.

Successful demultiplexing relies on the fact that each constituent data source has an identifiable unique parameter; that is each data stream is orthogonal to all others. The unique parameter may be:

- Time (as in *time division multiplexing*).
- Frequency (as in *frequency division multiplexing*).
- The modulating code (as in *code division multiplexing*).

The use of code division multiplexing is restricted mainly to specialist high security applications. Frequency division multiplexing is the longest established method; its use in analogue communication systems has already been mentioned. It also has application in some optical fibre systems where it is known as wavelength division multiplexing. For data transmission, time division multiplexing is currently the most common choice and forms the basis for many signal formats throughout the rest of this book.

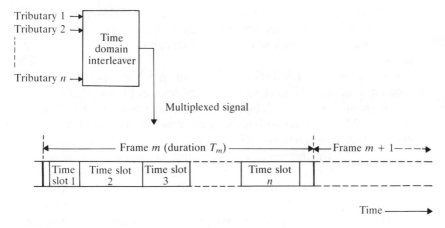

Figure 2.12 Time division multiplexing

2.3.1 Time division multiplexing

A very generalized view of a time division multiplexed frame structure is shown in Figure 2.12. Here, frame m is seen to contain n *time slots*. The distinction between byte- and bit-oriented protocols has already been made, thus a time slot may contain a single word as in Figure 2.3 or, more usually in modern systems, simply provides a transmission opportunity which may be unrelated to the word structure of the incoming data stream or *tributary*. Thus, n active tributaries can each be given a time slot in which to transmit their information. If all information arriving from the tributaries in time T_m is to be accommodated within frame m, it follows that it must support a bit rate that at least equals the aggregate bit rate of all the tributaries. The bit rate of a particular tributary will determine the duration of its corresponding time slot.

We note that time slots do not necessarily fill the whole of the frame. Spare message capacity may be required to accommodate variable tributary rates. Capacity will also be required to provide frame alignment and time slot identification information.

If all tributaries have an identical format, such as in 24-channel and 30-channel pulse code modulation systems, subsequent frames will usually have a similar format. However, if tributary rates can vary (such as in a packet network) then time slot allocations and their durations may differ from frame to frame.

2.3.2 Time division multiple access

Time division multiple access (TDMA) can be considered as an extension of the above, where there are more potential tributaries than time slots available. An *access protocol* must be used to determine the allocation of time slots to active tributaries. Thus a TDMA equipment acts as an adaptive multiplexer or traffic *concentrator* by allocating time slots in accordance with each tributary's demand for transmission capacity.

For example, the TASI (time assignment speech interpolation) system used on some long distance submarine cables uses the fact that in a two-way conversation each talker is active, on average, for less than half the time. When a talker pauses his or her time slot may be reallocated to another talker. When the original talker resumes, a new time slot must be found. Clearly, there is a delicate balance between the number of potential tributaries and the number of time slots available on the transmission link. If a time slot is not available when it is needed the system exhibits *overload* and information is either lost or is subjected to a delay before transmission. For speech a small amount of loss may be permissible (in TASI the beginnings of a speech burst will be lost if a time slot is not immediately available). For data, however, if no information is to be lost, packets will suffer a waiting-time delay. There will usually be a limit on how much delay can be tolerated.

These concepts of data concentration and message/packet switching using TDMA protocol techniques will be seen to be a recurring theme throughout this book.

2.4 Broadcast communications

Broadcast communication systems are those in which a message can be sent to more than one destination. The number and the location of destinations may or may not be known to the sender. Many radio systems (terrestrial and via satellite) fall naturally into this broadcast category in that a single transmitted message can be received in several places. Likewise, message sources connected to a common ring or bus, such as a local area network, may also be within the broadcast category. Figure 2.13 shows these network types, all of which are identifiable within the introductory drawing of Figure 2.1.

The time division multiplexing concepts of the previous section, when used in conjunction with a broadcast channel, have led to some important opportunities for the data network designer. Thus, multiple users generating packetized data are able to share a common broadcast channel. This requires an extension to the basic concept of TDMA introduced earlier. Now, not only has the activity of an incoming tributary to be taken into account, but also the activity of existing traffic on the broadcast channel. If a data tributary accesses the broadcast channel in an uncontrolled manner, *collisions* may occur which will result in a loss of information. This *contention* for use of the broadcast channel requires an access protocol which can resolve any problems. This can be achieved, with an increasing degree of sophistication, by:

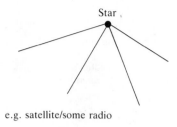

e.g. satellite/some radio

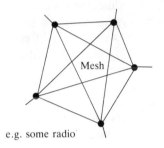

e.g. some radio

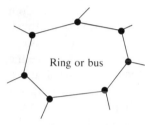

e.g. some local area networks

Figure 2.13   Broadcast network types

1   Allowing collisions to occur and then ensuring that lost packets are retransmitted.

2   Sensing the activity on the channel and only adding packets from waiting tributaries when there is unlikely to be a collision.

3   Avoiding collisions by waiting for a *token* which gives permission to transmit.

Such protocols, and combinations of them, can become quite complicated, as will be seen later. The reader is also referred to the comprehensive review and list of references in Sachs (1988). Here, we focus on basic concepts by considering an early access protocol system for packet transmission from which many recent developments have emerged. (More detailed treatment and comparison with alternatives are to be found in texts such as Schwartz, 1987 and Tanenbaum, 1988.)

2.4.1 The ALOHA system

This early technique, which enables uncoordinated users (message sources) to share a common broadcast channel, was originally invented for a packet radio system connecting remote stations to a central

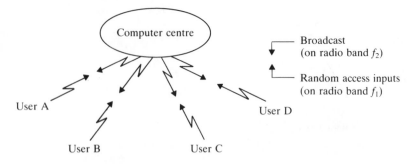

Figure 2.14 The ALOHA system

computer (Abramson, 1973, 1977). Figure 2.14 outlines the system. Users are permitted random access to a common radio frequency band (f_1). The computer centre rebroadcasts all received signals on a different frequency band (f_2); this enables users to monitor packet collisions. If a collision has taken place the whole packet is retransmitted after a random delay time. The system works quite well for low density traffic with packets originating at variable intervals.

A simple arrangement uses fixed duration packets and arbitrary transmission times and is known as *pure ALOHA*. This is shown in Figure 2.15 where it is seen that some packets suffer collisions and so are destroyed. If all packets have duration τ, then a given packet X will suffer a collision if another user starts to transmit at any time from τ before until τ after the start of packet X. That is, for a given packet, there is a *vulnerable time* of 2τ.

Throughput for an ALOHA access protocol

Let there be N stations contending for the common channel each generating on average λ packets/s, where all packets are of duration τ. Let S be the traffic throughput, that is the effective fractional utilization

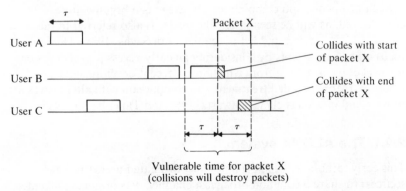

Figure 2.15 Pure ALOHA—packets transmitted at arbitrary times

of the channel (i.e. relative to a maximum possible throughput of 1). We assume all packets generated succeed eventually; thus:

$$S = N\lambda\tau \tag{2.1}$$

In practice collisions will take place that will result in retransmission attempts. Let the average number of *transmitted* packets per user per second (newly arriving packets and retransmitted ones) be $\lambda' > \lambda$. Thus, the total number of transmission attempts (per packet time) is:

$$G = N\lambda'\tau \tag{2.2}$$

Now, any packet of length τ will suffer a collision if another packet overlaps with it at any point over a time interval 2τ (Figure 2.15). For contending packets with Poisson distributed transmission times, the probability that no other packets are generated during this vulnerable time is approximately (for large N):

$$e^{-N\lambda'(2\tau)} = e^{-2G} \tag{2.3}$$

It follows that this is the probability that no collision occurs and is equal to the fraction of messages transmitted successfully, S/G. Thus,

$$\frac{S}{G} = e^{-2G}, \qquad \text{i.e. } S = Ge^{-2G} \tag{2.4}$$

This gives us the pure ALOHA curve in Figure 2.16 where it is seen that the best throughput achievable is $1/2e$ (18 per cent) when 0.5 attempts per packet time are made. If fewer attempts are made channel

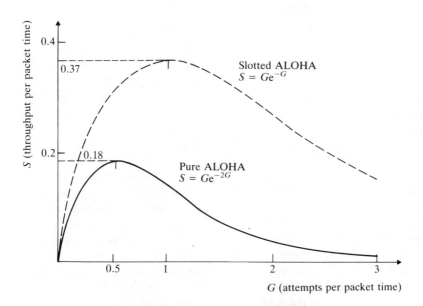

Figure 2.16 Throughput for the ALOHA protocols

capacity is underused. If more attempts are made too many collisions occur and so many retransmission attempts become necessary.

It should be noted that transmission attempts, that is both newly arriving packets and any associated retransmissions, have been assumed to have a Poisson distribution. This is not strictly true but is a reasonable approximation if there are a large number of users and if the random delay before retransmission is relatively long.

The probability of collision can be reduced and so the efficiency of the protocol increased if *slotted ALOHA* is adopted. This requires that users synchronize their packet transmission attempts within a τ spaced time slot framework. Now, collisions can only occur when there is a complete overlap of contending packets; that is the vulnerable time has been reduced to τ. Thus, equation (2.3) becomes:

$$e^{-N\lambda'(\tau)} = e^{-G}$$

leading to

$$S = Ge^{-G} \qquad\qquad (2.5)$$

This is also plotted in Figure 2.16 where it is seen that a maximum efficiency of $1/e$ (37 per cent) is now achievable. The system works well if transmission delays are small relative to τ so that slot alignment is maintained over the whole network. If transmission delays are large, then each user must have knowledge of its own round-trip delay so that it can time the transmission of its packets accordingly.

Packet transmission delay

Maximizing the packet throughput (S) is not the only parameter that must be considered when selecting a satisfactory protocol. As the proportion of transmission attempts increases, so also will the probability of collision. This leads to an increase in the number of retransmission attempts, that is to an increasing delay in the arrival of packets at the receiver. For interactive users in particular, this can cause problems.

To minimize contention and so control delay, the algorithm that determines what happens after a collision is crucial. If users involved in a collision try again at the same time, then another collision will occur. Such repeated collisions can be reduced by, for example, randomizing the waiting times before each user attempts a retransmission. A simple arrangement would be for a packet in contention to be retransmitted with equal probability in any one of the next K packet slots. This would continue, with further retransmission attempts as necessary, until the packet is transmitted successfully. For this arrangement, an equation for the mean packet transmission delay can be derived as follows.

Let the round-trip propagation delay time of the system be R (measured in packet times). A transmitter will not know the result of a

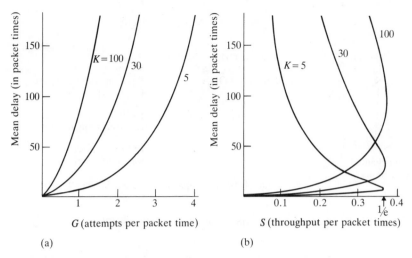

Figure 2.17   Mean delay against G and S for slotted ALOHA

transmission attempt (success or collision) and so will be unable to respond accordingly until $R + 1$ packet times later. If a collision has taken place, then on average, a packet will have to wait a further time of $K/2$ before a retransmission is attempted. Now, if the average number of retransmission attempts per packet time is E, then the mean packet transmission delay d (again in packet times) is

$$d = R + 1 + E\left(R + 1 + \frac{K}{2}\right) \qquad (2.6)$$

where

$$E = \frac{G}{S} - 1$$

In fact, when K is small there will be more collisions and so more retransmissions, which in turn will affect the value of G. Thus, the above is only an approximation, but it is found to be quite accurate for $K > 5$. Further discussion of this is to be found in Schwartz (1987) and a detailed analysis in Kleinrock and Lam (1975).

For slotted ALOHA, equation (2.6) is plotted as a function of G and S in Figure 2.17. Zero propagation delay is assumed, that is the delay arises only from the need to retransmit mutilated packets. It is seen that K has a significant effect on this delay. We also note, from Figure 2.17(b), that for a given throughput S there are two possible values of mean delay. Ideally, we wish K to be small so that mutilated packets are quickly retransmitted. However, if traffic on the system is heavy, this will cause contention to build up as increasing numbers of unsuccessful packets make retransmission attempts; that is the delay will increase, leading to possible instability. Under such circumstances a subsequent increase in K would spread out retransmission attempts and so tend to restabilize the system, but of course with attendant implications on the

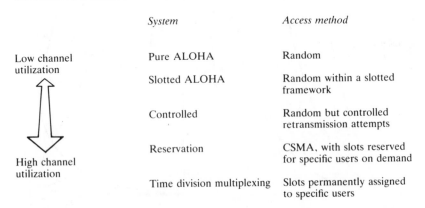

System	Access method
Pure ALOHA	Random
Slotted ALOHA	Random within a slotted framework
Controlled	Random but controlled retransmission attempts
Reservation	CSMA, with slots reserved for specific users on demand
Time division multiplexing	Slots permanently assigned to specific users

Low channel utilization ⬆ / High channel utilization ⬇

Figure 2.18   Access protocols discussed in this chapter

mean delay. This leads to the concept of *dynamically controlled ALOHA* where *K* is selected in accordance with traffic conditions with the objective of maintaining stability while attempting to optimize the delay–throughput relationship.

Reservation ALOHA

Finally, we note a system for users who require a packet slot at fairly regular intervals for delay critical signals such as speech. If several users are offering this type of traffic it is inefficient for each packet to 'fight it out' for channel capacity. A *reservation* system entitles successful users to another time slot after a predetermined time, and so on until they no longer need transmission capacity. The slot then becomes available for open contention by other users, as described previously. We thus have an adaptive variation on the pre-assigned time division multiplexing mentioned earlier in conjunction with Figure 2.12. Some rules are of course necessary. Reserved slots must be respected by other users; that is all users must sense the activity on the channel before they decide whether to make a transmission attempt. There must also be rules to prevent users from being too greedy and trying to acquire too much transmission capacity.

Figure 2.18 summarizes the range of access protocols discussed in this chapter. For low density traffic the basic ALOHA system (pure or slotted) may be adequate. These involve transmitting at random times and only sensing the channel *after transmission* to determine if collisions have taken place. For higher density traffic, where a more effective throughput–delay optimization is required, then a more sophisticated access algorithm is necessary. This will require that the activity on the channel is sensed, usually *before transmission*, and the appropriate algorithm initiated. (The sensing of a channel before transmission is known as *carrier sense multiple access* (*CSMA*); it is an important concept treated in later chapters.) For even higher density traffic, reservation systems can achieve up to full channel utilization but, as in the extreme case of time division multiplexing, this removes flexibility by requiring that time slots are assigned to specific users.

The access protocol selected in practice will thus depend upon the characteristics of the user data (continuous or bursty), whether access delay is important, the desired channel utilization and the permitted degree of sophistication in the protocol. These matters are considered in more detail in later chapters.

References and further reading

References

Abramson, N. (1973) 'The Aloha system', in *Computer-Communication Networks*, N. Abramson and Kuo (eds), Prentice-Hall, Englewood Cliffs, New Jersey.

Abramson, N. (1977) 'The throughput of packet broadcasting channels', *IEEE Trans. Commun.*, **COM-25**, 117–128.

Alisouskas, V. and W. Tomasi (1985) *Digital and Data Communications*, Prentice-Hall, Englewood Cliffs, New Jersey.

Beauchamp, K. G. (1987) *Computer Communications*, Van Nostrand-Reinhold, London, UK.

Bell Technical Staff (1982) *Transmission Systems for Communications*, 5th edition, AT&T Bell Laboratories, Holmdel, New Jersey.

Berlekamp, E. R., R. E. Peile and S. P. Pope (1987) 'The application of error control to communications', *IEEE Commun. Mag.*, **25**, April, 44–57.

Hamming, R. W. (1950) 'Error detecting and error correcting codes', *Bell System Tech. J.*, **26**, 147–160.

International Journal of Electronics (1983) 'Special issue on line codes', **55** (1).

Kleinrock, L. and S. S. Lam (1975) 'Packet switching in a multiaccess broadcast channel: performance evaluation' *IEEE Trans. Commun.*, **33** (4), 410–422.

Nyquist, H. (1928) 'Certain topics on telegraph transmission theory', *Trans. Am. Inst. Electl. Engr.*, **47**, 617–644.

Owen, F. (1982) *PCM and Digital Transmission Systems*, McGraw-Hill, New York.

Pahlavan, K. and J. L. Holsinger (1988) 'Voice-band data communication modems—a historical review', *IEEE Commun. Mag.*, **26**, January, 16–27.

Sachs, S. R. (1988) 'Alternative local area network access protocols', *IEEE Commun. Mag.*, **26**, March, 25–45.

Schwartz, M. (1980) *Information Transmission, Modulation and Noise*, 3rd edition, McGraw-Hill, Singapore.

Schwartz, M. (1987) *Telecommunication Networks: Protocols, Modeling and Analysis*, Addison-Wesley, Reading, Mass.

Shannon, C. E. (1948) 'A mathematical theory of communication', *Bell System Tech. J.*, **24**, 379–423 and 623–657. Also 'Communication in the presence of noise', *Proc. IRE*, **27**, 10–21, 1949.

Sklar, B. (1988) *Digital Communications*, Prentice-Hall, Englewood Cliffs, New Jersey.

Tanenbaum, A. S. (1988) *Computer Networks*, 2nd edition, Prentice-Hall, Englewood Cliffs, New Jersey.

Further reading

For more detailed treatment of the topics covered in this introductory chapter, the reader is referred to the many good books on the subject, for example:

On digital transmission principles and practice, Schwartz (1980), Sklar (1988), Bell Technical Staff (1982), Owen (1982) and Alisouskas and Tomasi (1985) are representative of texts ranging from the more theoretical to the more practical.

On computer and communication network analysis, see Tanenbaum (1988), Schwartz (1987) and Beauchamp (1987).

3 Wide area network design issues

STEPHEN HALL AND GILL WATERS

This chapter covers some of the basic concepts involved in the design of computer networks. In particular, techniques for ensuring that networks achieve efficient and reliable transmission of data between end users are considered. We concentrate here on *wide* area networks; issues specific to *local* area networks are left to Chapters 7 and 8. Throughout the chapter, the term *node* is used for a packet switching exchange; a *host* is a user system attached to the network.

3.1 Addressing

Every packet travelling across a network must contain sufficient information to enable the network to send it to the correct destination. It is also usual to include the address of the source of the packet, so that retransmission can be requested if necessary. The nature of the addressing scheme can have a considerable effect on both the efficiency of the packetization process and the flexibility of the network.

For *connectionless* working, every packet (i.e. datagram) must contain the full address of its destination. However, for *connection-oriented* working it is necessary to specify the full destination address only once, when the call is set up. At this stage a *logical channel number* is assigned to the call, and subsequent packets in the call use this number instead of an address. Each network node maintains a look-up table which it uses to translate the logical channel numbers of received packets into addressing information. Since the number of virtual circuits existing at any one time is likely to be far smaller than the total number of possible addresses, logical channel numbers require fewer bits in each packet header than full addresses, and therefore provide increased efficiency.

3.1.1 Address hierarchy

Address spaces can be either hierarchical or flat. In *flat addressing* every possible destination is assigned a unique number. When new hosts are added to the network, they must be given new addresses within the allowed range. This is rather like assigning pay numbers to the employees of a firm and giving each person to arrive a new number. Initially, the numbers may be assigned in alphabetical order, but the ordering soon deteriorates as people join and others leave.

In networking terms, flat addresses cannot be used to indicate where nodes are located in a network, as it is not possible to leave the correct number of gaps in the right places to ensure that spatial meaning is retained as new nodes are added. This means that flat addressing makes it more difficult for a switching node to decide where to send a packet it has received. On the other hand, flat addressing has the advantage that if the host moves it can retain its unique address. For example, all Ethernet addresses (which are 48 bits long) are allocated centrally, blocks of addresses being apportioned to manufacturers, so that it can be guaranteed that no two devices in the world will have the same address.

In *hierarchical addressing* each address consists of a number of fields; as each field is inspected, the packet is taken nearer its destination. This can be compared to reading an address on a letter, such as:

Mr. F. Smith
394 Roman Avenue
Colchester
Essex
Angleterre

Starting from the last line and working upwards, the letter is sent first to England, then to Essex, then to Colchester, where it is sorted on the name of the road. Each sorting office needs to understand only that part of the address which is relevant to it, so that it is not necessary for a sorter in France to understand the portion of the address which is in English, nor for a sorter in Essex to be able to read French.

A similar technique is used in hierarchical network addressing. If a

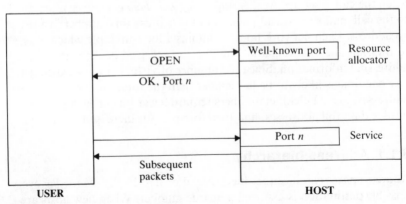

Figure 3.1 Addressing services via a well-known port

packet passes through many networks, each can use its knowledge of its part of the address in order to direct the packet. Network nodes need know nothing about the structure or coding of fields of the address that do not relate to them. Within each address field, the allocation of addresses to individual hosts is left to the body responsible for administering the associated network.

A disadvantage of hierarchical addressing is that if a host moves then it needs to have a new address allocated to it, in the same way that our addresses change when we move house. However, a significant advantage is that it is possible to relate a hierarchical address structure to the topology of the network, so that routing is simplified.

3.1.2 Naming

Network addresses are usually just strings of bits and are difficult to remember. It is therefore common practice to provide a list of names (text strings) which map onto addresses. This can be provided as a facility accessible throughout a network or locally for those services that one is most likely to need. The translation facility is often called a 'name look-up service' or *nameserver*. If a host is moved, its name may then remain the same even though its address may have changed.

Where users are relying on name translation, it is essential that the tables are kept up to date, and if several nameservers contain a common set of names, they must be kept consistent. It is also important that the address of the nameserver does not change and that the nameserver is extremely reliable. (Needham and Herbert, 1982).

In the DARPA Internet protocol suite, a system that does not know the mapping between a local and Internet (fully specified) address can determine this mapping by the address resolution protocol (Leffler *et al.*, 1983) (see also Chapter 8). This uses a broadcast request which receives a response from any system that knows the appropriate mapping. In this case the address of the responding system need not be known.

3.1.3 Addressing services within a host

Addresses are used to get to a particular host, but the host may provide several services. These are normally addressed as 'subaddresses' or *ports*. If a service can accept many remote users it is wasteful to set up too many service ports listening for incoming calls, and in any case the user may not know which ports are available. To get around this problem, the service machine can listen on one *well-known port* and allocate resources and a new port for the rest of the conversation when a connection request is made. The well-known port can then be identified by a single service name in the nameserver. This technique is illustrated in Figure 3.1.

3.2 Routing

For all but the simplest network topologies, there will be alternative paths which can be taken to route a packet from its source to its destination, and a *routing algorithm* is therefore required to decide which path should be taken. Typical goals of this algorithm include the avoidance of network congestion and the minimization of some cost function, such as packet delay. It is also important to ensure that packets do not go round in circles within the network, and that they are recognized at their correct destinations and accepted. Routing and addressing must therefore be seen as dependent networking issues.

For connectionless working, each packet from a particular host is routed independently. By contrast, a single route is usually set up for virtual circuit calls, so that in this case all packets follow the same path, with the advantage that they cannot get out of sequence. Nevertheless, the problems involved in setting up a route for a virtual circuit are similar to those for individual datagrams. There are two main approaches: either the packet can be routed down the shortest path or routing of packets can be varied to make the best use of the network facilities as a whole.

3.2.1 Static routing

Static routing is simple to implement and efficient. In this technique, each node has a table that indicates the best paths to any possible destination, in order of priority. Figure 3.2 shows an example network, with part of the *routing table* for node J. If a packet destined for node A is received at node J, then the first row of the table is used to determine the link on which the packet will be sent. The simplest approach is to use the link with the highest priority (link 1 in this case), assuming that it is available. However, a more balanced traffic distribution is obtained if the route is calculated on a statistical basis. For example, a random number between 0 and 1 is generated, and if it is less than 0.6, link 1 is used. If it is between 0.6 and 0.9, link 2 is used, otherwise link 3 is used. Static routing is straightforward to implement, as the tables change infrequently and can therefore be managed easily.

3.2.2 Centralized routing

Centralized routing is similar to static routing, except that the tables are updated periodically from a central node, the *network control centre*. Nodes inform the centre of the states of their local connections, so that the centre can then send them tables which provide a better view of the network as a whole. A disadvantage of this scheme is that it is vulnerable to failure of the network control centre. Furthermore, it leads to an excess of packets in the network near the control centre.

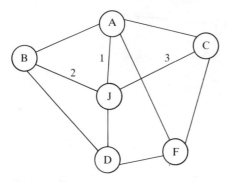

(a) An example network

To node	Via node	Priority	Via node	Priority	Via node	Priority	
A	A	0.6	B	0.3	C	0.1	
B	B	0.6	A	0.2	C	0.2	
F	D	0.5	C	0.4	A	0.1	

(b) Part of the routing table for node J

Figure 3.2   Static routing

3.2.3 Isolated routing

In this method, each node routes packets according to its own view of the network at a particular time. For example, it could forward each packet on the route that has the shortest queue, regardless of where that link goes. This technique is called the *hot potato* algorithm. Another possibility is to use *backwards learning*, where each data packet carries additional information about the source node and the number of hops travelled. However, both of these techniques are better when combined with some knowledge of preferred routes, as is provided by static routing.

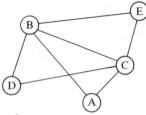

(a) An example network

To node	Delay (ms)
A	5
B	0
C	10
D	8
E	11

From B

To node	Delay (ms)
A	3
B	10
C	0
D	12
E	8

From C

To node	Via node	Delay (ms)
A	–	0
B	B	7
C	C	5
D	B	15
E	C	13

From A

Measured by A directly using echo packets

Calculated by A using information obtained from B and C

(b) Information obtained by A from B (c) Information obtained by A from C (d) Resulting estimates of delays from A

Figure 3.3 Distributed routing

3.2.4 Distributed routing

Instead of trying to obtain a view of the network conditions in an isolated manner, or from a central node, it is possible for nodes to exchange routing information with their immediate neighbours. Appropriate information would be an estimate of the delay involved in reaching a particular node. This could be obtained using special 'echo' packets, which the receiver time-stamps and sends back as fast as possible. However, as with centralized routing, it is important not to saturate the network with information packets to the detriment of data packets.

An example of distributed routing is shown in Figure 3.3. Using the delay information obtained from nodes B and C, node A can calculate new estimates of the delay to each other node in the network, for a number of alternative paths. It then enters the lowest delay estimates into its routing table, together with the first node on the corresponding path.

For example, consider the two routes from A to D via B and C, respectively. Node A measures the delay from itself to node B as 7 ms, using an echo packet. From the information supplied to it by node B, it finds that the delay from B to D is 8 ms, implying a total delay of 15 ms. Similarly, the route via C is estimated to have a total delay of 17 ms. As this is longer than the delay via B, the first route is preferred. The original ARPANET routing scheme was similar to this. A main drawback was that because a node's view of the network was limited, some packets were trapped in loops in the network. In the present scheme, each node has a view of the entire network including delays between every pair of nodes. This consistency reduces the number of looping packets.

3.2.5 Flooding

Flooding is a rather drastic approach to routing in which all packets that arrive at a switching node are forwarded on all the onward links. It results in large numbers of replicated packets, and at least two problems: how can the process be stopped and how are all the duplicate packets discarded at the destination? One solution is to limit each packet to a maximum number of hops, so that it is discarded when this count is exhausted. Alternatively, sequence numbers can be used, which, when taken together with the address of the source node, will indicate duplicates. However, flooding is only practical where there is a high probability of links or switching nodes failing, such as in military applications.

3.3 Congestion control

In circuit switched networks, congestion results in *blocking* when an attempt is made to set up a call. By contrast, congestion may be manifest in a packet switched network *after* a call (virtual circuit) has been set up, as a result of a sudden increase in the instantaneous traffic load. Due to the statistical nature of many traffic sources, it is usually difficult to anticipate such congestion at call setup time.

As the load offered to a network rises, the throughput rises at first, because the links are being used more efficiently. However, when the processing and transmission resources of the network start to become saturated, queues form and packets may be discarded. This results in retransmitted packets which add to the load, so that the throughput decreases, as shown in Figure 3.4. Important network design goals are therefore to control the onset of congestion, to maintain throughput under high loads and not to let the network get into a *deadlock* situation where no traffic can move at all.

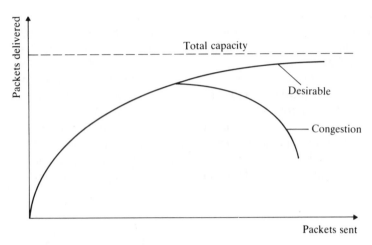

Figure 3.4 Performance degradation due to congestion

3.3.1 Deadlock

Two neighbouring nodes may get into a deadlock situation if each is allowed to use all its buffer space for queuing packets destined for the other. If the queues fill up at both nodes, neither node can accept any more packets, and a deadlock ensues (Figure 3.5(a)). A similar deadlock situation can be caused by the absence of receive buffer storage in a number of nodes which form a *cycle* within the network, as shown in Figure 3.5(b). In this case each node may not realize that it is contributing to the deadlock problem.

A third deadlock situation can occur in a node that *reassembles* subpackets before forwarding them. If at a given time a number of subpackets are required to make up complete packets, but there is no free buffer space left, then no packet can be completed and no buffer space can be released. An example of this problem is shown in Figure 3.5(c), where each packet consists of five subpackets and there are only sufficient buffers for 10 subpackets.

Avoiding deadlock situations amounts to preventing all the buffers in a node from becoming full. One solution would be to allow buffer space to be freed for high priority incoming traffic by discarding lower priority traffic already held in the node. More complex schemes are described in Tanenbaum (1988).

3.3.2 Congestion control techniques

The choice of a congestion control technique is essentially a trade-off between the amount of buffer memory in the nodes and the efficiency with which the network links are used. It is one of the major problems in designing node software.

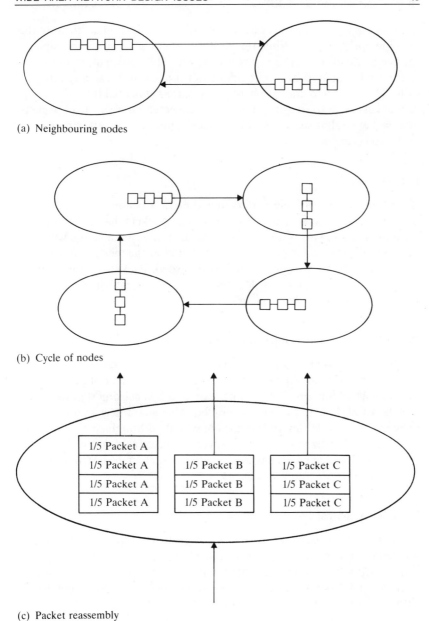

(a) Neighbouring nodes

(b) Cycle of nodes

(c) Packet reassembly

Figure 3.5   Deadlock situations

Pre-allocation of resources

In the case of connection-oriented working, it may be possible to allocate one buffer in each node to a particular virtual circuit when a call is set up. Then if each node only acknowledges a packet once it has been forwarded to the next one, there will always be a place to store the next incoming packet. While this means that congestion is impossible, the use made of network resources is potentially inefficient, as buffer space not being used by a particular call is not available to any other calls. This is a good choice where packets are segmented into subpackets. The source can indicate the expected number of subpackets in a packet so that sufficient buffer space could be reserved in advance at the receiver.

Packet discarding

If the number of available buffers in a node falls below a certain level, it may be desirable to simply discard incoming packets that are destined for long queues. This is not as drastic as it sounds, as timeouts will ensure that the packets are retransmitted, and in the meantime the packets already in the queues can be forwarded. Packets destined for shorter queues are not discarded, thereby ensuring that the total throughput of the node is maximized.

Restricting the number of packets in the subnet

If a node is allowed to transmit a packet only when it has acquired a special 'permit' from the network, then the total number of packets in the network at any one time can be restricted by ensuring that the number of permits in existence is limited. This guarantees that the network as a whole cannot become congested, although congestion can still occur in localized regions. However, a considerable difficulty involved in this scheme is ensuring the fair and efficient distribution of the permits throughout the network.

Use of flow control

Flow control (see below) can be used for ensuring that nodes are not flooded by information from their immediate neighbours. However, being essentially a point-to-point technique, it is usually insufficient for controlling network-wide congestion, and is therefore best combined with one of the other techniques described above.

Choke packets

Another means of controlling congestion requires each node to monitor the state of its output queues and to send a 'choke packet' back to the source of the traffic if the queues become too long. Upon receipt of this packet, the host is required to reduce the rate at which packets are sent to the associated destination. However, a problem with this scheme is that it may take a considerable length of time for the choke packet to

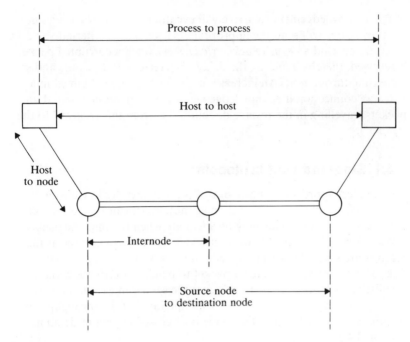

Figure 3.6   The uses of flow control in a network

travel back to the source if it is geographically remote, so that packet discarding may need to be used in the meantime.

3.4 Flow control

The fundamental purpose of flow control is to prevent a fast sender from swamping a slow receiver with packets. The sender and receiver are often network nodes at either end of a single link, but may in general be any pair of points within the network, as illustrated in Figure 3.6. It is also important that a flow control protocol does not break down in the face of misordered packets, duplicate packets, and corrupted or lost packets. Flow control is typically implemented using some form of feedback from the receiver to the sender, such as a positive *acknowledgement* which indicates that a packet of data has been received and that another may be sent. In some cases an acknowledgement may apply to more than one packet. (An example is HDLC, described in Chapter 4.)

An important application of flow control is in data link layer protocols, which control data transfer over a physical link and are discussed in Chapter 4. In that context, flow control is usually combined with *error control,* by specifying that the receiver will not acknowledge a packet (strictly, a frame) that has been corrupted in transit. The failure of the acknowledgement to arrive will eventually cause a *timeout*, and the transmitter will send the appropriate packet again. Alternatively, the receiver can send a *negative acknowledgement* when a corrupted packet is received, thereby avoiding the delay associated with timeouts and so increasing throughput. An increase in efficiency can be obtained in a two-way connection if acknowledgements are *piggy-backed* onto data packets travelling in the reverse direction (i.e. from the 'receiver' to the 'sender').

3.4.1 Stop-and-wait protocols

The simplest approach to flow control is for the sender to transmit a single packet and then wait until an acknowledgement is received before sending the next one. This may be adequate when the time the packet takes to travel between the two nodes and the turnaround time at the receiver are small. However, consider a network involving a satellite link, where the time taken for a packet to reach its destination via the satellite is typically 250 ms. In this case the sender must wait at least 500 ms between packet transmissions, implying that the throughput of the link is severely limited. This issue is discussed in greater detail in Section 3.4.3.

Another important issue is ensuring that the flow control protocol is robust to channel errors. For example, consider the case where a packet is transmitted and received correctly, but the acknowledgement is corrupted before it reaches the sender. This causes a timeout at the sender, which retransmits the packet. The receiver then thinks that the repeated packet is a new packet, and accepts it, which is clearly an error. This problem can be avoided if each packet is given a 1-bit *sequence number*, which alternates between 0 and 1 for successive non-repeated packets. If a lost acknowledgement causes the duplication of a packet, then the fact that both packets have the same sequence number will indicate to the receiver that the second one should be discarded.

3.4.2 Pipeline protocols

A means of increasing the throughput of a flow control protocol when operating over a link with a long delay is to allow *several* packets to be sent before a response to the first one is expected. This is known as *pipelining*, because it is like filling up a pipe between the sender and the receiver with information (Figure 3.7). However, in this case the fact that a number of packets may be unacknowledged at any one time means that a simple 1-bit sequence number is not sufficient to identify duplicate packets, so that the range of possible sequence numbers must be extended.

At the sender, a subset of the range of sequence numbers is contained

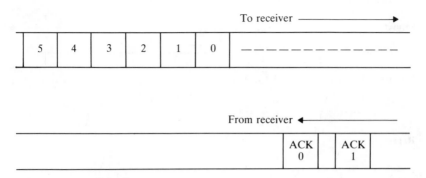

Figure 3.7 Pipelining

in a transmit *window*, and these represent the packets for which acknowledgements have not yet been received. The window *slides* along as sequence numbers are acknowledged, enabling more packets to be sent. The size of the transmit window is determined by the number of packets that may be 'in the pipeline' at any one time, which is in turn dependent on the length of the link. The receiver also has a window, which specifies the sequence numbers of packets that the receiver is prepared to accept and buffer.

The 'go-back-N' and 'selective repeat' protocols make use of this *sliding window* principle. Representative examples of these protocols which use positive acknowledgements and timeouts are given below, although it should be noted that a number of variants of each type exist; in particular, protocols may rely on a combination of positive and negative acknowledgements.

Go-back-N

The go-back-N protocol uses a receive window size of 1. If an error is detected in a packet, all subsequent packets are discarded by the receiver, until a new version of the packet is received. When the sender realizes that it is not going to receive an acknowledgement for the corrupted packet, it sends it again, *as well as* all the succeeding unacknowledged packets, as shown in Figure 3.8. While this approach has the advantage that only a single packet needs to be buffered at the receiver at any one time, it has the disadvantage that some packets are retransmitted unnecessarily, thus wasting bandwidth.

For a given transmit window size w, it is necessary to find the smallest range of sequence numbers 0,1,. . .,maxseq that can be used, so that the size of the field required to contain the sequence number can be minimized. In order to do this, consider the following example, where $w = 4$ and maxseq $= 3$. The transmitter sends packets with sequence numbers 0,1,2,3 and gets an acknowledgement for number 3, which in this case also acknowledges numbers 0,1 and 2. The transmitter then

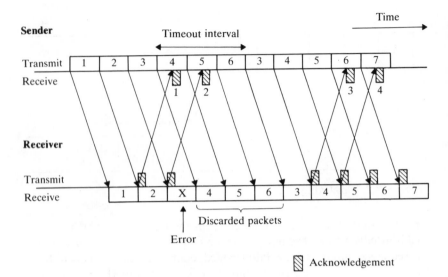

Figure 3.8 Go-back-*N* protocol

sends new packets with sequence numbers 0,1,2,3, and again receives an acknowledgement for number 3. However, it cannot know for sure whether the second acknowledgement relates to the first or second batch of packets, as the second batch of packets may all have been lost, and the second acknowledgement may be a duplicate of the first. The range of sequence numbers must therefore be *one greater* than the window size, in order to avoid this ambiguity (i.e. maxseq = *w* if the range of sequence numbers starts at 0).

Selective repeat

The distinguishing feature of the selective repeat protocol is that its receive window is *larger* than 1, and may even be as large as the transmit window. This means that an incoming packet which has any sequence number within the receive window is accepted, even if it is out of order. If a corrupted packet is received, it is discarded, but this does not have any effect on the acceptance of subsequent packets. When the transmitter realizes that something is wrong, it simply retransmits the appropriate packet, as shown in Figure 3.9. This technique makes efficient use of the channel bandwidth, but requires more buffering in the receiver than the go-back-*N* scheme.

Once again, the minimum range of sequence numbers required can be illustrated by means of an example, with *w* = 4 and maxseq = 3. The transmitter sends packets with sequence numbers 0,1,2,3. These are received correctly and acknowledged by the receiver, which then advances its window to expect new packets with sequence numbers 0,1,2,3. However, the acknowledgements are corrupted before they get

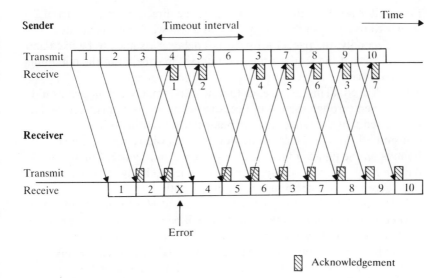

Figure 3.9 Selective repeat protocol

back to the transmitter. The transmitter therefore sends the first batch of packets again, and because they have sequence numbers that fall inside the receive window, they are once again accepted at the receiver. This is clearly incorrect, as the same batch of packets has been accepted twice. The solution to the problem is to ensure that when the receive window is advanced, there is no overlap with the previous receive window. This requires that the range of sequence numbers must be at least twice the window size, i.e. maxseq = $2w - 1$.

3.4.3 Protocol performance

In the absence of a flow control protocol, the sender transmits packets as fast as it can, regardless of what is happening at the receiver. However, if one of the protocols described above is introduced, then the packet rate is *reduced* when necessary and corrupted packets are *retransmitted*, implying that the rate at which data is transferred over the link will be reduced to some extent. The *link utilization*, which is the ratio of the *useful* data rate to the link capacity, is thus a reasonable measure of the performance of a flow control protocol. (Another possible performance measure is the mean packet delay introduced by the protocol.)

Link utilization

For the purposes of the discussion below, a number of simplifying assumptions are made about the flow control protocol and the link.

Firstly, it is assumed that data travels in one direction only, i.e. from sender to receiver, so that piggy-backed acknowledgements are not used. Secondly, the packet propagation delay is assumed to be fixed and known, so that the sender's timeout interval may be set to the minimum time required for a packet to travel to the receiver and be acknowledged. (This also means that no performance advantage would be obtained through the use of negative acknowledgements.) Thirdly, it is assumed that the sender always has a packet waiting to be transmitted, so that the calculated utilizations will represent upper bounds on the figures that would be obtained in practice. Finally, an infinite range of sequence numbers is assumed to be available, implying that there are no restrictions on window length.

If D is the length of the packet data field and H is the length of the packet header, then the link utilization will be proportional to the factor $D/(H+D)$ for all flow control protocols. Furthermore, if the probability of corruption of a packet is denoted by P, then the link utilization will be proportional to the probability of successful transmission $(1-P)$, assuming that acknowledgements are sufficiently short as to suffer a negligible probability of corruption. For many protocols, the link utilization will also be dependent on the delay before acknowledgements are received, and in this connection it is useful to define another parameter, A, as follows:

$$A = \frac{T_1 + 2T_P + T_{PROC} + T_S}{T_1} \tag{3.1}$$

where T_I is the time to transmit a packet;

T_p is the propagation delay;

T_{PROC} is the processing time required by the receiver to generate an acknowledgement for an incoming packet;

T_S is the time to transmit the acknowledgement.

The numerator in equation (3.1) is equal to the time between the transmission of a packet and the receipt of the acknowledgement and can be seen in Figure 3.10. The denominator is simply the time to

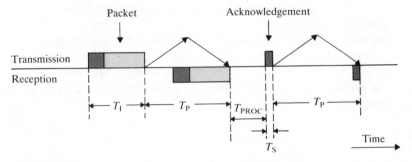

Figure 3.10 Time taken to transmit a packet using the stop-and-wait protocol

transmit the packet. Ideally, the propagation delay and 'turnaround' time at the receiver would tend to zero, so that A would tend to unity. In practice, however, A may be significantly greater than unity, and it is important to consider the effect of this on the throughput of the flow control protocol.

It can be shown that an upper bound on the link utilization for the stop-and-wait protocol is given by Schwartz (1987):

$$U = \frac{D}{H + D} (1 - P) \frac{1}{A} \tag{3.2}$$

The utilization is thus inversely proportional to A in this case, implying that the throughput of the protocol decreases as the link gets longer. This makes the stop-and-wait protocol unsuitable for very long links, e.g. satellite links, as mentioned previously.

A similar expression can be found for the go-back-N protocol (Schwartz, 1987):

$$U = \frac{D}{H + D} (1 - P) \frac{1}{1 + (A - 1) P} \tag{3.3}$$

In this case the more complex denominator in the third term reflects the fact that more than one packet is retransmitted when an error occurs, and these packets may themselves be corrupted. However, it is clear that the utilization is less sensitive to A than in the previous case, and hence to the distance between the sender and receiver.

For the selective repeat protocol (Schwartz, 1987):

$$U = \frac{D}{H + D} (1 - P) \tag{3.4}$$

In this case the link utilization is *independent* of the parameter A, and hence of the length of the link, which is clearly desirable. Furthermore, the sensitivity to link errors is lower than for the go-back-N scheme. It can therefore be concluded that the selective repeat flow control protocol will give the best link utilization over a wide range of parameters, although at the expense of extra buffering in the receiver.

Optimum packet length

As the ratio of the data field length D to the header length H for a packet is increased, the factor $D/(H + D)$ in equations (3.2) to (3.4) tends to unity, seeming to imply that the utilization increases. However, increasing D also increases the probability that a packet will be corrupted, in which case it will have to be retransmitted (as well as a number of other packets in the case of the go-back-N protocol).

Figures 3.11 and 3.12 are plotted using equations (3.2), (3.3) and (3.4). In Figure 3.11 the utilization is plotted against the total packet size for a 64 kbit/s satellite link with a one-way propagation delay of 250 ms. A bit error rate of 1 in 10^5 is assumed. Other relevant

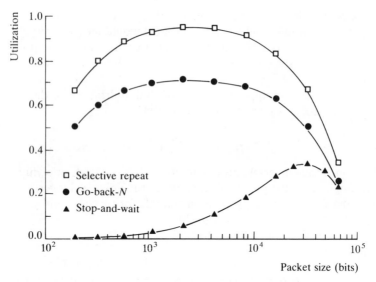

Figure 3.11 Link utilization with a one-way propagation delay of 250 ms

parameters are the packet processing time, which is 10 ms, the header length, which is 64 bits, and the length of acknowledgements, which is also 64 bits. The optimum packet size (i.e. header and data) is about 2560 bits and is similar for both the selective repeat and the go-back-N protocols. This graph clearly illustrates the advantage of selective repeat when the delay is long.

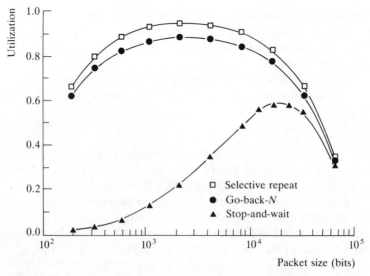

Figure 3.12 Link utilization with a one-way propagation delay of 50 ms

In Figure 3.12 a one-way propagation delay of 50 ms is assumed, otherwise the parameter values are those used in Figure 3.11. Optimum packet sizes are similar to the previous case for selective repeat and go-back-N protocols. In the case of the stop-and-wait protocol, the figure illustrates that packet size selection is even more critical, and a longer packet size is needed for optimum link utilization to compensate for the time spent waiting.

3.5 Network topology

The advantages and disadvantages of various network topologies were mentioned in Chapter 1. However, when designing a large network it is not always clear where nodes should be located and what capacity the individual links should have to achieve the best trade-off between cost and performance. Some theoretical approaches to this design problem are discussed briefly here. However, it is important to note that in practice the growth of many networks is influenced more by the constraints of existing equipment, lines and cable ducts, than by theoretical planning.

3.5.1 Backbone design

The first problem is to decide where to site the nodes of the network. A recent review paper on communication network optimization (Grout and Sanders, 1988) indicates that there are many approaches to this problem, several of which start with large numbers of nodes and apply some reduction algorithm.

Assuming the number of nodes in the network is n, then there are $n(n\text{-}1)/2$ possible connections between nodes, and since each connection may or may not be present, the number of distinct topologies is therefore $2^{n(n-1)/2}$. It is thus clearly not feasible to try out each topology when the network has more than about five nodes. The approach often taken in practice is to choose a likely configuration and then calculate whether it performs within a desired set of constraints. If so, then this design can be used as it stands, or alternatively it can be refined by making some small modification to reach a better design. The process of refinement may be repeated as many times as necessary, or until the available computing time has expired.

Typical design constraints that might be considered are:

1 The network is cheap to install.

2 Packets will be subject to minimal delays.

3 The network offers good throughput.

4 Running costs are low.

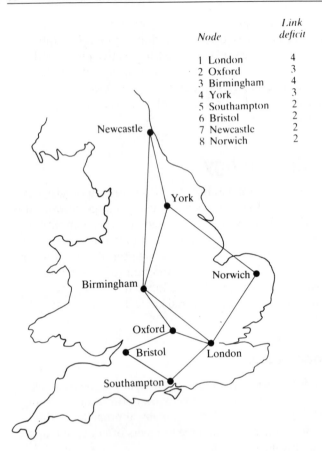

Node	Link deficit
1 London	4
2 Oxford	3
3 Birmingham	4
4 York	3
5 Southampton	2
6 Bristol	2
7 Newcastle	2
8 Norwich	2

Figure 3.13 Choosing an initial topology

5 The network is reliable in that it provides a sufficient number of alternative paths.

In practice it will be necessary to hold some factors constant while designing for others.

The main steps in the design procedure are:

1 Choose an initial topology.

2 Assign expected traffic flows and capacities, and see if the design meets the constraints.

3 Generate a slightly modified network and compare it with the previous best solution.

Choosing an initial topology

A suitable technique due to Steiglitz *et al.* (1969) is to choose a starting topology by considering the degree of connectivity required and the 'distance' between nodes, which may actually be geographical distance, cost, or some other metric. It is obviously desirable that the initial topology does not have too many links and that they are not too long. One way of generating a good initial topology is to use the method described below. This makes use of a variable known as the *link deficit* which is an estimate, at each iteration of the procedure, of the further links required at each node.

1   Number the nodes at random.

2   For each node, set the link deficit to the number of links which it is expected that the node will require.

3   Find the node with the highest deficit; in the case of a tie, choose the lowest-numbered node.

4   Link the node found in step 3 with another node with the highest deficit, provided that a connection does not already exist between these two nodes. If there are a number of candidate nodes, choose the 'nearest' one.

5   Repeat steps 2 to 4 until the link deficit at each node is zero (or less than zero).

As an example, consider the eight nodes in Figure 3.13, numbered from 1 to 8 as shown. An initial link deficit is assigned to each node which is also shown in the figures. Here, the nodes expected to carry more traffic have been given higher initial link deficits. The links might be inserted as follows:

First round:	(London, Birmingham)
Second round:	(London, Oxford), (Birmingham, York)
Third round:	(London, Southampton), (Oxford, Bristol),
	(Birmingham, Newcastle), (York, Norwich)
Fourth round:	(London, Norwich), (Oxford, Birmingham), (York, Newcastle),
	(Southampton, Bristol)

After the fourth round all nodes have link deficits of zero and the connections are as shown in Figure 3.13.

Assigning flow and capacity

The load borne by each link in a network will depend on the end-to-end offered load and on the routing algorithm used. Dynamic routing algorithms are usually very difficult to model, so a static algorithm based on the shortest path between two points is typically assumed. This might not correspond exactly with the real situation, but is a reasonable approximation where traffic is uniformly distributed throughout the network. If we are optimizing cost, it is then possible to

Figure 3.14 Topology modified by slight perturbation

calculate the total cost of the network as the sum of the usage-related and fixed costs for all the links, subject to given constraints such as packet delay. An analytical approach to this problem is described in Tanenbaum (1981). It is also common to apply computer simulation techniques in this area, and a number of suitable commercial packages are available. Optimization of channel capacities is discussed in detail in Seidler's (1983) book on network design.

Generating a modified network

In order to check whether the network design can be improved further, it is necessary to alter the existing one slightly (i.e. perturb it). One possibility is to remove or add a single link. Another is to choose two links, remove them and then add two new links, using another combination of the four nodes affected. For example, in Figure 3.13 the links (York, Birmingham) and (London, Norwich) could be replaced by (York, London) and (Birmingham, Norwich), to give the new topology shown in Figure 3.14.

Other perturbations would be to replicate links on heavily used routes or to increase their bandwidth, to add an extra node or to add a microwave or satellite link.

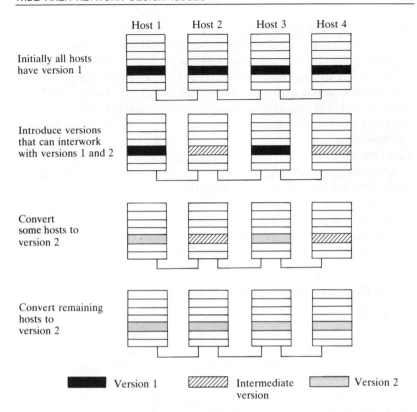

Figure 3.15   Network change control

3.6 Network maintenance and evolution

3.6.1 Changes to the network

There are a number of problems associated with making changes to an existing network. If physical links are added or moved, then name tables and routing tables will have to be changed and kept up to date. If a change is made to a protocol, perhaps to bring it into line with standards, it is essential that the change takes place in a phased manner, such that all systems can still communicate while some systems have been changed and some have not. Protocol layering helps in this respect, but a change to just one layer can be quite complicated, and software should be designed in a manner such that changes can be made easily and safely. Figure 3.15 illustrates a simple example of change control, where one layer of protocol is to be changed in four interconnected systems. The change is achieved by introducing an intermediate version at some locations which can work with earlier and later versions.

The recent agreements on standards for networking or open systems interconnection (OSI) (discussed in Chapters 4 and 5) have resulted in a migration problem for existing network providers. In the United Kingdom, the Joint Academic Network has extensive plans for migration from the existing suite of protocols to OSI standards. A key aspect of this strategy is to provide converters which will act as translators enabling existing and new versions to communicate with each other. Users requiring access to a service consult a central data base called the name registration scheme (NRS). The data base knows the version of both the user's and service's software and the location of the converters; thus the NRS itself can route the user's call to the appropriate converter. Full details of the transition plan can be found in Joint Network Team (1987). The migration plans for the European academic research network are described in Bryant (1987). The problems of migration from the widely used TCP/IP suite of protocols used in the DARPA Internet in the United States (see Chapter 8) to the OSI standards are discussed in Rose (1990).

3.6.2 Performance monitoring

An important requirement for networks is that statistics are kept so that overload problems can be detected and understood, and better planning can ensue. Normally, each node collects statistics about the data passing through it, and this can then be sent through the network to a collection centre where analysis can be performed. Suitable statistics might be:

1   Traffic levels in and out of a node.

2   Average and maximum sizes of queues.

3   Use of buffer space in the node.

4   Availability of links, including scheduled and unscheduled breaks.

A system to which such information is sent can act as a *network diagnostic centre* if necessary, so that not only does it record statistics but it also instructs nodes on any actions they should take to avoid temporary problems elsewhere in the network. In a network that is required to be highly reliable, it may be necessary to install two diagnostic centres situated far apart, in order to provide cover if one of them fails.

Performance monitoring is just one of several aspects of network management, which is the subject of Chapter 13.

3.7 Special network facilities

The inherent intelligence of packet switched networks enables special features to be offered to users, protocol implementers and network installers.

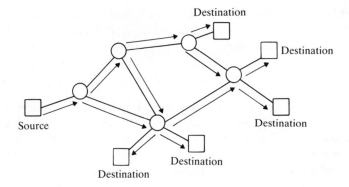

(a) Broadcasting facility

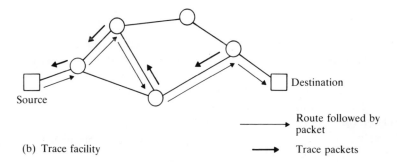

(b) Trace facility

Route followed by packet

Trace packets

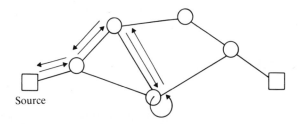

(c) Echo facility

Figure 3.16   Special network facilities

3.7.1 User facilities

Delivery confirmation

When delivery confirmation is supplied, an indication is sent to an originator when a particular packet has been passed all the way through the network to its destination. This indication can originate at the final node in the network which sends the delivery confirmation as soon as the packet has been delivered to the destination host's interface, or it can be a facility provided within the destination host, separate from normal flow control procedures.

Closed user group

An authority providing a private network can restrict access to those people with permission to use it. A similar facility can be provided on a public packet switched network. The network nodes are informed which host systems belong to a group. At each host in the group, packets may be sent to and received from other members of that group, but may not be sent to or received from other hosts on the network. The network provider must prevent access from hosts outside the group to those within it, if necessary using cryptographic means, as discussed in detail in Chapters 10 and 11.

Broadcasting

A packet that is broadcast is sent to every host attached to the network (see Figure 3.16(a)). This facility is not commonly found on wide area networks, but on a LAN of simple topology such as a bus or a ring it is easy to provide. It is also an inherent property of packet satellite and packet radio networks, where a single instance of the packet need be sent. Even on these networks the facility may be made available exclusively to privileged users, as receiving and discarding broadcast packets can represent a high and unnecessary burden for the network hosts. A better alternative is *multicasting*, in which the same packet is delivered to a group of users; provided the group has been appropriately defined, the burden of reception falls to the hosts that need to receive the packet and so is not excessive.

3.7.2 Network test facilities

Trace packets

For this facility, packets that need to be traced are marked in a way that can be interpreted by the nodes as the packet travels through the network. Each node sends its identity back to the source host in a trace packet (see Figure 3.16(b)), so that the originator can determine the path followed by the packet to a particular destination. This facility is unlikely to be made available to users because of the potentially large amount of traffic produced.

Echo

This is a useful facility when testing or updating protocol implementations. The echo facility can either be provided within a network node or by a remote host. All packets sent to the echo facility will be returned, the source and destination addresses having been swapped, and will have been subject to the normal handling and protocol procedures of the network (see Figure 3.16(c)). Thus a considerable amount of debugging can be carried out from a single site, without the need for time to be scheduled on some other system or with some other person.

Drop

This is another facility that may be required when debugging protocols. The packets are sent through the network, but are then discarded at a special remote destination, which could be either a host or a network node. Thus the ability to transmit a packet can be tested.

3.8 Conclusion

This chapter has shown solutions to some of the major problems involved in constructing, running and maintaining computer communication networks; it has also illustrated the complexities of network provision. Examples of many of the techniques discussed here will be found in the following chapters. Network users may not be aware of many of the problems discussed in this chapter, the details of which are often hidden from them. As we shall see, some standards define the interface between a host and the nearest network exchange and are not concerned with how the network is organized internally.

References

Bryant, P. (1987) 'The migration of EARN to use ISO protocols', *Computer Networks and ISDN Systems*, **13**(3), 201–203.

Grout, V. M. and P. W. Sanders (1988) 'Communication network optimization', *Computer Commun.*, **11** (5), October, 281–287.

Joint Network Team (1987) 'Transition to OSI standards', Final report of the Academic Community OSI Transition Group, the Joint Network Team, Rutherford Appleton Laboratory, Chilton, Didcot, UK.

Leffler, S. J., W. N. Joy and R. S. Fabry (1983) '4.2 BSD networking implementation notes', Computer Systems Research Group, University of California, Berkeley, July 1983.

Needham, R. M. and A. J. Herbert (1982) *The Cambridge distributed computing system*, Addison-Wesley, Wokingham, UK.

Rose, M. T. (1990) 'Transition and coexistence strategies for TCP/IP to OSI', *IEEE J. Selected Areas in Commun.*, **8** (1), January, 57–66.

Schwartz, M. (1987) *Telecommunication Networks*, Addison-Wesley, Reading, Mass.

Seidler, J. (1983) *Principles of Computer Communication Network Design*, Ellis Horwood Ltd, Chichester, UK.

Steiglitz, K., P. Weiner and D. J. Kleitman (1969) 'The design of minimum-cost survivable networks', *IEEE Trans. Circuit Theory*, **CT-16**, November, 455–460.

Tanenbaum, A. S. (1981) *Computer Networks*, Prentice-Hall, Englewood Cliffs, New Jersey.

Tanenbaum, A. S. (1988) *Computer Networks*, 2nd edition, Prentice-Hall, Englewood Cliffs, New Jersey.

4 Open systems interconnection: concepts and the lowest four layers

GILL WATERS

4.1 The need for standards

If all systems connected to a network are of a single type, it is possible to specify a set of private protocols for the network. Such a system is *closed* because it allows communication only within a closed group of users. Where there is a requirement for communication with systems outside the closed group, protocol conversion is necessary. Networks that offer support for communication between many types of system without protocol conversion are *open* because they allow open access to many other facilities. A system connected to such a network is called an *open system.*

Much of the motivation for open interconnection has come from computer users, because they did not wish to be restricted to a particular manufacturer's products for the whole of their computing resources. Manufacturers of intermediate size were also keen to see standards introduced as this would make their products acceptable to a wider range of customers. The complexities inherent in communication protocols made the task of matching up different protocol architectures (suites of protocols) a time consuming and error-prone task which was found to be uneconomic. When networks first became a possibility, it was not clear which were the best protocols to use. Users took the easiest option which was to use an appropriate conversion package to satisfy their short-term needs.

As the need for greater interworking between many machines of different manufacture became apparent, a programme of protocol standardization got underway. Manufacturers and suppliers were under pressure to produce software aligned to the standards because of users' procurement policies, and it is rapidly becoming necessary for manufacturers to follow standards policy in order to sell their products to a community eager for the benefits of open interconnection.

An additional advantage of standards is that they can form the basis for a public service. A public network that adheres to the standards can be accessed by any system that also adheres to the standards.

In this chapter and the next, we discuss a layered model for the connection of open systems—the *open systems interconnection* (*OSI*) model.

4.1.1 Standards bodies

Standards work has been carried out in many countries and by various standards bodies. The people involved in standards work have often already gained experience of the problems of designing and using protocols in prototype or early production networks. It is to their credit that a full set of quite complex standards is now available.

Two study committees of the International Standards Organization (ISO) were responsible for coordinating work on the various layers of the OSI models. The International Electrotechnical Commission (IEC) and ISO are now combining their activities in the area of data processing standards. The resultant joint technical committees working on OSI are ISO/IEC JTC1/SC6 for the lower layers discussed in this chapter and ISO/IEC JTC1/SC21 for the OSI model, the higher layers discussed in Chapter 5 and other related work. ISO receives input from national members such as the British Standards Institution (BSI) and the American National Standards Institution (ANSI).

The International Consultative Committee on Telegraphy and Telephony (CCITT) has been in close liaison with ISO and several of the standards are published as CCITT recommendations. Comments have also been received from organizations representing manufacturers such as the European Computer Manufacturers' Association (ECMA). Also, the Institute of Electrical and Electronic Engineers (IEEE) has been highly active in the development of LAN standards.

4.1.2 Reference material

Standards document references published by ISO and CCITT are listed at the end of this chapter. A number of books describe the lower layers of the OSI model including Tanenbaum (1988), Halsall (1988), and Marsden (1986). Additionally, papers have been published to explain the individual standards and these are referenced at the beginning of the appropriate section. Two issues of IEEE journals are particularly useful (*IEEE Proceedings*, 1983; *IEEE Transactions on Communications*, 1980).

4.2 OSI concepts

(ISO 7498; CCITT X.200; Day and Zimmermann, 1983; Linington, 1983)

OSI stands for open systems interconnection. A particularly important specification is the OSI *reference model* which is an abstract framework that can then be used for further specifications. Individual specifications deal with the procedures to be carried out in order to enable systems to interwork with other systems. We can regard a *system* as a computer or a set of computers with associated peripherals and an *open system* as one that can interwork with other such systems using OSI standards.

This chapter and Chapter 5 provide an informal introduction to the standards. (Standards documents are notoriously difficult to read.) They should not be taken as definitive; when writing software or making decisions about whether designs meet the specification the appropriate standards documents should always be consulted. Each standard consists of several aspects each of which must be carefully defined. Hence a standard will normally contain time sequence diagrams, state diagrams, packet or frame formats, definitions of terms and procedural rules. In the examples that follow I have used some, but not all, descriptive techniques for each layer and have also used informal textual descriptions, so that the reader can become familiar with the various methods of description encountered in a standards document, while maintaining an overview of the subject.

4.2.1. Layering

As discussed in Chapter 1, the many problems of protocol definitions become manageable by dividing the protocol definitions into layers. OSI has a precise terminology for how a layer fits into the model and communicates with other layers. The relationship between adjacent layers and with the peer layer (equivalent layer at the same level) at the other side of the network can be seen in Figure 4.1.

Note that the relationship is logical not physical. Here layer $(N-1)$ offers a *service* to layer N. Layer N (also called the *N-layer*) offers an enhanced service to layer $(N+1)$; this enhanced service is called the *N-service*. The *N*-layer *protocol* provides the enhancement; the protocol is interpreted and used by the corresponding layer at the other side of the network. The ways in which adjacent layers communicate determine an *interface* between the layers.

The service provided by the lowest layer is achieved by physical transmission across the network. If we consider the logical progression up the layers within each open system, each layer acts as a provider of extra facilities to the layer immediately above it.

The *N*-service is accessed at the *N-service access point* (*NSAP*). The

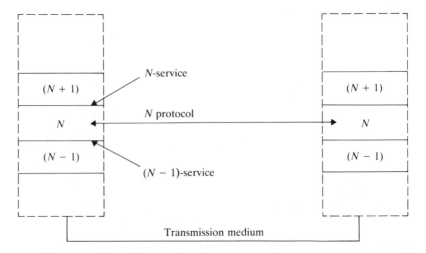

Figure 4.1 Relationship between layers in the OSI model

NSAP is given an address called the *N-address*. The access interface can be used by several $(N+1)$ users simultaneously.

Note that the standards are concerned with the external behaviour of the elements in the model; they are not concerned with their internal structure; that is the standards say what services should be provided, what protocol formats are valid and define terms to enable this information to be interpreted logically, but do not dictate how this will be implemented on any particular system.

4.2.2. Connection-oriented and connectionless approaches

In the first OSI reference model definitions, the *N*-service of each layer was connection oriented; that is communication between a pair of NSAPs was on a logical connection which is set up at the start of communication and cleared afterwards (a virtual circuit).

Connectionless options have now been added. In the connectionless network service, each item of data is sent between NSAPs independently of other items of data (in the form of datagrams) (Chapin, 1983). The protocol to provide these facilities (ISO 8473) is based on those of the Internet protocol (IP) of the ARPANET which is described in Chapter 8. The advantage of connectionless working is its simplicity and it is particularly useful where transmission errors can be tolerated or where communication between the two parties is very brief (e.g. a data base enquiry). It also makes interconnection of LANs and WANs simpler, as we discuss in Chapter 9.

Standards have now been introduced not only for a connectionless network service but also for connectionless transport, session and

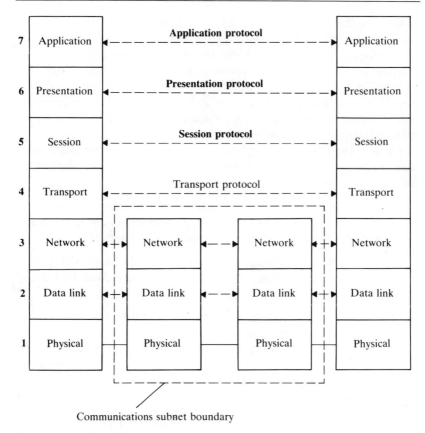

Communications subnet boundary

Figure 4.2 The OSI seven-layer reference model (after CCITT X.200)

presentation services. These are published as addenda to the corresponding connection-oriented service documents. In this chapter and in Chapter 5 we concentrate on the connection oriented services for the following reasons: they were the first to be standardized and most standard applications (user facilities) are based on them, they offer reliability for long duration applications such as file transfer and they reduce the addressing overhead once a connection is set up. Also, once an understanding of the connection-oriented approach is gained, it is a simple matter to understand the connectionless approach.

4.2.3 The seven-layer model

The seven-layer reference model is illustrated in Figure 4.2. The lower three layers are network dependent and the protocol correspondence at these layers is with the same layer in the nearest system in the network. It is possible for intermediate systems to act as *relays* which forward data received from one system to another system; the highest layer at which this can be done is layer 3.

Why are there *seven* layers in the model? Layers must be chosen so that changes can be made to one layer without affecting the others. The actual division into seven layers has resulted from experience of using existing networks. Important criteria included the grouping of logically related functions, the separation of distinct logical functions and the possibility of a layer providing a useful function in the future. Seven was seen as a number that satisfied these requirements and was not too many to describe and implement.

The purpose of the seven layers of the model is summarized here and we shall be looking into each layer in more detail, starting at the bottom and working up:

1 The *physical layer* defines the electrical and mechanical control required to transmit data bits onto the communications medium and procedures to make and release the physical connection.

2 The *data link layer* controls data transfer over the physical link. It includes frame delimiting, error detection and (optionally) error correction.

3 The *network layer* provides the means to establish, maintain and terminate connections, and offers a service that is independent of the workings of the underlying network such as routing. Note that the way in which the lowest three layers are implemented will depend on the type of network and will differ considerably, for example, between wide and local area networks.

4 The *transport layer* provides a service of a given quality to the session layer. Thus the service offered by the transport layer must respond to session layer requirements by making appropriate use of network service facilities. Any protocol defined at this layer has end to end significance.

5 The *session layer* controls the period over which data is exchanged. It includes facilities for delimiting parts of the conversation and synchronizing the dialogue between both ends of the conversation.

6 The *presentation layer* hides from the application the differences in representation of information, for example by providing character code conversion. The presentation layer is not concerned with the meaning of the information.

7 The *application layer* provides the user with the required access to the OSI environment. This may either be supplied in a user program or may be a facility provided by the system. Examples of applications are file transfer, messaging and virtual terminal (remote login) facilities.

We can now give an example of the terminology introduced at the beginning of this section. The *network layer* offers the *network service* to

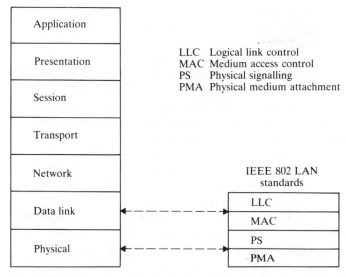

Figure 4.3 Relationship between LAN standards and the OSI model

the *transport layer*. This is enhanced by the *transport protocol* which is interpreted and used by the corresponding transport layer at the other side of the network. The enhanced service is offered as the *transport service* to the *session layer*.

4.2.4 Standardization for local area networks

The OSI model is applicable to all networks; it does not dictate what sort of network the higher layers must use. Local area networks have different properties to WANs; they are faster and largely error free. Standards for local area networks have developed from work done in the IEEE 802 committee, now published by the IEEE. The standards relate to the lower two layers of the OSI model as shown in Figure 4.3. Protocols up to and including the medium access control are specific to the LAN technology used. The logical link control (LLC) sublayer is the same for all LAN technologies. Because LANs including LAN standards are discussed at length in Chapters 7 and 8, in this chapter we shall be concentrating on the OSI model as applied to wide area packet switched networks.

4.2.5 Integrated services digital network (ISDN)

In order to meet some of the needs of a wide range of new services (voice and non-voice) and to provide a common network on which to carry them, the PTTs have developed recommendations for an integrated services digital network (ISDN). To minimize the amount of effort expended in developing a variety of new terminal equipment, users of ISDN have access to a limited set of standard customer interfaces. Services that can be carried by ISDN include telephony, circuit switched data, private circuit data, facsimile, teletex photovideotex, slow scan television and access to packet switched networks. It is basically a bearer network with the objective of carrying digital traffic truly transparently at any of a limited number of access speeds which are multiples of 64 kbit/s. Control and signalling are provided on completely separate channels. ISDN is specified in the CCITT I-series recommendations; it is now coming into service in a number of countries, and is likely to have an important impact on all communication systems including computer communication networks. However it is accepted that there are many issues requiring further work and there will be a long transition period. In this book we are concentrating on packet networks; readers interested in the basically circuit switched ISDN may wish to refer to Ronayne (1987).

The ISDN recommendations use the OSI model but, because user information and signalling flow through separate channels in ISDN, the model is extended to enable the seven layers to be applied separately both to control and to user information.

4.2.6 Status of open systems interconnection

Pouzin (1986) comments on the progress and outstanding issues in OSI. The models, services and protocols are mostly agreed now, and implementation is starting, with some manufacturers already offering conformance at a number of layers. It is important to note that the abstractions and the reference model can be applied to a wider range of networks than those discussed in detail in this book, for example ISDN.

When considering implementing standards, manufacturers and network providers are faced with a large selection of options. Interworking can take place only if selected options are compatible, and a number of bodies both nationally and internationally (for example the Commission of European Communities—Information Technology and Telecommunications Task Force) have taken on the job of promoting standards, recommending selected options and providing conformance testing.

MAP and TOP

The standardization of protocols is being given added momentum by two important groups of users in the United States. Each recommends the specification of certain standards for equipment procurement. The manufacturers' automation protocol (MAP) project is aimed at the purchase of networking equipment suitable for manufacturing; this has been led by General Motors. A sister project is the technical and office protocol (TOP) which is led by Boeing and specifies protocols for use in office systems. Both MAP and TOP use the connectionless network service beneath a connection oriented transport service (class 4—see Section 4.8).

4.2.7 Outstanding problems for open systems interconnection and continuing standards work

There are a number of problems that are still being resolved. The first is network management. Network systems must be evaluated to see if designs meet demand or need to be expanded. Each layer will need some form of local management to ensure that charges can be made appropriately and statistics can be collected. Network management is covered in more detail by Alwyn Langsford in Chapter 13.

The second main problem is in interworking between connection-oriented and connectionless protocols. Suggested solutions exist, but have not yet been agreed. There are also problems with addressing which have not yet been fully agreed. Another problem that needs to be tackled is refinement of the protocols with a view to greater efficiency. Work is also proceeding on the way in which packet switching protocols can be used in ISDN. (For more information refer to CCITT recommendation X.31.)

Standards will be needed to allow for novel physical networks or new applications. There will also be new types of protocols and new styles of working (for example for satellite transmissions, the use of broadcast or multicast at the application level, or solutions for distributed computing). Current research into ways of accommodating new services within networks, such as real time voice and image traffic, is another area that will require standardization when techniques are sufficiently robust.

4.3. CCITT recommendation X.25

(CCITT X.25; Deasington, 1988; Moudiotis 1986)

The OSI model does not dictate which protocols are to be used to provide the network service. Currently, most public and many private packet switched networks use the CCITT X.25 recommendation for call setup, operation and termination. Since X.25 was devised before the OSI model, it does not provide the full network service laid down in the OSI model. However, the majority of required facilities are available. ISO 8878 and CCITT X.223 describe the use of X.25 to provide a connection-oriented network service. X.25 has three levels which broadly correspond to layers 1, 2 and 3 of the OSI model. X.25 level 1 is the physical layer; X.25 level 2 is the data link layer and the unit of transmission is a *frame*; X.25 level 3 is part of the network layer and the unit of transmission is a *packet*.

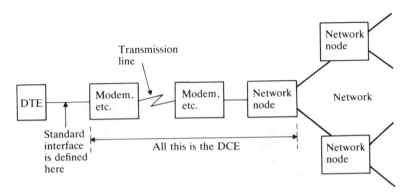

Figure 4.4 Relationship between DTE and DCE

Because of the importance of X.25 throughout the networking community we shall be using it as an example of the lower three layers of the OSI model for wide area networks.

All layers of X.25 specify the interface between a data terminal equipment (DTE) and a data circuit-terminating equipment (DCE). The DCE is the access point into a packet switched network. The DTE is the host wishing to communicate through the network. The relationship between DTEs and DCEs is illustrated in Figure 4.4.

4.4 The physical layer

(CCITT X.25; McClelland, 1983; Bertine, 1980)

The physical layer is concerned with the electrical and mechanical characteristics of the link. Our example protocol, X.25, specifies two different protocols at the physical layer—one for digital circuits and one for analogue circuits. These are described in CCITT recommendations X.21 and X.21bis respectively. X.21bis is equivalent to a subset of CCITT recommendation V.24 which has been in use for point to point analogue circuits for many years. A brief description of X.21 follows.

The X.21 digital interface defines the names and functions of eight interchange circuits (signal lines). (The physical connector has 15 pins but these are not all used.) The interchange circuits are illustrated in Figure 4.5. The T and C lines are used to transmit data and control information from the DTE. The R and I lines are used to transmit data and control information from the DCE. The S (signal) line provides bit timing. The B (byte timing) line is optional and can be used to group sequences of 8 bits into bytes. Whether or not byte timing is provided, all sequences must be preceded by two SYN (synchronization) characters. Ga is a reference ground and G is a protective ground.

X.21 provides for either point to point or circuit switched operation. We shall first discuss point-to-point operation suitable for leased lines,

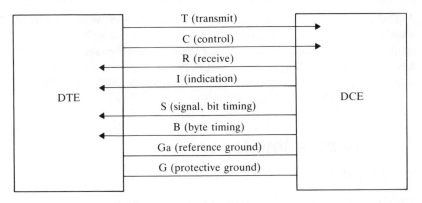

Figure 4.5 The X.21 digital interface

which is the one to which the X.25 specification refers. In this case the default state of the line is the data transfer state with the C and I lines held ON. If a DTE wishes to terminate the connection, it signals continuous binary 1 on the T line and sets C OFF. The data transfer state is reentered by the opposite procedure: setting binary 0 on the T line and setting C ON. Because data is transmitted transparently any synchronous link controls can be used; thus the procedure for the physical layer in no way restricts the possibilities for higher layer protocols.

In the quiescent phase before the data transfer phase is entered, a number of states can occur as shown below.

State	Signals	Meaning
DTE Ready	T = 1, C = OFF	DTE ready to go to data phase
DTE Uncontrolled not ready	T = 0, C = OFF	Abnormal DTE condition
DTE Controlled not ready	T = 010101.., C = OFF	DTE temporarily not ready
DCE Ready	R = 1, I = OFF	Able to go to data phase
DCE Not Ready	R = 0, I = OFF	Not ready

4.4.1 Circuit switched X.21

For circuit switching, a sequence of events must occur in order to start up and clear down a call on the line. The call may be set up from either end, involving slightly different procedures. A signal that a call is about to be made is given using the T and C lines (for a DTE) or the R and I lines (for the DCE). Information required to set up the call is transmitted on the T and R lines using sequences of International

Alphabet 5 (IA5) characters. The sequences can be seen as analogous to similar procedures on analogue telephone lines, for example the Bell symbol is sent to indicate ringing. The use of symbols enables a very flexible means of coding allowing a variety of responses to a call start-up request; X.21 circuit switching also provides high reliability and fast call establishment.

4.5 The data link layer

(CCITT X.25; Carlson, 1980; Conard, 1983; Brodd, 1983)

The data link layer of X.25 is based on the ISO high level data link control procedure (HDLC). HDLC is a bit-oriented protocol and is closely related to IBM's synchronous data link control (SDLC). HDLC is a family of protocols supporting a number of modes including multipoint polling, master-to-slave point-to-point working and peer-to-peer point-to-point working. Versions are differentiated by the link access procedure (LAP); for example LAPD is used for signalling in ISDN (see Section 4.5.4). X.25 level 2 allows the use of either LAP or LAPB (link access procedure balanced), but LAPB is preferred. LAPB covers either single links or multiple physical links between a DTE and a DCE.

The basic function of the layer is to divide the information into *frames* to achieve the multiplexing advantages of packet switching (see Chapter 1). Each frame includes a sequence number and check bits so that flow control, error detection and recovery procedures can be carried out (see Chapter 3). Two versions of HDLC LAPB are available: the basic version uses modulo 8 frame sequence numbers and the extended version uses modulo 128 sequence numbers, and is only required on lines where there is an exceptionally long delay.

4.5.1 HDLC framing

All bits transmitted by the HDLC protocol are grouped together into frames. A frame is delimited by the bit sequence 01111110 which is called a *flag*. In order to enable any bit pattern to be sent and yet not be misinterpreted as a flag, the flag pattern is not permitted within the frame, so that whenever a sequence of five '1' bits is encountered, an extra '0' bit must be inserted. This technique is called *bit stuffing*. An example for the three IA5 characters 'Eh?' is shown in Figure 4.6.

Any data code can be transmitted by a bit-oriented protocol such as HDLC and it is very easy to implement in hardware. A disadvantage is that because of bit stuffing, the length of the frame is not known before transmission (in fact there is a maximum overhead of 20 per cent on the bit stream contained within the opening and closing flags).

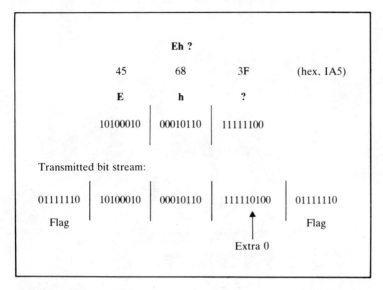

Figure 4.6   Example of bit stuffing to transmit the sequence 'Eh?'

4.5.2 HDLC frame types

The HDLC frame format is illustrated in Figure 4.7. The fields in the frame are defined as follows:

Address octet

This may take one of two values on a point-to-point link, depending on whether the frame is a command or a response; this can be determined by the frame type and context which is determined by the control field as shown in Figure 4.8.

Information bits

In an information frame, this field carries the information bits passed from a higher layer of protocol; for most control frames no information field is needed. The maximum length of information will depend on the

FLAG	ADDRESS	CONTROL	INFORMATION	FRAME CHECK SEQUENCE	FLAG
01111110	8 bits	8 bits		16 bits	01111110

Figure 4.7   HDLC frame format (after CCITT X.25)

Format	Type	Control field (bits)				Command/response
		8 7 6	5	4 3 2	1	
Information	(I)	N(R)	P	N(S)	0	Command
Supervisory	Receive ready (RR)	N(R)	P/F	000	1	Either
	Receive not ready (RNR)	N(R)	P/F	010	1	Either
	Reject (REJ)	N(R)	P/F	100	1	Either
Unnumbered	Set asynchronous balanced mode (SABM)	001	P	111	1	Command
	Disconnect (DISC)	010	P	001	1	Command
	Disconnect mode (DM)	000	F	111	1	Response
	Unnumbered acknowledge (UA)	011	F	001	1	Response
	Frame reject (FRMR)	011	F	011	1	Response

Figure 4.8 Coding of control byte for HDLC (after CCITT X.25)

network; the minimum value which a DCE must handle on a public
network is 259 octets.

Control field

This octet specifies the type of frame together with other information.
Three main types of frames are defined: unnumbered (containing no
sequence numbers), information and supervisory (which both include
sequence numbers). The control field illustrated is for modulo 8
sequence numbers; the extended sequence range requires a longer
control field. Individual bit(s) in the control field are shown in
Figure 4.8.

- *P/F* the poll/final bit—set to 1 to indicate *poll* when used as a
  command or 0 to indicate *final* when used as a response.
- *N(S)* is the sequence number of a transmitted information frame.
- *N(R)* is the receive sequence number—it indicates the N(S) of the
  next expected information frame.

The recommendation says that the low order bit of control octets is
transmitted first; for the information field the bit order is not specified.
Bit and byte ordering must always be carefully defined for any
communications link, and entrenched views on the matter have had to
be resolved. An interesting discussion on the different bit order
conventions in links and within computers is given by Cohen (1981).

Frame check sequence

A 16-bit cyclic redundancy check, using the polynomial
$x^{16} + x^{12} + x^5 + 1$. (Error detection coding is discussed in Chapter 2.)

4.5.3 HDLC phases and procedures for information transfer

The LAPB information transfer procedures have three main phases:
initial link setup, information transfer and finally link disconnect.

Link setup starts with the command SABM (set asynchronous
balanced mode); the response is UA (unnumbered acknowledge) if the
remote end agrees, or DM (disconnect mode) if the remote processing
equipment is logically disconnected from the link.

Link disconnect normally consists of one exchange of DISC
(disconnect) and UA.

During the *information transfer* phase, only information (I) and
supervisory frames (RR, RNR and REJ) are exchanged. Each
information frame has a send sequence number N(S), and all
information and supervisory frames contain an indication of the next
expected received information frame N(R). This enables flow control

using a sliding window and a go-back-N strategy independently for each direction. Each end stores two state transmission variables V(S) and V(R). V(S) is the next sequence number to be sent; it is incremented immediately after an information frame has been sent. V(R) is the next expected sequence number in a received information frame. This is incremented when a frame arrives without errors and in correct sequence. When sending an I frame the field N(S) is copied from the state variable V(S), and N(R) is copied from V(R). This enables acknowledgements to be piggy-backed onto information frames.

The supervisory frames *receive ready* (RR), *receive not ready* (RNR) and *reject* (REJ) are also used for flow control and error detection.

The *receive ready* (RR) frame is a positive acknowledgement, and acknowledges all frames up to and including N(R) − 1. Using the last received value of N(R), the transmitter may send frames with sequence numbers

$$V(S),..., N(R) + k - 1 \text{ (modulo 8 or 128)}$$

where k is the agreed window size for the link ($k \leqslant 7$ for modulo 8 mode). A poll bit setting of 1 is used to solicit an immediate response in order to determine the remote status.

The *receive not ready* (RNR) frame indicates a busy condition; following reception of RNR no further information frames may be sent until an RR or REJ frame is received.

The *reject* (REJ) frame indicates that a frame has arrived out of sequence. The transmitter should retransmit information frames again starting with N(S) equal to the value of the N(R) in the received REJ frame. The receiver will discard any information frames until the correct sequence number is received.

When an invalid frame is received, for example a frame containing a frame check sequence (FCS) error, it will be discarded, and recovery is by timeout and retransmission from the sending DTE or DCE. Timer T1 (which may be different in DTE and DCE) is started by the sending end when an information frame is sent, and must be longer than the minimum time taken for the information frame to arrive at the receiving end and for an acknowledgement to be given and received by the sender. This time must include processing and propagation delays. In order to keep this within reasonable bounds another timer T2 (with T2 < T1) is the longest time allowed for the receiving end to initiate sending an acknowledgement after it has received an information frame. Only one T1 timer is active at any time. A third timer (T3) in the DCE detects a long idle time and informs the packet level. When a frame has no bit errors but does not properly obey the protocol (for example the information field is too long) the FRMR (frame reject) frame is sent which results in a reset of the link.

Figures 4.9 and 4.10 illustrate typical exchanges of frames between

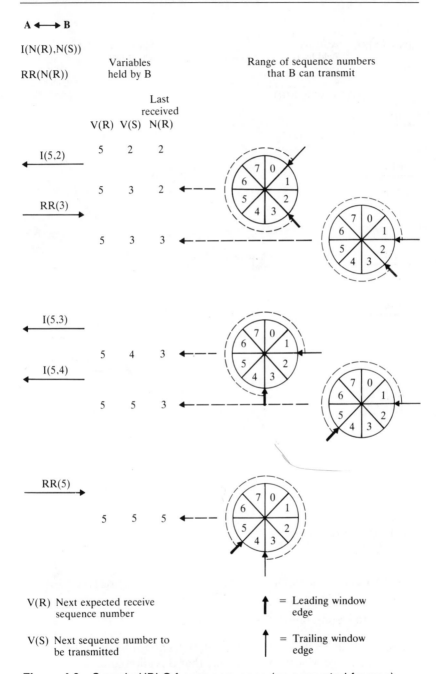

Figure 4.9 Sample HDLC frame sequence (no corrupted frames)

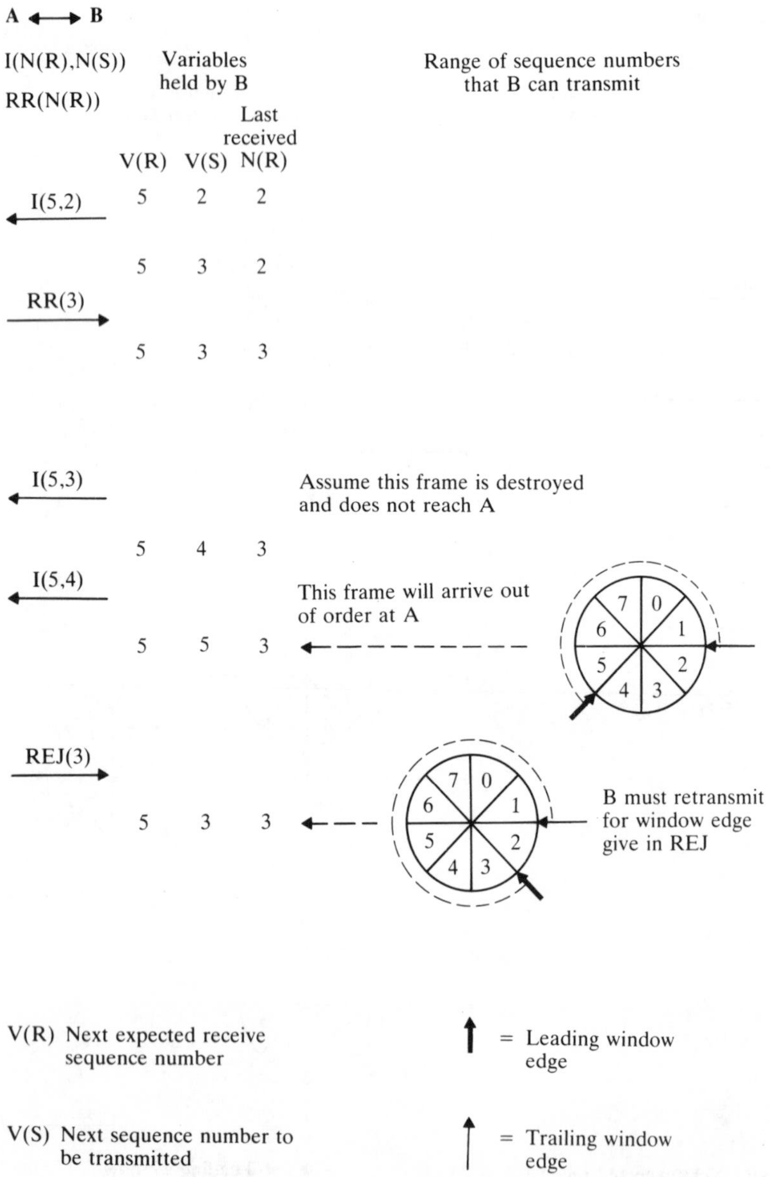

A ◄──► B

I(N(R),N(S))	Variables held by B			Range of sequence numbers that B can transmit
RR(N(R))				
		Last received		
	V(R)	V(S)	N(R)	

I(5,2)

5　2　2

5　3　2

RR(3)

5　3　3

I(5,3)　　　　　　Assume this frame is destroyed and does not reach A

5　4　3

I(5,4)　　　　　　This frame will arrive out of order at A

5　5　3　◄ ─ ─ ─ ─ ─ ─ ─ ─

REJ(3)

5　3　3　◄ ─ ─ 　B must retransmit for window edge give in REJ

V(R)　Next expected receive sequence number　　　↑ = Leading window edge

V(S)　Next sequence number to be transmitted　　　↑ = Trailing window edge

Figure 4.10　Sample HDLC sequence showing use of REJ frame

endpoints A and B during the information transfer phase. In the figures, the range of sequence numbers that B is allowed to transmit is illustrated by following clockwise round each disk from the leading window edge arrow until the trailing window edge arrow is reached. The state variables V(S) and V(R) and the last received N(R) value are shown for B immediately following transmission or reception of the corresponding frame. In Figure 4.9, V(S) is initially 2, so the first information frame sent I(5,2) has the sequence number 2. (The N(R) value of 5 is the next expected receive sequence number from A and does not change throughout the example.) Following the transmission of the first information frame, V(S) is increased by 1, and B may now transmit frames with sequence numbers 3,4,5,6,7 and 0. (Note that the trailing window edge can be calculated as the last N(R) received + window size -1 (modulo 8) $= 2+7-1$(modulo 8) $= 0$.) A's response is to acknowledge the frame with an RR frame with N(R) set to 3. The trailing window edge now progresses by 1 and becomes 1. The figure also indicates how more than one information frame can be sent by B before an acknowledgement is received.

Figure 4.10 shows what happens if sequence numbers arrive out of order at A. Here the second information frame (with N(S) set to 3) is lost and although A is expecting to receive frame 3, it actually receives frame 4. A must then issue an REJ frame with N(R) set to 3—the next expected receive sequence number. The effect of this is to bring the leading window edge back to 3, so that B must retransmit frames 3 and 4 (using the go-back-N technique).

4.5.4 Link access procedure for ISDN

The data link layer for use on the ISDN D-channel (for packet transfer and signalling) is another variant of HDLC called LAPD. It is described in CCITT recommendation I.441 and the major differences between LAPD and LAPB are summarized here.

The LAPD frame has two octets for the address, offering up to 64 service access point addresses and containing a subfield to specify the specific terminal endpoints. A broadcast group consisting of all the terminal endpoints can also be addressed. The FCS is slightly more complex than for LAPB.

Three modes of operation are provided: multiple frame operation which is very similar to LAPB, single frame operation which uses modulo 2 sequence numbers and unacknowledged operation which offers a single frame type with no sequence numbering and no error recovery. For single frame operation, only two frame types are used—called SI0 and SI1. These may or may not carry user information and are also used for acknowledgements and error reporting.

The recommendation also defines primitives for the data link layer's interfaces with layer 3 and with a management entity, and shows how these interactions are related to the peer-to-peer link procedures.

4.6 The network layer

(CCITT X.25; ISO 8878; Ware, 1983; Deaton and Hippert, 1983)

The network layer provides the network service to the transport layer.

Bits

8 7 6 5 4 3 2 1

Call request/incoming call	0 0 0 0 1 0 1 1
Call accepted/call connected	0 0 0 0 1 1 1 1
Clear request/clear indication	0 0 0 1 0 0 1 1
Clear confirmation	0 0 0 1 0 1 1 1
Data	x x x x x x x 0
Receiver ready	x x x 0 0 0 0 1
Receiver not ready	x x x 0 0 1 0 1
Reject	x x x 0 1 0 0 1
Interrupt	0 0 1 0 0 0 1 1
Interrupt confirmation	0 0 1 0 0 1 1 1
Reset request/reset indication	0 0 0 1 1 0 1 1
Reset confirmation	0 0 0 1 1 1 1 1
Restart request/restart indication	1 1 1 1 1 0 1 1
Restart confirmation	1 1 1 1 1 1 1 1
Diagnostic	1 1 1 1 0 0 0 1

Note. Bits marked x may be 0 or 1 as discussed in the text.

Figure 4.11　X.25 level 3: packet type identifiers (after CCITT X.25)

This service transfers data transparently from end to end between service users and makes the means by which this is accomplished invisible to the users. It must offer unambiguous addressing and the ability to select a quality of service. The service enables connections to be established at an agreed quality of service and to be reset or disconnected and provides a means to transfer data. It offers flow control and, optionally, expedited transfer (which enables data to overtake other data already queueing) and reception confirmation. Examples of quality of service include throughput, transit delay and the proportion of data units suffering corruption, duplication and loss.

The network service must be provided by the facilities of a physical network and for many networks the current choice is X.25 level 3. This is a reasonable choice for WANs because most Western countries have a packet switching network with an X.25 interface. For LANs, X.25 is inefficient for large systems and unsuitable for microcomputers. For high speed LANs, implementation is almost impossible.

X.25 level 3 recommends how *packets* should be exchanged between a DTE and a DCE. Each packet is contained in the information field of an X.25 level 2 information frame. A virtual circuit is a correspondence between two DTEs for the transmission of logically related packets across the network. All packets transmitted on a single virtual circuit are identified by a *logical channel* number, which distinguishes them from packets belonging to other virtual circuits on

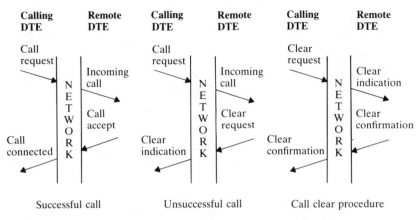

Figure 4.12 Time sequence diagrams for call connection and call clearing

the same link. A single link may support many virtual circuits at any one time.

A switched virtual circuit (SVC) is set up when required; a permanent virtual circuit (PVC) is always available and does not need a call setup phase. In order to set up and clear SVCs and to maintain SVCs and PVCs a number of different packets are provided. The packet type identifiers are shown in Figure 4.11. The data passed to layer 3 from higher layers is transmitted in data packets. All other types of packet control the progress of calls.

4.6.1 Call setup and clearing

When a new virtual circuit is required, a call request packet is sent from the DTE. It is transmitted across the network to the remote DTE and presented as an incoming call packet (which uses the same coding of the type field) (Figure 4.12). If the remote DTE accepts the call it will reply with a call accept packet and this is delivered to the originating DTE as a call connected packet. A call that is rejected results in a clear indication packet being sent to the originating DTE. The packets involved in clearing a call are also indicated in Figure 4.12.

The call request packet has the most complex of all the packet formats and is illustrated in Figure 4.13. The full DTE network addresses of both the called and calling DTEs are included (coded in binary coded decimal) in order to specify the destination of the call. A logical channel is chosen by the DTE for subsequent interaction with its local DCE, and another logical channel is chosen by the remote DCE for presentation to the remote DTE. The correspondence between logical channel numbers used on the same virtual circuit across the network must be kept by the network exchanges. The facilities field is of variable length and will specify such options as a closed user group or a

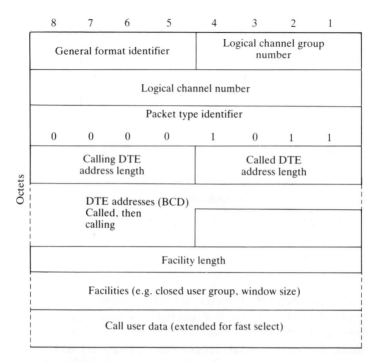

Figure 4.13 X.25 level 3: call request and incoming call packet format (after CCITT X.25)

non-standard packet or window size. (Some options can be can be negotiated, but only downwards. For example, if the call request packet specifies a packet length of 256 bytes, the call accept packet may agree to 256 bytes or request a reduction to 128 bytes, but may not increase the packet length.)

Figure 4.14 shows the state transition diagram for call setup. Note that, normally, the end that initiates the call proceeds to a waiting state until the call is accepted. For example, if the DTE issues a call request, this results in a transition from the *ready* state to the *DTE waiting* state. When the DCE responds with call connected a transition is made to the *data transfer* state. Once the call setup procedure is complete, data and flow control packets can be exchanged, and this is then subject to another state transition diagram (not illustrated). If both DTE and DCE were to try to initiate a call using the same logical channel number, a *call collision* would occur. Only one of the two call requests can be successful, and the call that is connected is the call requested by the DTE; the incoming call will be cancelled by the DCE. To avoid this situation, LCNs are generally allocated from different subsets of the available range, or starting at the low end and working up from the DTE and starting at the high end and working down from the DCE.

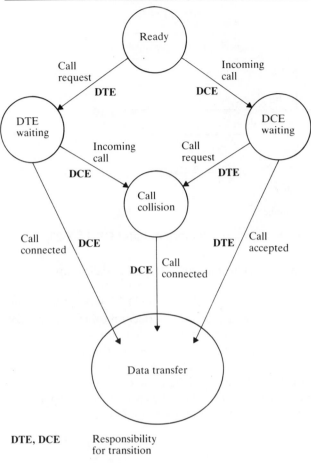

DTE, DCE   Responsibility
                for transition

Call connected,   Packet transferred
call request, etc.

Figure 4.14   X.25 level 3: call setup state diagram (after CCITT X.25)

4.6.2 The data phase

Once the call is established, it enters the data phase during which data, RR and RNR packets are exchanged. REJ is not normally required because the data link layer will have discarded any frames in error. The format of a data packet is illustrated in Figure 4.15. P(R) is the acknowledgement sequence number and P(S) is the sequence number of the packet. These are used in the same way as for the layer 2 frame sequence numbers.

The qualifier bit Q may be used to distinguish between different sorts of information contained in the packet. (An example is given in Section 4.7.3.) The more data bit M, when set to 1, indicates that the packet is a fragment of a larger message, and that this is not the last fragment of the message. The D bit indicates a requirement for delivery

8	7	6	5	4	3	2	1
Q	D	0	1	Logical channel group number			
Logical channel number							
P(R)			M	P(S)			0
User data							

Q   Qualifier bit
D   Delivery bit
M   More data bit
P(S) Send sequence number
P(R) Next expected
      received sequence number

Figure 4.15   X.25 level 3: data packet format (after CCITT X.25)

confirmation of the packet. (This may also be used in call setup packets.)

An interrupt packet can be sent to the remote DTE without following the normal flow control procedure; it has no effect on other packets in transit for that call, and receipt is confirmed with an interrupt confirmation packet. A reset packet may be issued during the data state and has the effect of removing all data and interrupt packets in the network for each direction of the call. Where there are more serious problems, the whole of the packet level can be reinitialized using a restart packet which clears all SVCs and resets all PVCs.

4.6.3 Optional user facilities

The X.25 recommendation details a wide range of options. These do not all have to be offered by every network provider. The variety of user-implementable options makes X.25 a flexible tool, but results in many different implementations of the same recommendation which can lead to problems of intercommunication.

The following are some of the user options:

- *Non-standard default window and packet size.* Each network controller may choose the maximum window and packet size. Within this maximum, DTEs may negotiate the packet size for a particular call. (Note that the standard window size is 2.)
- *Default throughput class assignment and negotiation.*
- *Reverse charging.*
- *Barring of incoming or outgoing calls.*
- *Closed user groups.* DTEs belonging to a closed user group are allowed to make calls only to other members of the group. DTEs outside the closed user group are barred from making calls to members of the group.

- *Choice of transit network* (for international working).
- *Fast select*. This facility is intended to reduce the number of packets exchanged over a virtual circuit if there is very little data to transmit. The call request packet has an extended user data field, and user data is also allowed in the call accept and clear request packets. This option is suitable for transaction processing where the complete call may consist of a simple request and a response. This option has also been used in networks where it has proved convenient to put extra information (for example subaddressing) inside the call request packet, which would not fit in a standard size call request packet. The full network service connection needs information which is not included in the standard X.25 call request packet; this can be achieved either by using fast select or by sending a data packet containing connection information once the X.25 call has been established.

4.6.4 Why offer flow control at both link and packet level?

At first sight the use of sequence numbers at both frame and packet levels seems redundant. At the packet layer the flow control refers to a single virtual call and can be used to control the buffering requirements for each call. At the data link layer, the sequence numbers apply to all the frames transmitted across the link, protecting the storage required for all calls between a DTE and a DCE. Note that at the packet level, sequence numbers are used for restraint (to avoid the recipient being swamped with packets), at the frame level they are used for restraint and for error recovery.

4.7 Packet assembler/disassembler

Although it does not strictly come in the sequence of the OSI model, it is convenient to discuss the concept of packet assemblers/disassemblers (PADs) at this point. A PAD enables terminals that have simple asynchronous interfaces (usually RS232C/V.24) to communicate with a remote computer across a packet network; it provides a buffering service between the *character oriented* terminals and the *block-oriented* packet switched network. A PAD system is interfaced to a number of terminals and is also connected as a DTE to a packet switched network (as illustrated in Figure 4.16). PADs are available either as systems in their own right or as a facility offered by a large system that also provides other services.

4.7.1 Recommendations for a PAD

The characteristics and procedures for connecting a PAD to an X.25 network are described in three CCITT recommendations collectively

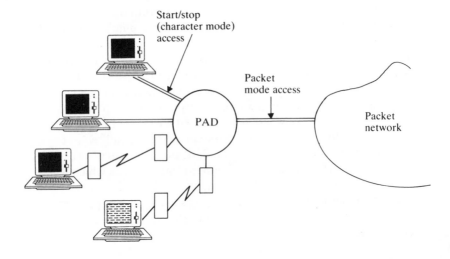

Figure 4.16 A packet assembler/disassembler (PAD)

known as *triple X*. The triple X recommendations are not part of OSI but are widely used. They will eventually be replaced by the application layer virtual terminal protocol (see Chapter 5). The three recommendations are:

X.3   The PAD parameters (terminal characteristics)
X.28   The terminal to PAD interface
X.29   The protocol between the PAD and the remote host

The relationship between the recommendations is shown in Figure 4.17.

4.7.2 X.3 parameters

The PAD parameters may be imposed by the host or set by the user. Most PADs offer a selection of facilities appropriate to a range of

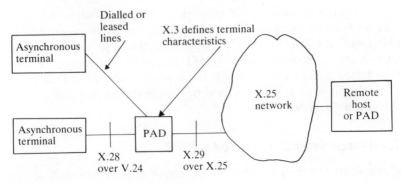

Figure 4.17 Triple X protocols

terminals, but a standard profile is also defined. Examples of PAD parameters are:

- Whether character echo is provided by the PAD or remotely.
- Specific action to be taken on receipt of a 'break' character from the terminal.
- Padding after a carriage return character (to allow for fly-back time on some terminals).
- Redefinition of the delete character.
- Terminal speed.

4.7.3 X.28: terminal to PAD interface

This recommendation defines how a terminal should be connected to a PAD service implemented by a public packet network provider. It includes specifications of the appropriate CCITT modem recommendations—V series for use over public switching telephone networks and X.20 or X.20bis over a public data switched network.

Also defined are a number of PAD commands for the terminal user, which include enabling the user to change the PAD parameter values, to check the status of the call or to send an interrupt packet.

4.7.4 X.29

X.29 defines the protocol to be used between the PAD and the remote service using X.25 calls; it also provides for interworking between PADs. When a user issues an appropriate command to start a call, the PAD issues an X.25 call request packet which includes the address given by the terminal user. Further information from the user is inserted in the call user data field and specifies such parameters as the host's high level protocol and other user data.

During data transfer, both packets containing characters of user messages and packets containing X.29 control messages are sent in X.25 data packets. They are distinguished by use of the qualifier bit. X.29 control messages indicate such actions as switching character echo off while a password is being entered. In order to make economic use of the network, packets are filled to the maximum packet size or until a data forwarding character is input at the terminal. (This character may differ for different terminals.) If a packet is filled before the data forwarding character is entered, the more data bit is set.

4.8 The transport layer

(ISO 8072/8073; CCITT X.214/X.224; Knightson, 1983; Stallings, 1984; Hunt, 1984)

Before describing the transport layer it is useful to recall the definitions of the OSI model. Because the details of the transport layer and higher

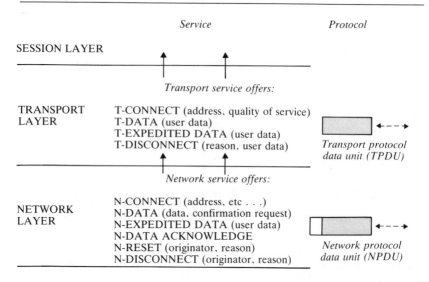

Figure 4.18 Services and protocols of the transport and network layers

layers have largely been decided following the precise definition of the seven–layer model, the standards documents carefully follow the separate definition of service and protocol for each layer, a distinction that was not so apparent in earlier interconnection documents.

We recall that each layer (except the application layer which is at the top of the model) offers a service to the next higher layer. The service provided by a layer is defined as a number of primitives with parameters. Primitives have names of the form x-FUNCTION where x represents the layer offering the service (P for presentation, S for session T for transport and N for network layers). Examples are T-CONNECT, T-DATA, N-CONNECT, N-EXPEDITED. Each block of information that forms an element of the protocol to be transmitted across the network contains control information related to that layer and may contain data passed transparently from the next higher layer; it is called a protocol data unit or xPDU. Examples are TPDU for the transport layer and NPDU for the network layer (see Figure 4.18).

4.8.1 Transport layer services

The transport layer relieves higher layers from the details of transporting data across the network. It hides the details of how the network is implemented. The transport service responds to the user's request for a particular *quality of service* (e.g. throughput, transit delay, residual error rate) by making the best use of the resources of the underlying network service and by enhancement to compensate for any deficiencies in the network service. For example, the required service may need to make use of several network service connections (e.g. to increase bandwidth) or it may be possible to map several transport service connections onto a single network service connection (e.g. to reduce network costs). Error recovery may need to be provided. User data may need to be segmented for transmission across the network and reassembled at the other end; or it may benefit from concatenation into single network data units and be separated again at the other end. The transport service is concerned with transporting user data between end systems and is not concerned with the routing or relaying of packets. All information from higher layers is transmitted transparently as octets and does not have to be interpreted by the transport layer.

Five classes of service are provided, which are selected depending on the reliability and other characteristics of the underlying network service. The class is negotiated when a transport service connection is established. All classes allow segmentation of blocks for transmission and reassembly at the other end. Expedited data can also be selected. The transport service classes (0 to 4) are as follows:

0 *Simple class.*   This adds very little to the network service and is for use in networks where there is a very low residual error rate, that is where the proportion of the total traffic sent which is incorrect, lost or duplicated is very low or where the packet loss rate or error rate can be tolerated.

1 *Basic error recovery class.*   This is for use where the network can inform the transport service that a disconnect or reset has occurred.

2 *Multiplexing class.*   Several transport connections are allowed to use a single network connection.

3 *Error recovery and multiplexing class.*   This class combines classes 1 and 2.

4 *Error detection and recovery class.*   This class has its own checksum in order both to detect and recover from errors, and is to compensate for low grade reliability networks. This class can also be used over a connectionless network service.

4.8.2 Transport protocol

The information passed across the network to the corresponding transport layer obeys the *transport protocol*. This is defined by a number of transport protocol data units (TPDUs). Figure 4.19 illustrates the connection request (CR) TPDU which is used to initiate a transport service connection. The *length indicator* gives the length of the TPDU; *destination* and *source references* are the transport layer addresses. The *class*, *options* and *parameters* specified in the CR TPDU are those that are preferred or proposed by the caller. These include transport class, transit delay, throughput, residual error rate, protection constraints and

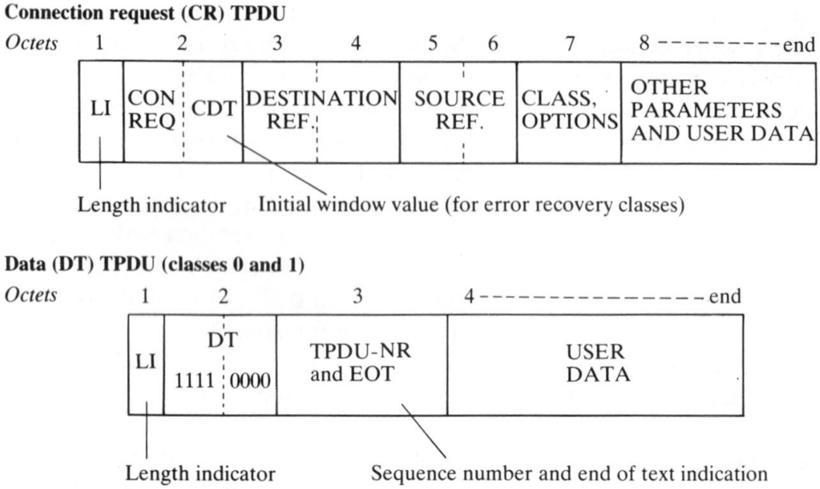

Connection request (CR) TPDU

Octets 1   2   3   4   5   6   7   8 --------- end

| LI | CON REQ | CDT | DESTINATION REF. | SOURCE REF. | CLASS, OPTIONS | OTHER PARAMETERS AND USER DATA |

Length indicator   Initial window value (for error recovery classes)

Data (DT) TPDU (classes 0 and 1)

Octets   1   2   3   4 -------------- end

| LI | DT 1111 : 0000 | TPDU-NR and EOT | USER DATA |

Length indicator   Sequence number and end of text indication

Other TPDUs

CC Connection confirmation   EA Expedited acknowledge
DR Disconnect request   RJ Reject
DC Disconnect confirmation   ER TPDU error
ED Expedited data   PI Transport protocol ID
AK Acknowledge

Figure 4.19 Transport protocol data units (after CCITT X.224)

TPDU size. If the remote session layer agrees to the connection, the transport service connection is confirmed with a connection confirmation (CC) TPDU. This contains the parameters that define the quality of service selected by the responding system from those proposed in the CR TPDU and may include changes to the class or option to be used depending on the responder's capabilities. Once the connection is established, information from the session layer is sent as *user data* in the data (DT) TPDU, which also includes a sequence number (for recovery classes) and an end of text (EOT) indication used to indicate the final segment if the data must be segmented into several TPDUs for transmission across the network. Details of the other TPDUs may be found in the standards documents.

4.8.3 Setting up a transport service connection

The example in Figure 4.20 illustrates the relationship between services, layers and protocols in the OSI model, showing how the transport service itself makes use of the network service offered by the network layer. The higher layer (the session layer) issues a service *request* to a lower layer (the transport layer), which makes use of the underlying services to transfer the request to the remote system. It is then presented

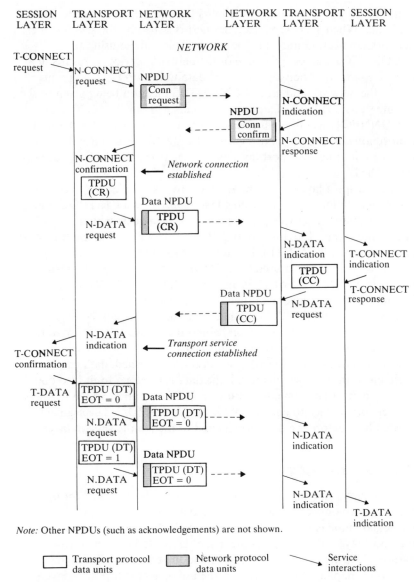

Note: Other NPDUs (such as acknowledgements) are not shown.

Figure 4.20 Examples of transport and network services and protocols

by the remote transport layer to the corresponding higher layer (session layer) as an *indication*. The *response* is then made from the higher layer and results are conveyed across the network, and presented as a *confirmation* passed from the transport service to the requesting session layer. These four terms—request, indication, response and confirmation—are used frequently at the service access points as defined in the seven-layer model.

In Figure 4.20, it is initially assumed that no network connection is available. When a T-CONNECT.request is issued by the session layer, a network connection must first be set up. This is done using the N-CONNECT.request primitive of the network service. It results in a connection request network protocol data unit (NPDU) being sent across the network and an N-CONNECT.indication being given to the remote transport layer. This layer responds with an N-CONNECT.response which is conveyed by the network layer as a connection confirmation NPDU. An N-CONNECT.confirmation then tells the initiating transport layer that the network connection has been established.

The transport layer can now use the network connection to convey its protocol data units. The transport layer constructs a CR TPDU and issues the network service primitive N-DATA.request which encapsulates the TPDU in a data NPDU. When the receiving transport layer interprets the CR TPDU it indicates to the remote session layer that a connection is required (T-CONNECT.indication). The response results in a CC TPDU which is also carried in a data NPDU. The initiating transport layer can then confirm that the transport service connection is established. Note that the initiating session layer is unaware of all the work going on behind the scenes and simply waits for the requested connection to be confirmed.

Once the transport service connection is established, data can be exchanged. In the case illustrated, the data is too long to fit into one TPDU so must be segmented into two TPDUs, each of which requires the network service to encapsulate it in a data NPDU. The segments are reassembled before the data is delivered to the remote session layer.

4.9 Summary

In this chapter, having described the OSI seven-layer model for open systems interconnection, we have looked at the lowest three layers with particular emphasis on public packet switched networks, and also at the transport layer which enables the higher layers to be independent of the nature of the network. Thus we have laid the groundwork for the following chapter which deals with the user oriented top three layers of the OSI model.

References

Standards documents

The definitive situation on current standards should be obtained from your national member of ISO (for example BSI in the United Kingdom and ANSI in the United States). A useful summary of the latest status of OSI and related standards appears periodically in *ACM Computer Communication Review*. Note that the CCITT plenary assembly makes revised recommendations every four years, the most recent at time of publication being 1988. The following list is a guide to some of the more important documents.

DIS = Draft International Standard, DP = Draft Proposal, DAD = Draft Addendum, AD = Addendum

Subject	ISO document	Related CCITT document
OSI reference model	7498–1	X.200
OSI reference model: connectionless data transmission	7498–1 AD 1	
OSI: service conventions	TR 8509	X.210
Physical service definitions	DP 10022	X.211
Physical layer: X.25 level 1		X.21, X.21bis
Data link service definition	DIS 8886	X.212
Data link layer: X.25 level 2		X.25
HDLC frame structure	3309	
HDLC elements of procedures	4335	
HDLC classes of procedures	7809	
HDLC: X.25 LAPB compatible DTE link layer procedures	7776	
Network layer: network service definition	8348	X.213
Protocol for providing the connectionless mode network service	8473	
X.25 packet level protocol	8208	X.25
Use of X.25 to provide connection-oriented network service	8878	X.223
Protocol combinations to provide and support OSI network service	8880	
Support of packet mode equipment by an ISDN		X.31
Interface for start/stop mode DTE to access PAD in public data network		X.28
Procedures for information exchange between PAD and packet mode DTE		X.29
PAD facility definition		X.3
Transport service definition	8072	X.214
Connection-oriented transport protocol specification	8073	X.224
Protocol for providing the connectionless mode transport service	8602	

Further references

Bertine, H. V. (1980) 'Physical level protocols', *IEEE Trans. Commun.*, **COM-28** (4), April, 433–444.

Brodd, W. D. (1983) 'HDLC, ADCCP & SDLC: What's the difference?', *Data Commun.*, **12** (8), August, 115–122.

Carlson, D. (1980) 'Bit oriented data link control procedures', *IEEE Trans. Commun.*, **COM-28** (4), April, 455–467.

CCITT X.200 (1984) *Red Book*, Vol. VIII, Fascicle VIII.5, International Telecommunication Union, Geneva.

CCITT X.25 (1984) *Red Book*, Vol. VIII, Fascicle VIII.3, International Telecommunication Union, Geneva.

CCITT X.224 (1988) *Blue Book*, Vol. VIII, Fascicle VIII.5, International Telecommunication Union, Geneva.

Chapin, A. L. (1983) 'Connections and connectionless data transactions', *Proc. IEEE*, **71** (12), December, 1365–1371.

Cohen, D. (1981) 'On Holy Wars and a plea for peace', *IEEE Computer*, **14** (10), October, 49–54.

Conard, J. W. (1983) 'Services and protocols of the data link layer', *Proc. IEEE*, **71** (12), December, 1378–1383.

Day, J. D. and H. Zimmermann (1983) 'The OSI reference model', *Proc. IEEE*, **71** (12), December, 1334–1340.

Deasington, R. J. (1988) *X.25 Explained: Protocols for Packet Switching Networks*, 2nd edition, Ellis Horwood, Chichester, UK.

Deaton, G. A. and R. O. Hippert (1983) 'X.25 and related recommendations in IBM products', *IBM Systems J.*, **22** (1/2), 11–20.

Halsall, F. (1988) *Introduction to Data Communications and Computer Networks*, 2nd edition, Addison-Wesley, Wokingham, UK.

Hunt, R. (1984) 'Open Systems Interconnection: the transport layer', *Computer Commun.*, **7** (4), 186–197.

IEEE Proceedings (1983) Special issue on 'OSI', **71** (12), December.

IEEE Transactions on Communications (1980) Special issue on 'Computer network architectures and protocols', **COM-28** (4), April.

Knightson, K. G. (1983) 'The transport layer standardization', *Proc. IEEE*, **71** (12), December, 1394–1396.

Linington, P. F. (1983) 'Fundamentals of the layer service definitions and protocol specifications', *Proc. IEEE*, **71** (12), December, 1341–1345.

McClelland, F. M. (1983) 'Services and protocols of the physical layer', *Proc. IEEE*, **71** (12), December, 1372–1377.

Marsden, B. W. (1986) *Communication Network Protocols*, 2nd edition, Chartwell Bratt, Kent, UK.

Moudiotis, G. (1986) 'Development of CCITT standards for packet switched data networks', *Br. Telecom Engineering*, **5**, October, 212–216.

Pouzin, L. (1986) 'OSI progress and issues', in *New Communication Services: a Challenge to Computer Technology*, *Proc. Int. Conf. on Computer Communication*, Munich, 1986, P. Kuhn (ed.), 154–158.

Ronayne, J. (1987) *The Integrated Digital Services Network—From Concept to Application*, Pitman, London.

Stallings, W. (1984) 'A primer: understanding transport protocols', *Data Commun.*, **13** (13), November, 201–215.

Tanenbaum, A. S. (1988) *Computer Networks*, 2nd edition, Prentice-Hall, Englewood Cliffs, New Jersey.

Ware, C. (1983) 'The OSI network layer: standards to cope with the real world', *Proc. IEEE*, **71** (12), December, 1384–1387.

5 Open systems interconnection: the application-oriented layers

GILL WATERS

5.1 Introduction to the higher layers of the OSI model

The top three layers of the OSI reference model (shown in Figure 4.2) are application oriented, so that the services provided by these layers offer useful facilities for applications and ultimately for the human user. The session layer includes synchronization facilities. The presentation layer ensures that despite differences in representation of the information exchanged, the information can be understood. The application layer also offers services that may be of interest to several applications. These include reliable transfer and concurrency control. Each specific application to achieve an objective such as file transfer or on-line use of a remote computer must also be defined.

5.1.1 Reference material

As in Chapter 4, standards document references published by ISO and CCITT are listed at the end of the chapter. A number of books describe these layers of the OSI model including Tanenbaum (1988), Halsall (1988), and Marsden (1986). Additionally, papers have been published to explain the individual standards and these are referenced at the beginning of the appropriate section.

5.2 The session layer

(ISO 8326/8327; CCITT X.215/X.225; Emmons and Chandler, 1983)

The session layer manages an open communication path on a channel between two application processes that are using the network to

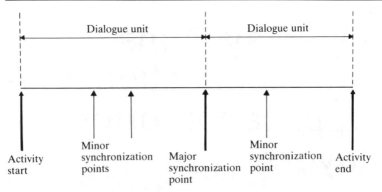

Figure 5.1 Synchronization points at the session layer (after CCITT X.215)

perform some activity. In fact it provides a service to the presentation layer, but since the presentation layer simply transforms the representation of the data, the session layer can be seen as being directly related to the timing and dialogue control requirements of the application. An example of a session would be the time between logging-in and logging-out using a terminal to access a multiuser system. The address for communication is passed from the application layer.

The session service provides a number of facilities including the following:

1 It enables connections to be established, data to be exchanged and connections to be released in an orderly manner (i.e. without the loss of any messages in transit).

2 It enables information to be exchanged either in a full duplex manner (both ways simultaneously) or in a half duplex manner (one way at a time).

3 It enables exceptions (error conditions) to be reported outside the normal information flow.

4 It enables the dialogue between the two users to be synchronized. Synchronization is carried out by labelling instants of time throughout the session which may be of three types: activity start or end, major synchronization points and minor synchronization points (see Figure 5.1).

An *activity* is an agreed programme of interchange between the two end users. During one session connection an activity may be interrupted, making way for a different activity, and then resumed when the interrupting activity is complete. For example, while logged-on remotely to a distant machine, it may be required to transfer a short file; the log-on activity would be interrupted, then the file transfer activity would take place and then the log-on activity would be resumed.

A *major synchronization point* divides an activity into *dialogue units.* Major synchronization points must always be confirmed, so that both end users agree exactly what message interchanges were completed prior to synchronization.

The *minor synchronization points* give an additional means of indicating that the communication has reached a certain stage, but are not confirmed and thus do not affect the information flow. Within an activity, each synchronization point is numbered in sequence. It is then possible to resynchronize either to the most recent major synchronization point or to any subsequent minor synchronization point. Note that the semantic meaning associated with the synchronization points and defined by the application layer is of no significance to the session layer, except in the choice of synchronization primitive (activity, major or minor synchronization point).

However, it may be helpful to give an arbitrary example. Suppose we are transferring all of the files in a user's directory. The *activity* consists of the entire directory transfer, major synchronization points could be taken between each file and minor synchronization points could be taken for every 10 blocks of each file transmitted. Thus at any point it would be possible to return to the beginning of the file currently being transferred or to any position in the file marked in multiples of 10 blocks.

To provide many of the services described above, the session layer defines the use of *tokens*, each of which may be held at any instant by one of the two session service users of a connection. There are four tokens as follows:

- *Data token*—enables half-duplex working, the end holding the token may transmit data.
- *Release token*—enables an orderly closing of the session, ensuring all data in transit has been received.
- *Minor synchronization token*—the end holding the token may define a minor synchronization point.
- *Major synchronization and activity token*—needed to issue start activity or end activity, or to take a major synchronization point.

If one end wishes to make use of a token, but does not possess it, it can issue an S-TOKEN-PLEASE request. For dialogue control, the use of a single token ensures that both end users agree on the identification of each synchronization point.

Table 5.1 lists the session layer service primitives. The recommendation defines three subsets of session services to aid with conformance by preferred selection of optional services, but other subsets may be defined if necessary. The three defined subsets are:

- The *basic combined subset (BCS)* offers connection, data transfer and orderly release, with a choice of half duplex or full duplex working.

Table 5.1 Session layer service primitives

Primitive	Function
S-CONNECT	Session connection and confirmation
S-DATA	Normal data transfer
S-EXPEDITED-DATA	Data transfer not subject to token and flow control restraints
S-TYPED-DATA	Data transfer ignoring token restrictions
S-CAPABILITY-DATA	Data exchange while not in an activity
S-TOKEN-GIVE S-TOKEN-PLEASE }	Token of any type is given to or requested by the other end
S-CONTROL-GIVE	Allows entire set of tokens to be passed across when no activity is in progress.
S-SYNC-MINOR S-SYNC-MAJOR }	Define synchronization points
S-RESYNCHRONIZE	Go to specified synchronization point
S-P-EXCEPTION-REPORT S-U-EXCEPTION-REPORT }	Exception reporting from service provider or user
S-ACTIVITY-START S-ACTIVITY-RESUME S-ACTIVITY-INTERRUPT S-ACTIVITY-DISCARD S-ACTIVITY-END }	Control for activities. Interrupt and resume allow a temporary pause in activity. Discard terminates (possibly cancelling) the current activity.
S-RELEASE	Orderly release
S-U-ABORT S-P-ABORT }	Early termination by service user or provider

- The *basic synchronized subset* augments the BCS with other services including major and minor synchronization points and resynchronization.
- The *basic activity subset* provides facilities for activity management and exception reporting.

5.3 The presentation layer

(ISO 8823/8824; CCITT X.216/X.226; Hollis 1983)

The presentation layer is rather different from the layers underneath it which are concerned with the orderly transmission of transparent information. The presentation layer is concerned with how information is represented and with preserving the meaning of that information, independent of its syntactic representation in the course of its transfer across the network. The layer therefore caters for the differences in

Character '5'
System A: ASCII, even parity, bit 0 least significant
System B: EBCDIC, bit 0 most significant
Network: ASCII, odd parity, least significant bit transmitted first

Figure 5.2 Example of representation differences for transfer between systems

representation of information in the systems in which the applications reside.

The principal choices in representation for system manufacturers or managers are the character code (e.g. EBCDIC or ASCII), character size, how numbers are represented (e.g. in ones or twos complement), the bit and byte ordering for integers and real numbers and the way in which characters are stored in strings. It is not always possible simply to change the position of high- and low-order bytes to convert to another manufacturer's representation, as this may work for integers but not for character strings. The scope for confusion is apparent (Cohen, 1981).

In any point-to-point transfer, there may be three different ways in which information is represented—one in each of the end systems (the *local syntax*) plus one while the information is in transit. A simple example is shown in Figure 5.2, in which the character '5' is represented in three different ways. To understand how the presentation layer tackles these problems, we must first introduce some definitions. Figure 5.3 shows the relationship between the various syntaxes involved. For the exchange of a particular sort of data, users must agree on an *abstract syntax* which describes the generic data requirements. For example, this might be a Baccus Naur form (BNF) description of the fields with their types for the records in a data base. The information to be transferred must be coded as a structured bit string; this concrete pattern to be transmitted is called the *transfer syntax*. For example a character string may be represented by an octet coded with 06_{16} followed by an octet containing the count of characters in the string, followed by an octet containing each character coded in ASCII without parity. In order to accomplish a transfer, a relationship called a *presentation context* must be defined between an abstract syntax and a transfer syntax. An application may need to use several presentation contexts to convey all the types of information it needs to transmit, and

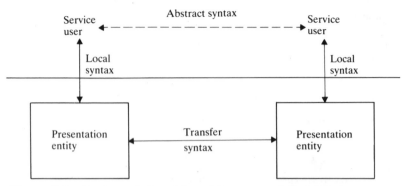

Figure 5.3 Syntax relationships at the presentation layer

the agreed set of such contexts in use at any one time is called a *context set*. The presentation layer negotiates transfer syntaxes on behalf of the user and transforms the user information into and out of the transfer syntax.

5.3.1 The presentation service

The presentation service offers all the services of the session layer—connection establishment and termination, information transfer and dialogue control—by invoking the appropriate session service primitives. In addition, it offers services for context management. The P-CONNECT primitive defines a context set at connection time, which is agreed by negotiation with the remote presentation service. The P-ALTER-CONTEXT primitive changes this set at a later time during the session by addition or deletion of presentation contexts. Data sent by the user must specify the particular presentation context to be used and is transformed into the transfer syntax by the presentation layer before the appropriate S-DATA primitive is invoked to transmit the information.

For implementation purposes the presentation layer may be treated as a set of facilities available to the application layer (for example a set of library routines).

5.3.2 Abstract syntax notations

There is no single common abstract syntax for all applications. A number of syntaxes to support specific applications have been defined, for example that for message handling (see Section 5.4.2). To aid in the definition of abstract syntaxes a notation for defining a number of simple types, and structures based on these types has been agreed. This is called abstract syntax notation one (ASN.1) (ISO 8824; CCITT X.208). The notation is rather like the type and structure definitions of a programming language such as Pascal or C. The notation has wider use

beyond a definition of presentation abstract syntaxes, and indeed
ASN.1 is now used as a tool to help define higher layer services and
protocols. The details are beyond the scope of this book, but a very
simple example is given below:

```
Student Record ::=  [APPLICATION 0] IMPLICIT SET
  {Name                                        ,
   dateOfBirth          [0] Date               ,
   married              [1] Married            ,
   labPartner           [2] Name               }

Name ::=  [APPLICATION 1] IMPLICIT SEQUENCE
  {givenName            VisibleString          ,
   familyName           VisibleString          }

Date ::= [APPLICATION 2] IMPLICIT
                         VisibleString -- YYYYMMDD

Married ::=  [APPLICATION 3] IMPLICIT BOOLEAN
```

All items described by the notation are distinguished by their *type*. In
this example the record format for a student is defined as being of type
StudentRecord. It consists of four fields: the student's name, date of
birth, whether he or she is married and the name of his or her
laboratory partner. Each of these fields is defined later, being of type
Name, Date or Married. An annotated version of the definition of
StudentRecord is shown in Figure 5.4.

A number of common type definitions together with the values that
may be assigned to them are included in the ASN.1 definition. For
example, BOOLEAN is defined as a type that can take the values true
or false. From these simple types, more complex types can be built up,
for example by defining a new type which is a SEQUENCE (ordered
list) of previously defined types. In Figure 5.4 'Name' is defined in this
way as a SEQUENCE of two character strings.

Each type definition is given a label called a *tag*. The tag consists of
two qualifiers—a class and a number. The class indicates the context in
which the tag will be used, the number enables different types to be
distinguished within that context. The class is one of the following:

- UNIVERSAL   Defined within the ASN.1 recommendation

- APPLICATION   Defined for a specific application e.g. message
                     handling

- PRIVATE   Defined by an enterprise (not part of ISO or
                     CCITT recommendations)

- Context specific (no   Tag numbers distinguish between different
  class name included   types in the context within which they are
  in tag definition)   used.

Both APPLICATION and context specific tags appear in Figure 5.4. When a type is defined in terms of a previously defined type, the correspondence may be designated IMPLICIT. If so, the correspondence does not have to be explicitly notified during transfer thus keeping the length of the encoded sequence to a minimum.

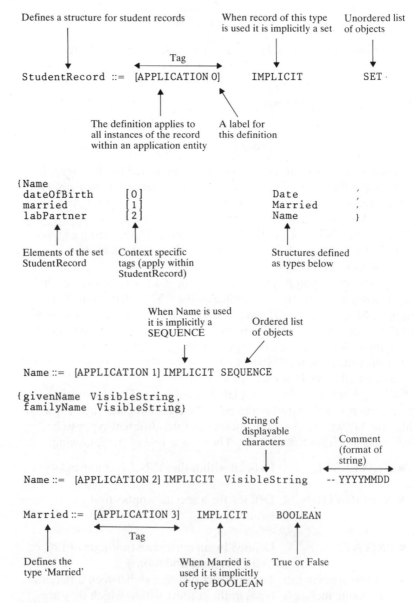

Figure 5.4 Annotated example of an ASN.1 definition

Using the above record structure, and assigning values to the types defined, an individual student might have a formal record of the form:

```
{{givenName ''Charles'',familyName ''Darwin''}         ,
  dateOfBirth ''19690104''                             ,
  married FALSE                                        ,
  labPartner {givenName ''Susan'', familyName ''Smith''}}
```

(Charles Darwin was born on 4 January 1969 and is not married. His laboratory partner is Susan Smith.)

This notation is not the transfer syntax. In order to transmit these values across the network, the information must be encoded into an appropriate bit sequence. A further definition (ISO 8825, CCITT X.209) specifies the basic encoding rules for ASN.1. Each type is encoded in three fields, the first indicates the type, the second is a count which is the number of octets in the contents field and the third field, the contents field, carries the represented information. The information field may contain further nested types.

5.3.3 Data compression and encryption

Before leaving the presentation layer, two other topics related to information coding should be mentioned—namely data compression and encryption. Data compression is a technique for reducing the amount of information transferred by replacing sequences of repetitive information by shorter coded values. See Black (1989) for a detailed description, which includes data compression for facsimile. Encryption also changes the coding of bit sequences (see Chapter 10). Both techniques can be applied after presentation layer conversion to an agreed transfer syntax, seeming to take a position intermediate between the presentation and the session layer.

Encryption could in principle be carried out at any of the layers in the OSI model (Price, 1985; Branstad, 1987) and work is proceeding to advise on the most appropriate position. Encryption at the data link layer, for example, could conceal the pattern, duration and quantity of information sent between two users. At a higher level, protection can be much more specific and therefore responsive to the user's needs, and this is seen as more appropriate in an open systems environment.

5.4 The application layer

(ISO 9545; Bartoli, 1983)

The application layer is concerned with the provision of user services (applications) such as file transfer or remote job submission. It is possible for users to define their own applications which use the rest of the OSI model, but obviously from the point of view of full interworking, standards are needed at this layer just as at all the lower layers. Work is still in progress on some of the applications, but implementations of others such as the CCITT X.400 series recommendations for message handling are now available.

Because several applications have common needs in the way that they access and use facilities of the lower layers, a number of standards have been drawn up to satisfy these needs. These are called *application service elements* (*ASEs*) and they form building blocks from which applications can be constructed. Different applications will require the use of a different selection of ASEs. Also some of the ASEs use the services of other ASEs.

We shall first briefly describe some of the application service elements and then look at some specific applications. The ASEs to be discussed are association control, reliable transfer, remote operations and commitment, concurrency and recovery. The five applications to be described are message handling, directory services, file transfer access and management, job transfer and manipulation and virtual terminal services.

5.4.1 Association control service element (ACSE)

(ISO 8649/8650; CCITT X.217/X.227)

The association control service element (ACSE) provides the basic facilities for an application to set up an association with the remote application entity and to release that association either normally or abnormally. The services offered are A-ASSOCIATE, A-RELEASE, A-U-ABORT and A-P-ABORT and these make use of P-CONNECT and the other corresponding presentation service primitives. All other services are provided directly to the application entity by the presentation service. The purpose of association is to pass additional information in the form of an application protocol data unit (APDU) which includes a protocol version number and an application context name which defines which of the application service elements are being used.

5.4.2 Reliable transfer service element (RTSE)

(ISO 9066; CCITT X.218/X.228)

The purpose of the reliable transfer service element is to help applications to transfer information and to recover from either communication failure or end-system failure, while minimizing the amount of retransmission. RTSE uses the services in ACSE. Reliability is achieved through use of confirmed transfer which can be initiated only by the end system possessing the 'Turn'. The other end system may request the 'Turn' in a similar manner to that used in the synchronization layer for tokens. Each reliable transfer of information is an *activity* and this is further delimited using the *minor synchronization points* of the session service.

5.4.3 Remote operations service element (ROSE)

(ISO 9072; CCITT X.219/X.229)

This service element is intended for users in an interactive communication situation. The idea is to provide the user with a view of remotely executed functions independent from the details of the OSI communications services (an abstraction similar to that of object oriented programming). The ROSE services are based on the request/ reply transaction concept, where the recipient of a request carries out some work and replies either that it has been carried out successfully or that an error has occurred.

The three major services of ROSE are to invoke an operation, to send a result or to indicate an error. These services can be requested only within an *application association*, i.e. once an association between two application entities has been established. The application association may be set up using either ACSE or RTSE.

Several optional classes are defined within ROSE based on:

1 The sort of reply expected.

2 Whether the initiator of an application association or the responder or both can invoke ROSE operations.

3 Whether the invoker of an operation must wait for a reply before invoking another operation (synchronous mode) or not (asynchronous mode).

The service definition document for ROSE uses ASN.1 to define the following macros for user facilities:

• BIND and UNBIND to set up and clear the application association. These are mapped directly onto A-ASSOCIATE and A-RELEASE in ACSE (or similar facilities in RTSE).
• OPERATION and ERROR which are mapped onto the ROSE services.

5.4.4 Commitment, concurrency and recovery (CCR) service element

(ISO 9804/9805)

The commitment, concurrency and recovery (CCR) service element allows a single application which is achieved through a number of individual connections, to coordinate its activities and ensures the reliability required for such multiparty communications. An example

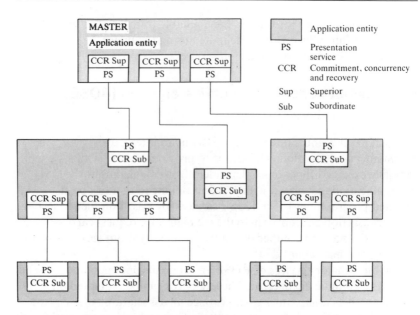

Figure 5.5 Tree of connections between application entities for CCR

would be where it was required to update the same record in several copies of a data base.

The basic concept on which CCR is based is the *atomic action*. An atomic action consists of a sequence of operations which either must all be completed, or none of which must be completed. Cooperating application entities must be arranged in a tree structure; this can be either a static arrangement agreed before any connections are made or a dynamic arrangement where new nodes can be added. Figure 5.5 illustrates the arrangement. How the tree is formed is a matter for each application and is not defined in CCR. An atomic action is controlled by a *master* application entity at the root of the tree, then the interactions take place between a *superior* entity which initiates the atomic action and *subordinate* entities which receive and carry out the atomic action. Note that the defined service is provided on top of the the presentation layer at each application entity, and operates between each (superior, subordinate) pair as shown in the Figure. Different application protocols may be used on different branches of the tree.

The primitives used in a successful atomic action are illustrated in Figure 5.6(a). This is a *two-phase commit* procedure, the first phase being to indicate some action to the subordinates and the second to ensure that all the subordinates carry out that action at the same time. In the first phase the atomic action is started with a C-BEGIN to each subordinate. Information is then passed between superior and subordinates using the normal facilities of the lower layers. The amount of data sent will depend on the particular application involved;

information about the action must be kept in *stable storage*, that is
somewhere where it could be recovered after a system crash. It is also
necessary for each subordinate to take a note of its initial position for
possible reinstatement. In order to prepare for commitment, the
superior sends C-PREPARE to each subordinate. Each subordinate
must respond either with C-READY or C-REFUSE. The atomic action
is completed only if all the subordinates are ready to go ahead. In phase
two, the superior sends a C-COMMIT to each of the subordinates, and
they must carry out the action completely, updating all relevant
information, before replying with an acknowledgement. Alternatively all
subordinates are instructed to roll back to the initial state and not
implement any changes (Figure 5.6(b)).

Concurrency is provided by the knowledge that a previous atomic
action has been completed before another atomic action which depends
on it can take place. Unlike some of the service options at lower layers
in the OSI model, all the services of CCR must be made available if it is
used, in order to ensure full reliability of the service.

5.5 Message handling

(ISO 10021; CCITT X.400; Malde, 1986; Kirstein, 1986)

In electronic messaging, a user enters a message indicating the subject
and the recipient's electronic address, the message is then forwarded
across the network and stored in the recipient's mailbox where it is
available for inspection at a later time. Such facilities have been offered
for some time by a variety of electronic mail systems which have
evolved based on ad hoc proprietary or emerging open communication
principles. The advantages of electronic messaging are: the ability to
send a message without the need for the direct services of other people
such as postal or facsimile operators, the speed at which the message is
carried and the asynchronous nature of the transfer, the recipient and
the sender each accessing the system at a time convenient to themselves
(unlike a telephone conversation). Scope for additional features includes
notification of delivery, message forwarding for mobile users,
distribution lists, message translation and electronic notice boards.

The CCITT X.400 series of recommendations provides a framework
for the preparation and transmission of messages of many types (e.g.
telex, teletex, facsimile, voice, IA5 characters), although the current
definitions deal in detail with character messages only. The X.400
recommendations were first published in 1984 and implementations for
character messages have followed swiftly. An individual demonstration
by Bull, ICL and Siemens in 1985 has been followed by demonstrations
involving 21 concerns including PTTs from Japan, North America and
Europe at the Telecom '87 exhibition in Geneva. This application has

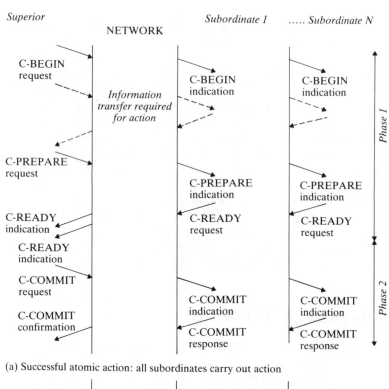

(a) Successful atomic action: all subordinates carry out action

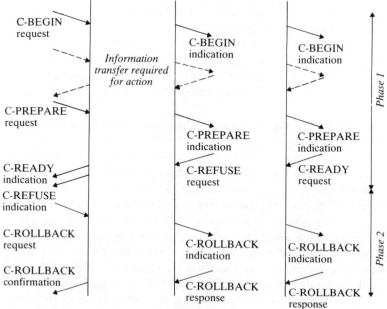

(b) Unsuccessful atomic action: no changes made

Figure 5.6 Two-phase commit in CCR

Table 5.2 X.400 recommendations

X.400	Message handling system model and service elements
X.401	Categorizes the basic service elements and optional user facilities
X.408	Conversion rules between different types of encoded messages
X.409	Presentation transfer syntax and notation for X.400 messaging
X.410	Remote operations including the use of lower layers in the OSI model and the reliable transfer service
X.411	Message transfer layer including the P1 and P3 protocol definitions
X.420	Interpersonal messaging user agent layer—defines the sublayer that provides user interaction
X.430	Access protocol for Teletex terminals

therefore shown the way in which OSI principles are taken up where there is a widespread need for a particular service.

The CCITT has worked closely with ISO's proposals for a message oriented text interchange system (MOTIS), and the two systems are now compatible. The 1988 (*Blue Book*) version of X.400 addresses such topics as security including authentication of users and proof of submission and delivery, distribution lists which may be used by simply giving a group name, a message store facility for use by systems that are not switched on for 24 hours every day and the provision of physical delivery (e.g. printed output). It also shows how the X.500 directory services (see Section 5.6) can be used to provide more user friendly access.

5.5.1 Overview of the X.400 recommendations

This overview covers the 1984 X.400 series of recommendations; each of the eight recommendations defines some aspect as shown in Table 5.2.

The message handling system model is depicted in Figure 5.7. Users of the model are either people or computer programs. The user interacts with a *user agent* (*UA*) which assists in the preparation and reception of messages. The transfer of messages takes place between *message transfer agents* (*MTAs*). Each message consists of an envelope and its contents; the information in the envelope is used to send the message across the network. Each message may either travel direct from the nearest message transfer agent to that used by the remote user agent, or be relayed via other MTAs. A wide variety of terminals is catered for; some typical physical configurations are shown in Figure 5.8.

The only user agent service defined in X.400 (1984) is the interpersonal messaging (IPM) service which enables a person to prepare messages and to submit them to and receive them from the

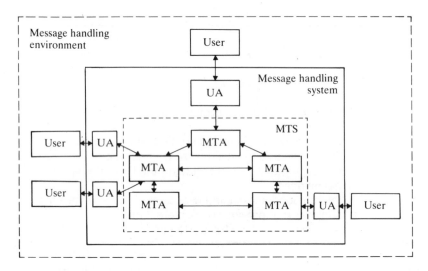

UA   User agent
MTA Message transfer agent
MTS Message transfer system

Figure 5.7   Message handling system model (after CCITT X.400)

message transfer system. It also offers cooperation with other user agents and message storage and manipulation functions.

5.5.2 Layered representation of the message handling system model

The message handling system uses the lower layers of the OSI model to establish connections between individual systems over different kinds of underlying networks and to signal to lower layers that it wishes to use the presentation syntax defined in X.409.

The message handling functions themselves are divided into two layers, the user agent layer and the message transfer layer. The protocols by which the various elements exchange information are illustrated in Figure 5.9. The entities shown in the figure are:

1   The *user agent entity* (*UAE*) provides interactions with the user; it can communicate with other UAEs and with the lower layers.

2   The *message transfer agent entity* (*MTAE*) coordinates with other MTAEs to ensure the transfer of messages.

3   The *submission and delivery entity* (*SDE*) enables the UAE to accomplish message transfer through the services of the message transfer layer which is located in another system.

The three entity types make use of three protocols—the message transfer protocol, the submission and delivery protocol and the cooperating user agent protocol.

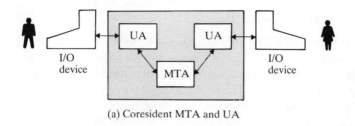

(a) Coresident MTA and UA

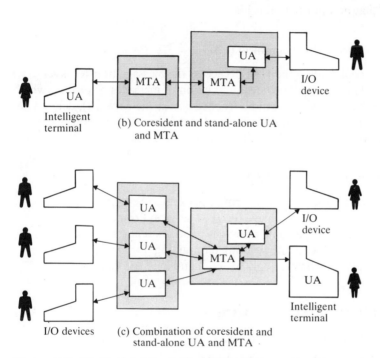

(b) Coresident and stand-alone UA and MTA

(c) Combination of coresident and stand-alone UA and MTA

Figure 5.8 Typical configurations for message handling systems (after CCITT X.400)

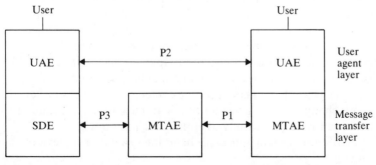

Figure 5.9 Protocol interaction in X.400 (after CCITT X.400)

Message transfer protocol (P1)

The P1 protocol describes how messages are relayed between message transfer agents. Each message consists of the message content plus envelope, the envelope providing all the information to route the message to its destination. The protocol has two types of elements—those carrying messages and those providing control between the MTAs. Transmission is accomplished by means of the reliable transfer server of X.410, which uses the session and other layers to ensure that no information is lost.

Submission and delivery protocol (P3)

This protocol enables the UAE to access the message transfer layer services. The protocol includes submission of messages or probes (which check that an individual can be reached across the network), delivery of messages to the SDE, notification and management operations to register with and control the other operations in use.

Co-operating UA protocol (P2)

The P2 protocol specifies the form of packets to be sent between the user agents in order to achieve the IPM service. Protocol data units contain either interpersonal messages or status reports. The contents of a message are divided into heading (containing, for example, subject, who to reply to, when to reply by, etc.) and message body.

5.6 Directory services

(ISO 9594; X.500; Patel and Ryan, 1988)

Computer network users need to know where services or other people are located, and should be able to specify the service or person using a 'user friendly' name which is easy to remember. A directory service can satisfy these needs, the two main provisions being user friendly naming and translation from a name to an address. A directory is normally implemented as a data base, possibly distributed over more than one system, and the directory service responds to user requests to extract information or to update items in the data base. A reasonable assumption is that enquiries occur more frequently than update requests. By allowing changes to be made to data base entries, the directory can ensure that a person can be contacted even if his or her location has changed. Also, given a suitable framework for the data base, names can refer to a variety of entities such as groups of users (e.g. for distribution lists in a message handling system) or to devices (e.g. printers).

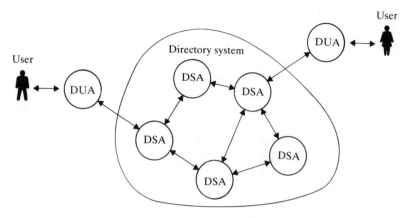

Figure 5.10   Environment for directory services

Since the availability of a directory is central to the ease of use of
network facilities there are several established solutions e.g. for the
ARPANET (Crocker, 1982) and the UK academic network JANET
(Larmouth, 1983). More recently the CCITT and OSI have worked
together to produce recommendations within the OSI framework for a
directory service (CCITT X.500, ISO 9594).

The X.500 series of recommendations defines the models, services and
protocols of a directory service and an authentication framework. It also
defines a number of useful classes of objects about which information
can be stored and the attributes (or properties) which comprise that
information. Examples of objects are person, organization, locality and
device; examples of attributes are countryName, postalCode,
internationalISDNNumber, presentationAddress and businessCategory.
The object class and attribute definitions of the recommendations may
be supplemented by other object class and attribute definitions defined
by an administrative authority.

The distributed data base is called the directory information base
(DIB); it may be located in one or more open system and is accessed
through a directory service agent (DSA). The DIB is a logical tree in
which every entry has a unique identifier. User access is through a
directory user agent (DUA) which communicates with a DSA (see
Figure 5.10). DSAs also need to communicate with each other in order
to maintain the data base and to forward user enquiries or results.

Users may interrogate or modify the data base through the DUA.
The interrogation options are to read a single entry, to compare a
stored value with a given value (suitable for password verification), to
list the immediate subordinates of an entry in the logical tree and to
search for all entries that have certain attributes (e.g. have the same
businessCategory). The facilities commonly found in paper 'Yellow
Pages' directories can be made available through a combination of the

list and search options. The DSA may respond with the information directly or it may suggest that the user consult an alternative DSA. Optionally, it may forward the request to other DSAs either by chaining (forwarding requests through intermediate DSAs) or by multicasting (sending the same request to several DSAs). Where the information cannot be found or the request was incomplete, a suitable error indication is returned. Modification options are to add, delete or modify entries in the DIB.

Two protocols are defined: the first between the DUA and the DSA and the second between DSAs. The protocols are defined by the mapping made onto other OSI services including ROSE, ACSE and presentation layer services. The operations of the directory service and the protocol definitions are formalized in the X.500 recommendations using the abstract syntax notation ASN.1.

5.7 File transfer, access and management

(ISO 8571; Lewan and Long, 1983; Linington,1984)

This is an extensive recommendation covering many options and responds to a wide variety of user needs, for example the use of remote data bases across a WAN or the implementation of a fileserver on a LAN. *Transfer* involves moving part of or the whole of a file from one place to another. *Access* allows the contents of a file to be read, modified or partially updated. *Management* allows files to be created or deleted, or changes to be made to file attributes such as the level of security for a particular access to be made. File transfer, access and management (FTAM) may also be used to pass information between application processes that do not use a real filestore.

The definitions of FTAM are:

- The virtual filestore—a well-defined file structure to which real file storage arrangements can be mapped.
- The file service—the user of the service may be either a person or another program.
- The protocol specification—between peer entities offering the file service.

FTAM makes use of a presentation service offering an appropriate transfer syntax.

5.7.1 The virtual filestore

To enable users with different local implementations of filestores to transfer and manipulate files, a common view of file structure and management must be agreed by those users. In FTAM, this is provided by the *virtual filestore*. Each application entity maps the real filestore to

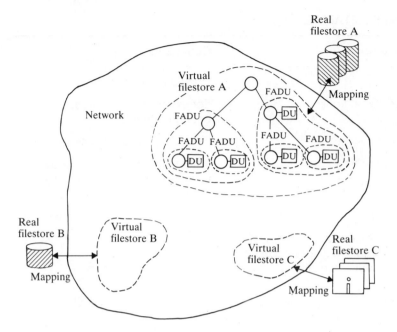

Figure 5.11   FTAM virtual filestore and hierarchical file structure

its virtual filestore, a form that can be understood and used by remote
application entities (see Figure 5.11). The virtual filestore's file structure
is a tree of arbitrary size and can be simplified to cover flat or
unstructured files depending on the 'access context' agreed between the
two application entities. Each subtree is called a file access data unit
(FADU) and may have a data unit (DU) containing information
associated with it. The properties of each file in the virtual filestore are
described by attributes; these are of two types—file attributes and
activity attributes.

File attributes

These attributes relate to each file in the filestore, each attribute of a file
has a unique value which is seen globally by all application entities. File
attributes are listed in Table 5.3.

Activity attributes

These attributes relate to the activity that takes place during a file
transfer or access (see Table 5.4). Some activity attributes (such as
access request) show the active state of the file attributes; others (such as
location) are concerned with the state of the exchanges taking place.

Table 5.3 FTAM file attributes

File attribute	Comments
Name	Unique to any filestore
Permitted actions	For example, insert, read, erase, change attribute
Order of access	For example, random, sequential
Access control	Whether checks are required on initiator identity or password for read, insert, replace, etc.
Account	Authority responsible for charging
Data and time of creation, last modification, last read access and last attribute modification	
Identity of creator, last modifier, last reader, and last attribute modifier	
File availability	Whether a delay is expected before a file can be opened
Contents type	Set of abstract syntaxes, etc., to be used by presentation service
Encryption name	Defines algorithm used
Size of file	Current and expected maximum size
Legal qualifications	Relate to national data privacy regulations
Private use	Access restrictions

Attribute groups

Not all of these attributes will be used during every FTAM connection. However, rather than offering a completely flexible choice of attributes to individual users, a number of sets of attributes are defined to encourage the use of FTAM and promote a high degree of interworking. These sets are defined as follows:

- *Kernel subset*—attributes that must be defined, e.g. filename, contents type, current processing mode.
- *Storage subset*—attributes relating to the storage of a file (e.g. account, creator identity, date and time of last modification, current access context).

Table 5.4 FTAM activity attributes

Activity attribute	Comments
Access request	For example, read, insert, etc., for checking access to file when it is opened
Initiator identity	Tells the responder who has initiated the action
Password	Separate values can be given for each access request attribute
Calling and responding applications' titles	Names of communicating application entities
Account	Responsible authority
Access context	View of file access structure (hierarchical, flat or unstructured)
Concurrency	Restrictions for shared or exclusive access
Location	Position within file
Processing mode	Current operation: read, write etc.

- *Security subset*—attributes relating to security and access control (e.g. access control, encryption name, current access password).

5.7.2 The file service

The service offered to the user (which may be a person or another program) by the application gives a choice of two levels of reliability:

- A *user correctable service* where the user will decide what action to take if errors occur.
- A *reliable service* which ensures that the file service handles all errors up to some agreed quality of service.

The large number of service options offered may be customized for a particular application by combinations of classes—again to encourage interworking. Four classes are available: file access, file management, file transfer and management, and the unconstrained class.

5.7.3 The FTAM protocol

The protocol used for communication between peer application entities is large and only a brief summary is given here. The file PDUs required

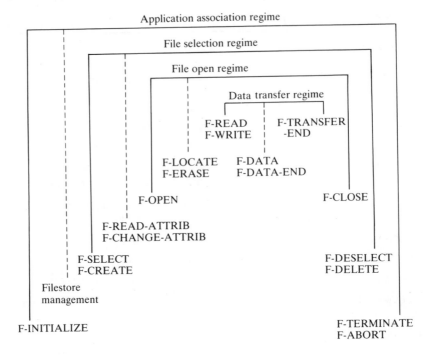

Figure 5.12 File service regimes

must support the various parameters of the FTAM service; that is they must provide the coding to indicate normal progress and error recovery, file selection, creation and deletion, reading and changing of attributes, reading and updating file entries, etc. An abstract syntax definition is used to define each file PDU as a structured data type.

FTAM makes full use of the lower layers. The file PDUs are carried either by application service element PDUs or by presentation data PDUs. The association control service element (see Section 5.4.1) is always used to set up associations between the two application systems. The CCR application service element may be used to support commitment control. Error recovery is provided by session layer synchronization points. A special presentation context is used to ensure that the FTAM PDUs are always identified as protocol units rather than user data.

Each of the FTAM activities can take place only at the appropriate time in the connection when all previously required services have been invoked. The connection time consists of a number of nested periods called *regimes* and Figure 5.12 indicates within which regimes the various service parameters may be used. For example, a file cannot be written to unless the FTAM activity has started, the file has been selected and opened and the appropriate record has been located.

5.7.4 Implementation

The standards documents as usual do not dictate how the implementation should be done, but it is intended that FTAM can be used between systems that are very different in terms of sophistication. For example, in a small single user system, FTAM may be provided as a utility program. In a multiaccess system, it may be provided as a permanently active subsystem that reacts to requests from the users.

5.8 Job transfer and manipulation

(ISO 8831/8832)

Job transfer and manipulation (JTM) provides the facilities for jobs of work to be specified, submitted and processed, and for the results to be transferred to the appropriate location. A job may need the services of several open systems as it progresses. JTM allows jobs to be controlled and manipulated as they pass through the various open systems involved. The processing to be carried out by the job is implementation dependent. Thus JTM is suitable not only for the execution of conventional computer jobs on a large batch processing machine, but also for other activities such as requesting a system (or human operator) to supply goods and to dispatch invoices and delivery notes.

JTM does not attempt to standardize job control language; instead it enables all necessary instructions and data to be transferred as documents to the system that carries out the processing.

JTM entities which cooperate in the progress of the job are called *agencies* (see Figure 5.13). The user issues a J-INITIALIZE-WORK request to an *initiating agency*, which creates a work specification including information such as: who submitted the job, a job reference name and where output documents are to be sent. A typical job will consist of a number of subjobs (the smallest part of a job known to JTM) and it is possible for further subjobs to be spawned by the system on which the job is processed. Once the initialization agency has responded to the user's J-INITIALIZE request, the user need have no further interaction with a job's progress.

Three other agencies are involved—these are source, sink and execution agencies. Whenever a document is required (such as data to carry out a subjob) a J-GIVE is issued to the *source agency* (for example a file system) which supplies the document. When output, such as a line printer or plotter output or a user message, is produced a J-DISPOSE is issued to transfer the output to an appropriate *sink agency*. The *execution agency* is the processing system and will carry out each of the job steps invoking J-GIVE and J-DISPOSE when appropriate. This procedure is also illustrated in Figure 5.11. Some of the agency functions may be contained within a single open system.

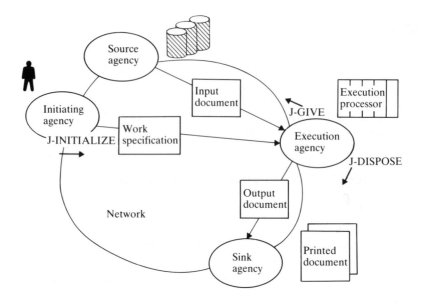

Figure 5.13   Role of agencies in JTM

There is a great deal of flexibility in the system, for example subjobs may be run on different open systems or may spawn further subjobs. JTM uses CCR to ensure that all systems have the same view of how the job has progressed. If for any reason a transfer cannot be carried out immediately, hold queues are provided by the agencies so that the job can be continued at a later time either by JTM or under the control of a user. This is particularly useful for mobile systems that may only be connected for short periods of time or where not all communicating systems are located in the same time zone. Job manipulation, that is the monitoring or changing of jobs, is achieved by using a different initiation primitive J-INITIALIZE-WORK-MAN which also involves passing documents between agencies. Results can either be returned as soon as possible or stored for later access.

JTM must ensure that its documents are obtained from, destined to and executed by the correct systems, so it offers authentication procedures that can check user and system names, audit traces and password encryption.

5.9 Virtual terminal services and protocol

(ISO 9040/9041; Lowe, 1983)

A person wishing to access the facilities of a remote machine, for example logging-on to and using a remote system will need virtual terminal facilities. For many years, triple X protocols have been used (see Chapter 4), but a more flexible and powerful standard has now been drawn up (ISO 9040). The application service and protocol to support virtual terminals is currently defined for what are called basic class terminals. The virtual terminal is agreed by viewing the characteristics of *a display object*, a one-, two- or three-dimensional array of elements, each of which may be represented by a graphic character. By so doing, interactive applications requiring terminal access can communicate using an agreed view of the same display object (or virtual terminal). The open system at one end maps the virtual terminal to a real display terminal and the associated open system across the network manipulates the information within the virtual display. Another type of object (a control object) defined in a similar way can be used for remote control or signalling, e.g. to sound a bell or light a lamp. Communication may be either in synchronous mode with a single dialogue providing alternate communication in each direction or in asynchronous mode consisting of two independent one-way dialogues.

Each element in the virtual terminal may be defined in any one of up to four character codes depending on the repertoire of the connection. Additional information which may be conveyed about each element includes font type, foreground and background colour, highlighting, etc., thus allowing both for very simple or very complex devices. Because it is necessary to agree the nature of the display object, its capabilities and how it should be controlled, the virtual terminal protocol is particularly rich in PDUs for negotiation; indeed the only options in the basic class protocol are in the way the negotiation takes place.

The two open systems use a conceptual communication area (CCA) containing one or more display objects whose state is changed by invoking the various virtual terminal services. This is illustrated in Figure 5.14.

5.10 UK coloured book protocols

The UK's joint academic network (JANET) has evolved over many years and offers facilities similar to those described in Chapter 4 and this chapter. The interim standards used for the network are described in a series of coloured books. They include a transport protocol to enhance X.25, a network independent file transfer and network mail (see the reference section). The coloured books have been a very important stepping stone not only for this community but also in encouraging manufacturers to realize the importance of open systems interconnection and take on the implementation task. Although they have been in use for many years they are now obsolescent, because of the more recently defined OSI standards, and are gradually being

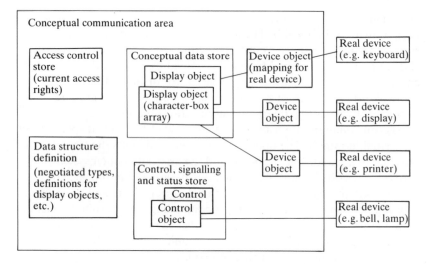

Figure 5.14 CCA partitioning for virtual terminal applications

phased out. A good descriptive overview of these protocols is given in Marsden (1986).

5.11 Summary

We have now reached the end of our explanation of the individual layers of the OSI model. In this chapter we have concentrated on the application-oriented protocols which assume that any differences in the physical network and its operation are hidden by the transport layer. At the application layer we have looked at the way problems common to several applications can be solved by application service elements, and have also looked at five examples of specific applications. Other applications are now being defined, and further applications can be expected when the full seven layers of OSI are widely implemented.

References

Standards documents

The latest situation on current standards should be obtained from your national member of ISO (for example BSI in the United Kingdom and ANSI in the United States). Note that the CCITT plenary assembly makes revised recommendations every four years, the most recent at time of publication being 1988. The following list giving status at 30 January 1990 is a guide to some of the more important documents.

Note that DIS = Draft International Standard, TR = Technical Report

Subject	ISO document	Related CCITT document
OSI reference model	7498	X.200
OSI: service conventions	TR 8509	X.210
Session service definition	8326	X.215
Basic connection-oriented session protocol specification	8327	X.225
Connectionless session protocol	DIS 9548	
Connection-oriented presentation service definition	8822	X.216
Connection-oriented presentation protocol specification	8823	X.226
Connectionless presentation protocol specification	DIS 9576	
Abstract syntax notation one (ASN.1)	8824	X.208
Basic encoding rules for ASN.1	8825	X.209
Application layer structure	9545	
Service definition for the association control service element (ACSE)	8649	X.217
Protocol specification for the association control service element (ACSE)	8650	X.227
Connectionless ACSE protocol specification	DIS 10035	
Service definition for the commitment, concurrency and recovery service element (CCRSE)	DIS 9804	
Protocol specification for the commitment, concurrency and recovery service element (CCRSE)	DIS 9805	
Reliable transfer Part 1: model and service definition	9066–1	X.218
Part 2: protocol specification	9066–2	X.228
Remote operations Part 1: model, notation and service definition	9072–1	X.219
Part 2: protocol specification	9072–2	X.229
Message handling	10021	X.400 series
The directory	DIS 9594	X.500 series
Office document architecture (ODA) and interchange format(ODIF); document transfer, access and manipulation (DTAM)	8613	T.400 series
File transfer, access and management (FTAM) Part 1: general introduction	8571–1	
FTAM Part 2: the virtual filestore definition	8571–2	
FTAM Part 3: the file service definition	8571–3	
FTAM Part 4: the file protocol specification	8571–4	
Job transfer and manipulation (JTM) concepts and services	8831	
Specification of the basic class protocol for JTM	8832	
Virtual terminal service: basic class	DIS 9040	
Virtual terminal protocol: basic class	DIS 9041	

Further references

Bartoli, P. D. (1983) 'The application layer of the reference model for the open systems interconnection', *Proc. IEEE*, **71** (12), December, 1404–1407.

Black, U. D. (1989) *Data Networks: Concepts, Theory and Practice*, Prentice-Hall International, London, UK.

Branstad, D. K. (1987) 'Considerations for security in the OSI architecture', *IEEE Network*, **1** (2), April, 34–39.

CCITT X.215 (1988) *Blue Book*, Vol. VIII, Fascicle VIII.4, International Telecommunication Union, Geneva.

CCITT X.400 (1984) *Red Book*, Vol. VIII, Fascicle VIII.7, International Telecommunication Union, Geneva.

Cohen, D. (1981) 'On Holy Wars and a plea for peace', *IEEE Computer*, **14** (10) October, 49–54.

Crocker, D. H. (1982) 'RFC 822: standard for the format of ARPA Internet text messages', University of Delaware, USA, August 1982 (see the end of Chapter 8 for details of obtaining RFCs).

Emmons, W. F and A. S. Chandler (1983) 'OSI session layer: services and protocols', *Proc. IEEE*, **71** (12), December, 1397–1400.

Halsall, F. (1988) *Introduction to Data Communications and Computer Networks*, 2nd edition, Addison-Wesley, Wokingham, UK.

Hollis, L. L. (1983) 'OSI Presentation layer activities', *Proc. IEEE*, **71** (12), December, 1401–1403.

Kirstein, P. T. (1986) 'Computer based message services', *Computer Commun.*, **9** (2), April, 60–66.

Larmouth, J. (1983) 'JNT name registration technical guide', Salford University Computer Centre, UK, April.

Lewan, D. and H. G. Long (1983) 'The OSI file service', *Proc. IEEE*, **71** (12), December, 1414–1419.

Linington, P. F. (1984) 'The virtual filestore concept', *Computer Networks*, **8** (1), 13–16.

Lowe, H. (1983) 'OSI virtual terminal service', *Proc. IEEE*, **71** (12), December, 1408–1413.

Malde, S., (1986) 'Message handling standards', *Computer Commun.*, **9** (2), April, 78–85.

Marsden, B. W. (1986) *Communication Network Protocols*, 2nd edition, Chartwell Bratt, Bromley, Kent, UK.

Patel, A. and V. Ryan (1988) 'Electronic directory services', *Computer Commun.*, **11**, (5), October, 239–244.

Price, W. (1985) 'Standards for data security—the state of the art', *Computer Commun.*, **8**, (5), 231–234.

Tanenbaum, A. S. (1988) *Computer Networks*, 2nd edition, Prentice-Hall, Englewood Cliffs, New Jersey.

UK academic community coloured books: interim standards

(obtainable from the Joint Network Team, Rutherford Appleton Laboratory, Chilton, Didcot, Oxfordshire, UK)

Yellow Book   Network independent transport service, prepared by Study Group 3 of the PSS User Forum, SG3/CP(80)2, 1980. (Concerned with both transport and network layers of OSI. Includes protocol for use over X.25.)

Blue Book   A network independent file transfer protocol (NIFTP), prepared by the File Transfer Protocol Implementors Group of the Data Communications Protocol Unit, FTP-B(80), 1981.

Red Book   A network independent job transfer and manipulation facility, prepared by the JTP Working Party of the Data Communications Protocol Unit DCPU/JTMP, 1981.

Grey Book   Joint Network Team mail protocol, C. J. Bennett, Department of Computer Science, University College London, 1982.

Green Book   Character terminal protocols on PSS: a recommendation on the use of X.3, X.28 and X.29, prepared by Study Group 3 of the British Telecom PSS User Forum, SG3/CP(81)/6, 1981. Includes TS29 as an annex.

6 Proprietary networking architectures

MARK CLARK

6.1 Introduction

In this chapter two of the most common and successful commercial network architectures, namely DECnet and SNA, are discussed and a comparison is made with the OSI model.

6.2 DECnet—digital network architecture

(DEC, 1982)

In the early 1970s, the Digital Equipment Corporation (DEC) was producing minicomputers and not mainframe systems. Each would support its own local terminals over standard communication lines such as V.24. DEC developed its networking product to support communication between minicomputer hosts that would be used within a company. There was no requirement at that time to support terminals since this was dealt with directly by the local host.

DEC announced its digital network architecture (DNA) in 1975. DNA was immediately conceived with sophisticated networking using distributed network management and control providing dynamic routing through meshed networks. DNA was designed to achieve the following goals (Wecker, 1980):

1   Create a common user interface hiding the network topology from the user.

2   Support a broad spectrum of applications which share resources, allow distributed computation and provide efficient communication with remote sites.

DNAnet is the family of communication products, both software and hardware, which implement the DNA architecture and supports synchronous and asynchronous, full and half duplex communication

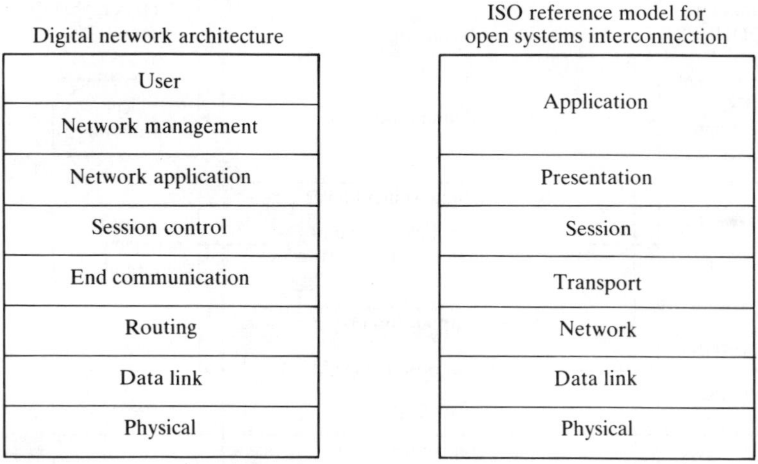

Digital network architecture

ISO reference model for
open systems interconnection

Digital network architecture
User
Network management
Network application
Session control
End communication
Routing
Data link
Physical

ISO reference model
Application
Presentation
Session
Transport
Network
Data link
Physical

Figure 6.1   Comparison of DECnet and the OSI reference model

devices. DNAnet networks support a variety of communication channels such as leased lines, satellite links, Ethernet local area networks, X.25 based packet switching networks and local links. DNAnet architecture allows a wide range of network topologies ranging from point-to-point, star or bus to hierarchical structures. The system is highly distributed and allows for security at several levels.

6.2.1 Comparison of DNAnet with the OSI reference model

DNA might be considered to have eight layers when compared to the OSI reference model. Some of the layers contain network control functions, providing more than is required to handle layer-to-layer communications; for example configuration services and logical connection services are included in the DNA layers. The network management layer might be thought of as a very privileged application which may make calls on all levels of the layered model (see Figure 6.1). However, the calls are direct access to the appropriate layer and not calls made through the layers to the required level.

DNA defines layers and protocols from the lowest to the highest for host-to-host communication, as seen in Figure 6.2. A user process on host 1 might make a request for communication with a process to be executed on host 2. The request will use symbolic names for the remote host which must be mapped by the session control module (SCM) to provide a node address on the network for host 2. The SCM will make a request upon the network services protocol (NSP) module for the establishment of a logical channel providing a full duplex logical circuit between the requesting and requested host processes. The NSP is also responsible for activating a timer which will monitor the establishment

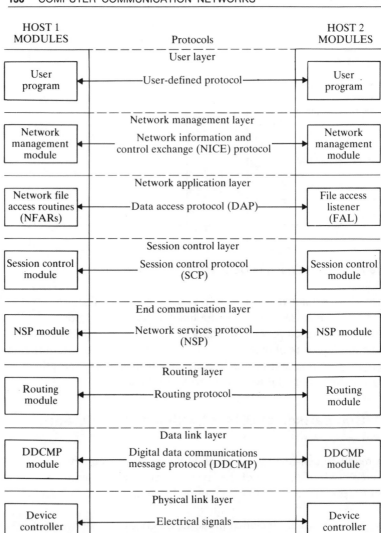

HOST 1 MODULES   Protocols   HOST 2 MODULES

User layer
User program ← User-defined protocol → User program

Network management layer
Network management module ← Network information and control exchange (NICE) protocol → Network management module

Network application layer
Network file access routines (NFARs) ← Data access protocol (DAP) → File access listener (FAL)

Session control layer
Session control module ← Session control protocol (SCP) → Session control module

End communication layer
NSP module ← Network services protocol (NSP) → NSP module

Routing layer
Routing module ← Routing protocol → Routing module

Data link layer
DDCMP module ← Digital data communications message protocol (DDCMP) → DDCMP module

Physical link layer
Device controller ← Electrical signals → Device controller

Communication lines forming physical connection

Figure 6.2 DECnet communication between hosts

of the connection. The NSP sends a request to the layers below, which provide a transport service, requesting a connection be initiated by sending a connect initiate request message. The routing layer will be responsible for choosing an optimum path from available paths. The request message is decoded to see from the destination address which link to use and hence put the request on the data link layer to be routed to the remote host. At host 2 the request message will be recognized as being for this host and not for forward routing; hence it is passed up to the SCM for validation. If the access permissions are validated the source address will be determined and the requested user process will be initiated and passed over the connection. At host 2 the SCM starts a timer to monitor the response of the process to the request for connection. Assuming all is well, the SCM formats an acceptance response to the request for connection which is passed to the NSP module and hence back to host 1, finally allowing the processes to exchange data.

If the host 1 process requires to access the data on host 2 it can be performed by either using the data access protocol (DAP) in the network application layer, described later, or file data may be obtained from host 2 by allowing the process that is communicated with on host 2 to do the file accesses and then send the data over the connection. In the latter case the network application layer performs no task and provides a null protocol layer.

Physical layer

DNA does not specifically define the physical layer which manages the transmission over a data channel. Many are supported by a DECnet such as RS232C, CCITT V.24 or Ethernet.

Data link layer and DDCMP

The data link layer creates a communication path between adjacent nodes and ensures the integrity and sequential order of the data passed over a channel between nodes. The protocol used at this layer is the DEC defined digital data communications message protocol (DDCMP) (DEC, 1980). DDCMP operates over both synchronous and asynchronous communication links. It is responsible for correct sequencing of data and for error control. In contrast to most network architectures which use a bit-oriented protocol such as HDLC, DDCMP is a byte-oriented protocol which includes a count of how many bytes will be sent in a data portion of a message. DDCMP assigns a number to each data message, incremented by one (module 256) for subsequent messages. It also places a 16-bit cyclic redundancy check error detection polynomial at the end of ech transmitted message.

The receiving DDCMP will check for errors and if none are found then it returns the message number with a positive acknowledgement of message reception. The receiver need not acknowledge all received packets since acknowledgement of message n implies all messages up to and including n have been correctly received. The ACKs may be carried by data messages received by the sender, hence providing a piggy-back of this acknowledgement, or by separate control messages. DDCMP is a positive acknowledgement retransmission protocol. Errors detected at the receiver use timeouts and control messages for resynchronization and retransmission. NAKs may be sent in a control message carrying the message number n of a message which was received in error, but in the case of a sequence error then a timeout at the transmitting station will occur and error recovery may commence. DDCMP supports five types of control message:

Message type	Function of message
STRT	Start message
STACK	Start acknowledge message
ACK	Acknowledge one or more messages, includes acknowledgement number
NACK	Negative acknowledgement
REP	Reply to message number n

Each data message within the protocol is given a message number n, which is contained in the message header. The transmitter sends the message with its count n in the header and a cyclic redundancy check (CRC), a CRC-16 polynomial block check, to the header and data. When the message is sent a timer is started at the transmitter. The receiver checks the CRC with the message data and if received without error then the message number is checked against the expected message number; if correct it may be passed on to a higher layer of protocol. The receiver then sends an ACK control message with the number of data messages received and increments its own expected message number. The transmitter receives the ACK and compares it with the number it was expecting; if similar, then it may discard all messages held for recovery purposes and it informs the higher layer of its success and increases its expected number.

If the transmitter fails to receive an ACK prior to the timer expiration, then a recovery procedure is started. There are of course several possible sources of error and recovery procedures are necessary to handle each of them. The transmitter might send an REP control message containing the message number n. If the receiver had discovered a transmission error resulting in an incorrect CRC for the data message itself then rather than wait for a timeout it could send the receiver a NACK control message with the message number n. If, however, the error was in the header itself then the receiver must do nothing, waiting for a timeout since it cannot assume that the received message was necessarily for itself.

In the event of a persistently erroneous link, after reinitialization an STRT control message is sent which makes the receiver notify its user. If in agreement, the user will have the receiver send an STRT control message and the transmitter will acknowledge with a start acknowledge, STACK, message which must in turn be ACK'd by the receiver. Thus a four-way sequence has both stations reset their message sequence numbers for the data messages and ensures all is correctly initialized.

The specification for DDCMP has features to allow the reporting and recording of errors which may be accessed by the network management function.

The routing or transport layer

The routing layer, called the transport layer prior to the introduction of OSI terminology, carries user data in messages from the end communication layer in the source node and routes these messages possibly through intermediate nodes to the destination node. These messages are effectively datagrams and routing implements a datagram service. In this context a node is any source, destination or intermediate host in the communication path through which the data passes.

This layer will select the route based on the network topology and circuit costs entered by the operator. There is a data base concerned with node-to-link combinations in order to reach the required destination and another data base concerning the number of hops and route cost for a packet to be sent on a particular route. This layer can adapt to changing network topologies in the event of a path failure and periodically updates other routing modules on other nodes.

The layer maintains counters and generally collects data for the network management modules. It also provides congestion control by rejecting packets from the local host in favour of those already en route and for which this host is acting as an intermediate node to an existing link. A packet can only pass through a limited number of nodes or it will be rejected by a loop detection module which reads information carried in the 'transport' header and compares it against a local data base. When no data is being transmitted a check is made to see if the link remains available and echo tests with local nodes are performed to check the link and status of hosts.

End communication layer and network services protocol

The end communication layer (previously called the network services layer) is responsible for the system-independent aspects of creating and managing logical links for network users. This layer allows two processes to exchange data both reliably and sequentially regardless of their network location. The communication is done on a connection basis where the connection between two processes is called a logical link. The protocol used at this layer is the network services protocol (NSP), and it uses three types of message—data, acknowledgement and control. The NSP provides the following functions:

1 To create, maintain and destroy logical links.

2 Data reception and transmission.

3 Flow control.

4 Data packetization.

5 Error control handling.

The SCM requests the NSP layer to create logical links and also to destroy them when communication between processes is closed. The NSP performs data flow control so the receiving process is not swamped with incoming data, end-to-end error control using the positive acknowledgement scheme, message segmentation into data units that can be handled by the transport service and subsequent reassembly in the correct sequence. It corresponds to the transport layer in the ISO reference model.

Session control layer

A process requiring another host to be accessed will request a virtual circuit. The session control layer will map a request for access to a remote host name into a node address and then request the NSP to create a logical link to it. The session control module maintains a node table which can map node names to addresses. Once the logical link is established data may be transferred with this layer acting as no more than a transporter of the data to the NSP from the user's process. When the connection is no longer required by the process this layer requests the destruction of the link by the NSP. This module may also create a new process or activate an existing process to hande incoming requests for service by a remote host requiring a specific application process to be accessible.

Network application layer

The network application layer provides generic services to the user layer for file access and remote file transfer, remote interactive terminal access, gateways to non-DNA systems and resource management programs. This layer is equivalent to both the application and presentation layers in the ISO reference model. The user process may, if it wishes, communicate directly with the session control layer.

One such generic service is the digital access protocol (DAP), which permits remote file access and file transfer in a manner that is independent of the structure of the operating system that is to be accessed. The cooperating processes exchange DAP messages such that the user I/O commands on the remote file are mapped into DAP messages and transmitted. At the remote node the server interprets the DAP messages and performs the actual file I/O required by the user. The use of these facilities greatly simplifies the task of the programmer who would otherwise have to write specific tasks to perform the required functions.

Network management layer

The network management layer provides user control of and access to operational parameters and data information counters in lower layers. It also provides down-line loading and up-line dumping, remote system control and test functions. It has facilities for network event logging and is the only layer that has direct access to each lower layer.

During communication between remote processes this layer will not be used at all. The user processes communicate directly with either the session control layer or the network application layer. The network management layer may be accessed by user programs or system processes that wish to collect statistics from a particular layer or perform a link test, etc. A network information and control exchange (NICE) protocol allows different nodes to exchange commands at the network management layer, allowing system managers to change routing parameters to avoid routes that are proving to be congested or to reconfigure network topologies.

Application or user layer

This is essentially the same as for the OSI reference model.

6.2.2 Observations on the comparison

The comparison of DNA to the OSI model is somewhat subjective. However, there are some clear conclusions that can be drawn.

The presentation layer of the OSI model maps to the network application layer of DNA but certainly does not encompass all those functions provided by DAP since it is concerned with data manipulation and transformation which is not evident or a major concern in DAP. Other modules could be provided at this layer that are more concerned with presentation aspects.

Figure 6.1 shows that the session control layer has only partial overlap with its OSI equivalent since some of the services associated with the OSI model at this layer are found in the lower layer in DNA. Messages passed over the network use a datagram service which means that they may follow different paths since the path will be optimized depending on algorithms implemented in the routing layer, and message delivery is on a best-effort basis. Thus the provision of a data stream to user processes has to be performed in the network services protocol layer.

The network management function is unusual in terms of the OSI model since it has hooks at all layers of the DNA. This is valuable to monitoring and management functions which do not have to suffer the overheads associated with access to data via the layers. It provides rapid reconfiguration as well as excellent tools for remote network management. It should be noted that DECnet is a homogeneous architectural network since it is supported by one manufacturer only and by a narrow range of computer system architectures. Many of the considerations necessary in a mixed architecture environment may be discounted. DECnet gives a good performance and excellent management facilities for the average company which supports only single vendor equipment on its network. Gateways to most major networks are available and supported by DEC.

6.3 Systems network architecture (SNA)

6.3.1 Introduction

The International Business Machine (IBM) Corporation's requirement for a networking architecture resulted from the growth in requirement for teleprocessing, the name given to the connection between a host computer and its terminals where the connections were made over telephone links or dedicated lines. The terminals were of a static nature and not programmable, as is the case for many modern varieties, thus the support for communications was fixed and no buffer storage for data existed. The supporting processor was required to deal with the communication requirements and because a variety of different types of terminals had to be used for different applications, several support packages could be required simultaneously. This gave IBM a difficult problem so that in 1973 IBM defined an interface and network that would have to be complied with by terminals wishing to communicate with the host. This architecture was the root of the IBM network architecture.

Systems network architecture (SNA), as introduced by the IBM Corporation in 1974, was, and still is, a design specification. SNA specifies the various data communication facilities and functions that need to be provided and performed by the products intended for operating either within, or supporting, a distributed data processing system. SNA is a large topic and a very detailed description is outside the range of this book; the reader requiring further information is recommended to read a text specifically on SNA (e.g. Meyers, 1987).

The SNA concept was made public through the announcement of a number of products in September 1974 (Schultz and Sundstorm, 1980). The products were built according to the SNA specification of which the principle elements were:

1 Virtual telecommunications access method (VTAM), an access mechanism for telecommunications circuits written to run in the host computer system, the IBM 370.

2 Network control program (NCP), which runs in the network communication controllers, IBM 3705.

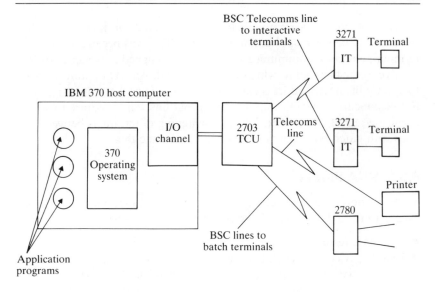

IT   Interactive terminal
TCU   Transmission control unit
BSC   Binary synchronous communications

Figure 6.3   Pre-SNA era IBM data processing environment

3   SNA versions of IBM's 3600 and 3650 cluster controllers used for banking and retail applications respectively.

Thus the first releases of SNA were designed to support terminal-to-host communications only, but provided the necessary facilities for extension to host-to-host communication. Each new release of SNA from version 0 (1974) through to version 4 has enhanced the control functions, the interconnection and data interchange capabilities and the configuration possibilities of SNA-based data communication environments.

6.3.2 Why was SNA required?

In the early 1970s IBM data processing systems provided an interactive service and a batch processing facility for remote locations. This would typically consist of the system 370 host computer with one or more input/output (I/O) communication channels, a $270x$ transmission control unit (TCU), supporting the necessary telecommunication links to remote locations with 2780 batch terminals or 3271 interactive terminals. Binary synchronous communications (BSC), a byte-oriented asynchronous data link protocol, was used for remote communication (see Figure 6.3), while an IBM channel protocol was used to communicate with local terminals.

Pre-SNA systems were unable to switch terminals or devices dynamically between application programs. This was because each terminal, device or communications line was assigned a 'device address' relative to the channel to which it was attached. All I/O operations had to specify the device address explicitly. Thus a user who required access to two concurrently active application programs would require two duplicate sets of equipment. Physical switching of terminals between lines with patch boards was performed to alleviate this.

6.3.3 SNA ideals

SNA was developed to achieve the more flexible communication necessary for a data processing environment; its aims were to:

- Facilitate resource sharing.
- Enable devices to switch between application programs.
- Enable mixed terminal/device types to coexist on the same communications link.

The first step was the introduction of a universally applicable communication protocol, synchronous data link control (SDLC) protocol, similar to HDLC (see Chapter 4). This was not itself part of the SNA but was a product facility announced at the same time. It enabled several lines to be multiplexed into one or more trunks and provided efficient and unrestricted interactions between application programs.

The data communication functions became distributed throughout the system. The communication facilities were provided by dedicated system control programs removing the requirement for this from the application programs.

6.3.4 The SNA specification

SNA as a specification identifies various entities in a data processing environment involved in data communication together with the necessary characteristics of the data transfer between nodes. It specifies:

1 The functions that must be provided to support data communication between the various entities that are served.

2 The protocol to be followed when two entities wish to communicate in an error-free manner and later wish to terminate their communication.

3 The commands associated with the communication procedures.

6.3.5 SNA

The SNA performs all communication between network addressable units (NAUs). All sources or destinations of information are 'end users', each of which access the network via a port called a logical unit (LU). Other NAUs are physical units (PU) which interface a particular piece of equipment to the network through their control functions and system services control points (SSCP), used in network management. All communication with NAUs occurs through messages called either request or response units, each of which has a header relevant to the layer or layers to which it is directed. The headers are of four types:

1 Link header, containing information concerning the messages on an identified link.

2 Transmission header, containing information to control the path control network.

3 Request and response headers, concerning control information regarding two communicating entities.

4 Function management header, to specify actions to be carried out on the data contained in the request or response unit (RU).

All communication must occur within what is termed a session, which may only be set up with the permission of an SSCP.

The LU acts as the intermediate between the transport network and the users, devices and application programs using or attached to the network. (Figure 6.6 shows a comparison to the OSI reference model and the position of the LU functions within the SNA protocols.) The LU is primarily concerned with session protocols for paired end users. The LUs serve as points for ultimate destinations of data and provide the end-to-end flow control, session cryptography, name-to-address translation using distributed directory services of the network and packaging for efficient transfer of messages.

The communications environment is accessed by an end user via an LU which should be thought of as a set of functions that are invoked by the end user to enable it to communicate with another end user, interconnected by a path control network (see Figure 6.4). The physical unit (PU) is a piece of software that controls physical entities and physical operations of a particular equipment type, for example the loading of software to control that equipment's operation. Every SNA network must have at least one system services control point (SSCP) to manage the resources of the SNA environment. The resources controlled by each SSCP are referred to as the domain of that SSCP. An SSCP will control the configuration of all resources within its domain, including the addition and removal of resources from its domain or the activation or deactivation of resources within its domain.

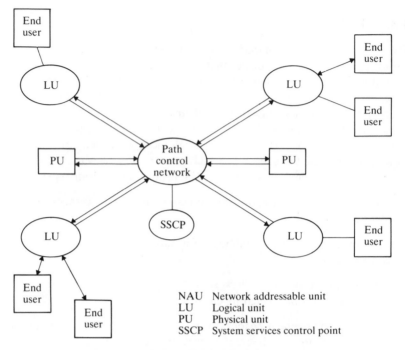

Figure 6.4 SNA for communicating NAUs

It may establish communication paths or test the resources of its domain.

6.3.6 Path control network

NAUs communicate by SNA specified message units, which in SNA are called request units. Responses to them are called response units; these are transmitted by the path control network. Headers relevant to a particular layer(s) are added to the request and response message units where the headers are of four types:

1 Request/response header containing control information relevant to the state of the communiction between LUs.

2 Transmission header containing control information related to the transport of messages through the PCU.

3 Link header containing messages relevant to one particular link.

4 Function management header which is part of the request unit to specify actions to be taken with data in the RU.

The path control network ensures that messages are delivered as specified and in the order in which they were sent. The path control network consists of path control and data link control components. The

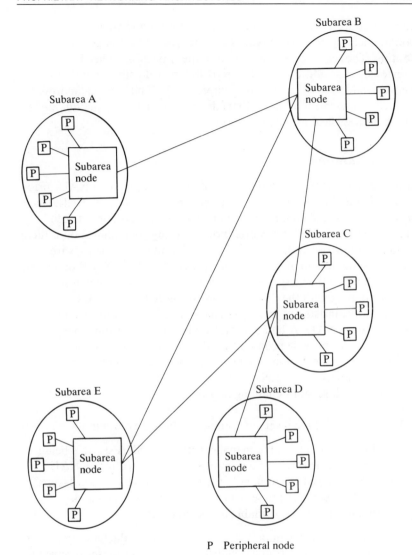

P   Peripheral node

Figure 6.5   SNA network topology

path control is responsible for inter-NAU routing and the data link control organizes the message units across each data link.

6.3.7 Addressing in the SNA

In the SNA network, the architecture consists of 'subarea' and 'peripheral' nodes, as seen in Figure 6.5. The subarea nodes are interconnected to form a spine or major network while the peripheral nodes attach to one subarea network. Every NAU has a network address consisting of two parts: a subarea and element address. The element address must be unique to each subarea and when communicating between subareas the subarea address is used for routing. These addresses are carried in the transmission header seen previously. The address may be up to 48 bits wide but currently the network address used is 23 bits, comprising an 8-bit subarea address and a 15-bit element address. Extended addressing using the full range of the address field may be supported in future.

6.3.8 Sessions

A formal correspondence must be established between two NAUs prior to a dialogue. The initiation and subsequent dialogue is referred to as a *session* in SNA. All communication is in terms of sessions between various types of NAU. SNA does not allow any communication if there is no session between the NAUs. The session must commence with the entity wishing to communicate using a session with the SSCP to request permission for other sessions. Thus, for example, an LU might have one session running with the SSCP and one or more to other LUs.

A session consists of two *half sessions*—one in the primary and one in the secondary NAU. A half session supports the set of functions used by the NAU to provide protocols needed to support communications within a particular session. Two half sessions are interconnected by the path control network. An NAU should be able to support multiple concurrent half sessions enabling multiple concurrent sessions with other NAUs.

Within an LU three layers of communication must occur:

1 At the highest layer is the management of the SNA services being offered to the end user and a set of services related to the NAU services management to handle the half sessions and lower layers. Some of the services are session specific and hence define the presentation services available to the end user.

2 The next layer is the data flow layer and handles the message exchange protocols. It must relate the message responses received to the relevant requests. It may perform data flow control by allowing or refusing message transfer in each half session as required.

3 At the lowest layer of the NAU function is the transmission control layer concerned with session maintenance and handling messages from the path control network to the NAU functions themselves. It builds the messages that are to be sent to the path contol network and in the case of a PU it would assist the handling of the physical equipment associated with it.

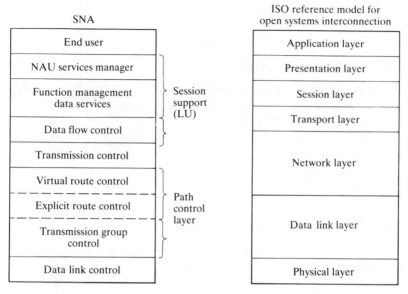

Figure 6.6 Comparison of the SNA layering and the OSI reference model

6.4 Comparison of SNA with the OSI reference model

SNA is a hierarchically layered architecture. The reference model for open systems interconnection is based, as has been seen previously, on seven functional layers and hence there is considerable similarity between SNA and the reference model (Corr and Neal, 1979). Figure 6.6 shows a comparison of SNA layers and the reference model. The function performed at each layer is described in the following sections.

6.4.1 The physical layer

The physical layer of the SNA does not explicitly address this aspect of an SNA link which is only referred to as being an IBM System/370 channel or a switched or non-switched telecommunications link, along with appropriate DCEs. Reference is made to existing international standards such as X.21 and V.24.

6.4.2 Data link control layer

The data link layer and the transmission group control sublayer of the path control together are similar to the functions provided in the ISO data link control layer. IBM's definition of SDLC (IBMa) was released prior to the ISO standard HDLC (ISO 3309,4335,7809) for commercial reasons and minor changes were made to make SDLC a subset of HDLC.

6.4.3 Path control layer

The path control layer overlaps OSI layers but is most readily mapped into the network control layer. It consists of virtual route control, explicit route control and transmission group control. This layer specifies SNA network addresses, the control commands required, formats of message units and the sequences of commands needed to establish, maintain an error-free, and disestablish a route for communication (Ahuja, 1979).

6.4.4 Transmission control layer

The transmission control layer is equivalent to the transport layer and relies upon the underlying layers to provide a logical route such that this layer may communicate through the two halves of a session. It is responsible for functions that establish sessions and perform error recovery as required.

6.4.5 Data flow control layer

The data flow control layer is mapped against the session layer and is responsible for establishing the session between two end users. Once the session is established it is responsible for data exchange.

6.4.6 Function management data services layer

The presentation layer of the OSI model cannot be mapped directly to an SNA layer but is in part the highest layer of the SNA in the form of the network addressable unit (NAU) services management layer and the function management data (FMD) services layer. The NAU services can be loosely classified as offering:

1 End users' services in the LU–LU sessions providing connectivity to end users.

2 Session network services that are used in sessions between SSCPs and SSCP to PUs or LUs. They interface users to functions controlling the network operation.

Data transformation in the form of compression and compaction facilities using string control byte (SCB) is provided in the FMD services layer and the data encryption facility in the transmission control layer. (Note that encryption is not regarded as a presentation layer facility in SNA but rather a security feature available to a session.)

The data handled at the FMD services layer may be of either EBCDIC or ASCII code format but no translation service is offered.

6.4.7 The end user layer

This is not defined in the SNA architecture but should be equivalent to the application layer of the OSI model. However, some of the functions seen in the NAU services layer concerned with network services are in many ways more closely related with the application layer than the presentation layer. Network management services are defined in SNA to assist the end user in the overall network management task which is usually a complicated and organizational process. Support for fault recognition and diagnosis are provided. Tools are provided to assist in the overall management of specific products to perform tasks such as performance and accounting, configuration management and change management.

6.5 X.25 and SNA

An alternative mechanism for routing information is to use the public packet network services. SNA therefore manages virtual circuits in a manner consistent with the management of real circuits, allowing the use of public service networks as available (Deaton and Hippert, 1983). SNA defines two types of connection for X.25 networks, SNA to SNA and SNA to non-SNA. Both support connections by permanent virtual circuits services or virtual call services or both. For SNA to non-SNA, connections are between SNA X.25 DTEs and X.25 DTEs. It should be noted here that SNA supports many alternative data transmission services and a connection between end users may traverse several data links and intermediate nodes.

6.6 X.25 for SNA to SNA connections

The data link control (DLC) elements provide the low level management of the link, for data transmission, and provide the functions such as link initialization and identification exchange between adjacent nodes. SNA X.25 DTEs use a logical link control (LLC) protocol to provide certain adjacent node services. A protocol called qualified logical link control (QLLC) is used to send control information for the data link between SNA nodes. The SDLC functions for exchange of identification and disconnect are among many sent through the public switched data network (PSDN) using qualified data packets. When the link is established then the path information units (PIUs) are transferred in unqualified data packets. If the PIU exceeds the maximum data size of the packet they are transferred as packet

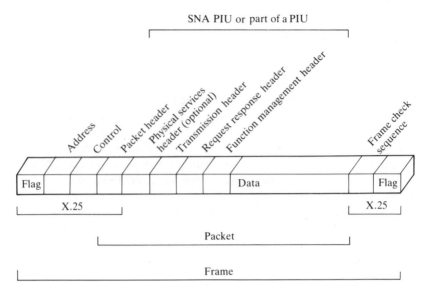

Figure 6.7 Packet and frame formats for SNA connections via X.25

sequences concatenated by the more data mark (M-bit). See Figure 6.7 for packet and frame formats.

6.7 SNA and OSI

IBM have support for OSI (Rutledge, 1982); there is open systems network support software (IBMb) which adds the necessary functionality to provide the OSI network service and on top of that the open systems transport and session support (IBMc) to provide an implementation of the OSI transport and session layers. Interconnection with the application level may be performed through an application bridge such as the IBM general teleprocessing monitor for OSI (IBMd).

Three approaches to SNA/OSI integration were possible and IBM has used each of them (Tillman and Yen, 1990):

1 Direct (proprietary).

2 Indirect (open systems).

3 Mixed (selective layering).

In a manner similar to DECnet, IBM has sought to provide compliance with OSI on a layer-by-layer basis. Direct conversion is dependent upon the availability of processors acting as protocol converters or gateways which provide the necessary protocol translations between SNA and OSI. The disadvantage is the high cost associated with the interfaces and processor required to perform the transformation of protocols. It is also true that the lower layers are far more compatible with each other than the upper layers, resulting in limited interoperability.

IBM has also tried to use indirect mechanisms by providing interworking with other vendors' networks and demonstrating interconnectability. Intermediate nodes can act as gateways from SNA to vendors' proprietary networks and hence through to OSI compliant networks. This has the advantage of lower costs since gateways to existing proprietary networks will usually be simpler and may be required anyway. DEC, for example, is committed to OSI compliance with its fifth generation offering (Edwards, 1988). A dedicated DECnet/SNA gateway provides the necessary link, allowing the users of either system to utilize the capabilities of both systems without being aware of the technical aspects of the connection. The disadvantage of this approach to OSI integration is the lost processing time spent in the vendor protocol system (gateway). There is also some loss of management control and a reduction in overall reliability due to the increased hardware and software.

The final approach has been to use the lower layers of OSI as a transport service for SNA sessions, a technique called selective layering. As discussed earlier, it is relatively easy to obtain compliance at the lower layers and, for example, it would be possible for SNA and OSI networks to run their upper layers independently while sharing an X.25 packet switched network.

IBM has committed itself to OSI and has made product announcements which will assist in all three approaches to OSI integration (Axner, 1988; Booker, 1988).

References

Ahuja, V. (1979) 'Routing and flow control in systems network architecture', *IBM Systems J.*, **18** (2), 298–314.

Axner, D. H. (1988) 'Packet switching offers improved connectivity', *TPT*, December, 40–44.

Booker, E. (1988) 'Latest IBM product avalanche embraces OSI compatibility', *Telephony*, September, 8.

Corr, F. P. and D. H. Neal (1979) SNA and emerging international standards, *IBM Systems J.*, **18** (2), 244–262

Deaton, G. A. and R. O. Hippert (1983) 'X.25 and related recommendations in IBM products', *IBM Systems J.*, **22** (1/2), 11–28.

DEC (1980) 'DNA digital data communications message protocol (DDCMP) functional specification', Version 4.1.0, Order AA–K175A–TK.

DEC (1982) 'Digital network architecture (phase IV)—general description', AA–N149A–TC, May.

Edwards, M. (1988) 'OSI standards are guiding computer vendors to compatibility of components and systems', *Commun. News*, April, 50–53.

IBM Corporation (IBMa) 'IBM synchronous data link control', General Information GA27–3093.

IBM Corporation (IBMb) 'Open systems network support', General Information GH12–5145.

IBM Corporation (IBMc) 'Open systems transport and session support', General Information GH12–5450.

IBM Corporation (IBMd) 'General teleprocessing monitor for open systems interconnection', General Information GB11–8201.

ISO 3309 Data Communication (1984) 'High-level data link control procedures—frame structure'.

ISO 4335 Data Communication (1987) 'High-level data link control procedures—consolidation of elements of procedures'.

ISO 7809 Data Communication (1984) 'High-level data link control procedures—consolidation of classes of procedures'.

Meyers, A. (1987) *A Systems Network Architecture: A Tutorial*, Pitman, London.

Rutledge, J. H. (1982) 'OSI and SNA: a perspective', *J. Telecommun. Networks*, **1**, Spring, 13–27.

Schultz, G. D. and R. J. Sundstorm (1980) 'SNA's first six years: 1974–80', *Proc. 5th Int. Conf. on Computer Communiction*, Atlanta, 1980.

Tillman, M. A. and D. Yen (1990) 'SNA and OSI: three strategies for interconnection', *Commun. ACM*, **33** (2), February, 214–224.

Wecker, S. (1980) 'DNA—the digital network architecture', *IEEE Trans. Commun.* **COM–28** (4), April, 510–526.

7 Local area networks: introduction and medium access techniques

SIMON JONES

7.1 Introduction to local area networks

[A local area network (LAN) is a computer network designed to operate within a limited physical area. Typically the number of devices to be interconnected is small (less than 500) and the maximum network length is less than 2500 metres.] An early introduction to LANs is given by Clark *et al.* (1978); a review of later systems is presented by Stallings (1984). [A LAN can be considered to consist of four elements. The three hardware elements are a transmission medium, a mechanism for controlling transmission over the medium and an interface to the network for the host computer. The fourth basic element is the set of software required in the host computer to implement the protocols which control the communication from one host computer to another. The software functions at different levels to provide a number of protocol layers ranging from packetization to the application level. This combination of layered hardware and software is a noticeable feature of local computer networks.]

Inexpensive hardware, such as a twisted pair or coaxial cable, is used to interconnect systems. The characteristics of LANs are high bandwidth (with data rates in the range 1–20 Mbit/s), low error rate and small delay. These characteristics enable LANs to interconnect processors to provide distributed multiprocessor systems which enable the sharing of resources such as secondary storage and other peripherals.

The LANs described in this chapter are principally used for data transmission. Other forms of information, such as voice and video, are time critical and will not tolerate the variable delays of LANs.

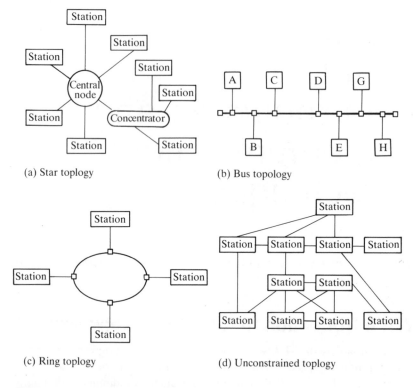

(a) Star toplogy

(b) Bus topology

(c) Ring toplogy

(d) Unconstrained toplogy

Figure 7.1 Local area network topologies

Techniques for handling integrated traffic on LANs and for the larger metropolitan area networks (MAN) are dealt with in Chapters 14 and 15.

7.1.1 Network topologies

There are a number of possible topologies for a computer network, the choice of which is usually determined by the purpose for which the network is installed and the physical location of the devices to be interconnected. The number of devices that are to be connected can dictate the choice of topology. If the number is very small, less than say 10 devices, then a point-to-point connection or polling system is possible, but as the number of devices to be interconnected increases then alternatives have to be considered.

The basic alternatives, shown in Figure 7.1, are: the star, where all devices are connected to a central point; the bus, where all devices are connected by a common link between them; the ring, where all devices are successively linked together and the ends of the connection are brought together again to complete a ring; and the unconstrained network, where links can exist between any one device and any number of other devices. (Note that a device connected to a local area network is often called a *station*.) There are variations of these basic topologies such as branching buses, trees, etc., but in general these can be considered to be extensions of the basic topologies in which the connected devices are components of another bus, star or ring.

7.1.2 Comparison of the basic topologies

The disadvantage of a star compared with a bus or a ring is that each device requires a connection to a central point which can become a bottleneck and, with a large number of connected devices, that the interconnection costs are comparatively high. The advantage of the star configuration is that the devices can use a simple point-to-point protocol rather than a common access protocol.

The interconnection cost disadvantage can be reduced by providing device concentrators located adjacent to a number of devices. This saves the cost of individual connections, distributes some of the control and also allows the possibility of continued operation of parts of the network in the event of the failure of the central point.

The advantage of the bus and the ring is that a simpler interconnection strategy can be used—that of single interconnection from one device to the next. A simple topology allows network configurations to be changed as the number of stations varies. With a bus, stations are simply connected to the bus cable. With a ring, the network is temporarily broken and a new station inserted between two existing stations.

Most LANs are based on either a bus or a ring topology with common access protocols. The unconstrained topology is uncommon due to the necessity to handle the routing of data between devices and the problem of locating a path between devices. The passage of information through a node will affect its performance and will generally require a processor system of similar performance to the host system.

7.2 The physical connection between devices

The physical connection between devices on a LAN is characterized by (Bertine, 1980):

- The nature of the medium, e.g. fibre optic, coaxial cable, current loop, etc.
- The method of transmission, e.g. broadband, baseband.
- The signal levels and the modulation schemes used.
- The methods of signal coding.
- The types of connectors used and their pin allocations.
- The logic levels of the signals.

- The use of point-to-point or broadcast techniques.
- The synchronization method used.

Physical level transmission techniques are described in Chapter 2.

Most LAN systems use synchronous schemes. The requirements of the link are specified in terms of data and error rates; this usually results in a compromise between performance and cost. Other important considerations in the selection of a physical medium are its integrity and reliability. These factors will seriously affect the performance of the network and its ease of use.

In this chapter we shall concentrate on baseband LANs. Some successful broadband LANs have also been implemented which carry information on channels of different frequencies. Often, one channel on a broadband network is shared using an access technique similar to those discussed below. For further discussion on broadband LANs see, for example, Hopkins (1979) or Pliner and Hunter (1982).

7.3 Medium access control techniques

The physical connection ensures that one device can communicate with another. In most LAN topologies part of the transmission system is shared, so some form of control is required to arbitrate between contending devices. This is called the *medium access control* (MAC). In this section, a broad outline is given of the variety of techniques available. In later sections, examples of these techniques are covered in more detail. The techniques to be discussed are token passing and contention for both rings and buses, and slotted and register insertion rings.

7.3.1 Common access control schemes: token passing

A token is a unique control message that conveys the right of access to the medium. The token is passed between devices; the device that received the token most recently is said to hold the token and can access the communication medium. Token passing can be used with both broadcast (token bus) and sequentially connected (token ring) networks.

A number of potential problems exist in token passing systems. The systems must be able to detect and recover from lost or duplicated tokens. These problems have different solutions for ring and bus networks.

In a token ring a unique marker is used to delimit the end of a transmission. When this marker is detected by the next device on the ring it looks for a pattern that represents the token and, if found, inserts its message into the ring. If this device does not wish to transmit, it passes the frame marker and token on to the next device.

A token bus transmits a unique message for the token and monitors the bus for activity that indicates that the token was successfully received. If this is not so then a token recovery procedure has to be started. For a detailed description see Section 7.7.2.

7.3.2 Common access control: contention

Contention control is a distributed control scheme where each device attempts to gain control of the transmission medium. Each device competes against the other devices, so, to ensure fairness, each device must follow a precise set of rules for accessing the medium. LAN contention schemes are derived from the ALOHA and slotted ALOHA approach to multiple access to broadcast networks, (see Chapter 2).

Contention control for network systems

Contention access control methods allow any device to attempt transmission and then take steps to resolve and limit the collisions that will occur. In a LAN, the prevention of collisions is attempted by sensing the activity on the bus before transmission is attempted. Any activity on the bus can be detected by looking for the presence of the carrier on which the data is encoded. This is known as *carrier sense*. The detection of a carrier can be used to prevent a device from corrupting the transmission in progress.

An algorithm is needed to specify what a station should do if the medium is found to be busy. Three approaches or persistence algorithms are possible:

1 *Non-persistent*. The station backs off a random amount of time and then senses the medium again.

2 *1-persistent*. The station continues to sense the medium until it is idle and then transmits.

3 *p-persistent*. The station continues to sense the medium until it is idle and then transmits with some probability p. Otherwise it backs off a fixed amount of time and then transmits with probability p or continues to back off with probability $(1 - p)$.

The non-persistent algorithm is effective in avoiding collisions; two stations wishing to transmit when the medium is busy are likely to back off for a different amount of time. The drawback is that there is likely to be wasted idle time following each transmission. In contrast, the 1-persistent algorithm attempts to reduce idle time by allowing a single waiting station to transmit immediately after another transmission. Unfortunately, if more than one station is waiting, a collision is guaranteed. The p-persistent algorithm is a compromise that attempts to minimize both collisions and idle time.

If two devices find that the bus is free then both may start to transmit at the same time and their messages will collide. This is detected by each station monitoring its signal on the bus. This is known as *collision detection*. To ensure that the collision is detected by all devices on the network, the devices must continue to transmit for a period after the detection of a collision. After the collision both devices will then wish to transmit. To prevent an immediate collision a random backoff delay is used. The system just described is called *carrier sense multiple access with collision detection* (CSMA/CD).

7.3.3 Other access control techniques for ring systems

A ring can be configured to contain a number of circulating data minipackets; this allows two methods of access. In the first method, if a device is able to determine if a packet is empty it can fill this with data. This is called a *slotted ring*; see Section 7.9 on the Cambridge ring. Alternatively, a device can detect the end of a packet and insert its own packet between two existing packets; this is known as *register insertion* (Hafner *et al.*, 1980).

7.3.4 Allocated control for buses

In a bus system the access to the bus can be controlled by allocating time slots to connected devices. The simplest form is to divide the time equally between each device using time division multiple access (TDMA) (Tanenbaum, 1981). This has the disadvantage that if a device does not wish to communicate then its allocated time is unused and resources are lost. Polling schemes can be used in which one device is allocated the role of bus master to handle requests. When a device wishes to transmit, it waits to be polled by the bus master and either transmits its message or declines the offer.

Both of the above schemes have high overheads when the network traffic is low. This has led to the development of combined contention and allocation schemes. In a combined scheme contention control is used when the load is light and the probability of collisions is low, resulting in low overheads. As the network utilization increases the chances of collisions increases and the throughput decreases. When this occurs an allocation scheme is utilized to ensure the maximum use of the network without contention. For example, see Kleinrock and Yemeni (1978).

7.4 Local area network standardization

Local area networks have developed rapidly over a short period of time, presenting the prospective user with a wide choice of systems. To help the prospective users and suppliers, national and international standards bodies have produced standards specifically for LANs, concentrating on the lower levels of the OSI model. The standards developed for wide area networks will also be applicable to the higher levels of LANs.

The IEEE is the organization principally involved with the standardization of the lower levels of LAN systems. The standardization was started by the IEEE Computer Society in 1980 by the formation of the IEEE Standards Project 802 whose role is to standardize the means of connecting digital equipment within a local area (Myers, 1980).

The project has noted that one single standard will not suffice and has accepted three main access methods. The first is the CSMA/CD bus (802.3) and the other two are token passing rings (802.5) and token passing buses (802.4). All three are designed to provide for the transmission of data over the network and are generally not suitable for the transmission of time critical information such as speech and video.

Standardization in metropolitan area networks is closely following that of LANs and will be important for LAN users wishing to interwork between LANs interconnected via gateways to the larger MANs. The IEEE 802.6 group is working in this field and this work is discussed in Chapter 15.

7.4.1 The IEEE 802 standardization

The IEEE 802 committee has developed an overall structure into which a family of substandards interrelate as shown in Figure 7.2. These standards have also been adopted by the International Standards Organization (ISO) with the series numbers 8802 (i.e. IEEE 802.2 is the same as ISO 8802/2). The individual IEEE standard numbers are:

802.1 Relationship between the 802 standards and the OSI model
802.2 Logical link control
802.3 Carrier sense multiple access with collision detection
802.4 Token passing bus access method
802.5 Token passing ring access method
802.6 Metropolitan area networks

In addition the following Technical Advisory Groups (TAG) have been initiated which may subsequently produce standards:

802.7 Broadband TAG
802.8 Fibre optic TAG
802.9 TAG concerned with the integration of LANs and digital telephony

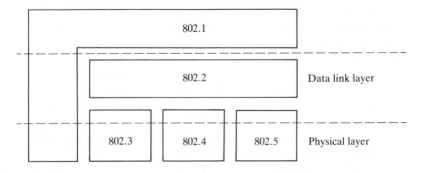

Figure 7.2 IEEE 802 standards family

7.4.2 IEEE 802.1 standard and the OSI model

The IEEE standard 802.1 describes the interrelationship between the members of the IEEE 802 standard family and their relationship to the ISO OSI reference model for LANs. Each of the access methods includes both physical layer components and a medium access control technique. The 802.2 logical link control (LLC) standard describes a network-independent protocol set used for passing data over any of the standardized systems 802.3–5; this is discussed in Section 8.3.2. The combination of the MAC and LLC layers can be considered to form layer 2 (the data link layer) of the OSI model (see Figure 7.3).

The principles adopted for local area computer network systems introduced in the previous sections have been incorporated into many experimental and commercial systems. As examples, we now describe the three IEEE 802 MAC techniques. Two other systems are also discussed: Ethernet, the precursor to IEEE 802.3, and the Cambridge ring, a slotted ring.

7.5 The Ethernet

The majority of installed LANs are based on Ethernet, which is a passive bus network that utilizes CSMA/CD. The system was developed in 1976 by Metcalfe and Boggs. The original system operated with a data rate of 3 Mbit/s connecting up to 256 stations with a maximum network length of 1 km. The system has evolved (Shoch *et al.*, 1982) into a commercially available system operating at 10 Mbit/s, connecting a maximum of 1024 stations over a branching bus with a maximum network length of 2.5 km.

Three of its major manufacturers, Xerox, Intel and DEC, have produced the industry standard definition which has been accepted by the IEEE. The advantages of the Ethernet system are its commercial standard and the fact that commercial interfaces are available for most

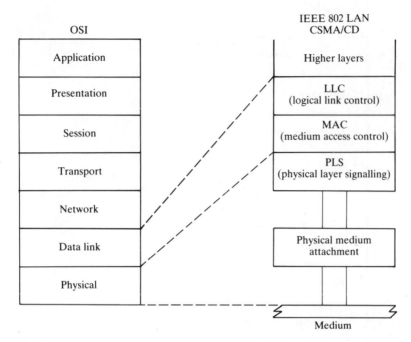

Figure 7.3 CSMA/CD system components

common computer buses. This standard formed the basis of the IEEE 802.3 system described in Section 7.6.

Two of the disadvantages of the full specification system are the cost per connection and the cabling costs. A number of manufacturers now provide lower cost connection and cabling systems called Cheapernet or thin Ethernet. This has a reduced specification for the number of devices and cabling lengths, but is very suitable for small-scale networks.

7.6 IEEE 802.3 CSMA/CD bus

The IEEE standard 802.3 describes a LAN system using a multiple access, broadcast, baseband transmission medium with CSMA/CD. The sections of the IEEE 802.3 system are shown in Figure 7.3.

7.6.1 Physical layer

The 802.3 standard specifies the physical layer of the system which has the following characteristics:

- *Transmission*. Baseband signalling over shielded coaxial cable.
- *Topology*. Branching non-rooted tree.
- *Network length*. Maximum cable segment length is 500 m, which may be interconnected via repeaters to a maximum system length of 2500 m.

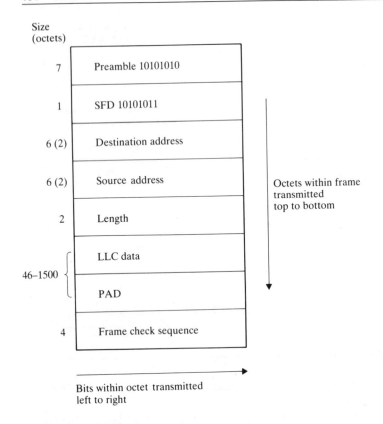

Figure 7.4   CSMA/CD frame format

- *Data rate.* 10 Mbit/s.
- *Maximum number of stations.* 1024.

7.6.2 Medium access control

IEEE 802.3 uses a CSMA/CD system as described in Section 7.3.2. To handle collisions the system uses a p-persistent algorithm with a variable back-off time.

Frame format

The format of the MAC frame is shown in Figure 7.4. When the frame is transmitted a minimum size limit of 46 data octets is imposed. Shorter blocks of information presented for transmission are padded to the limit using the PAD field. This is to ensure that a frame will have been sensed by all stations on a maximum sized network before the transmission is completed. The time taken by any transmission must be longer than the *slot time*, that is the maximum propagation time from one end of the network to the other and back. This enables the deference and collision detection procedures to operate correctly.

The *preamble* is a 7-octet field with a bit pattern that results in a transmitted signal of half the transmission frequency; this has a polarity transition at the centre of each symbol time which simplifies synchronization. The *SFD* is the start of frame delimiter and signifies the start of the frame. The *destination* and *source address* fields are either 2 or 6 octets in length. The address length is constant for a network implementation. The *length* field contains the number of octets in the data field. The *LLCData* field contains the *LLC* information and information from higher layers. The *FCS* (frame check sequence) field contains a 32-bit cyclic redundancy check (CRC) computed on both address fields, length, LLCData and PAD fields.

Transmission without contention

Transmission is initiated by the MAC on reception of data from the LLC. The MAC constructs a frame, to the above format, with the data provided by the LLC. The MAC monitors the medium for the presence of a carrier, indicating that a transmission in progress, and defers to passing traffic. When the medium is clear transmission is started.

Reception without contention

All stations listen to the network; a reception is begun by the detection of the preamble which is used for clock synchronization. The incoming bit stream up to the end of the start delimiter is discarded and then the frame is reconstructed. The destination address of the incoming frame is checked to see if it should be received by the station, if not it is discarded. If the address matches and the reception was completed without error, this is signalled by the MAC to the LLC and the address and data fields are passed to the higher layer.

Access contention

If more than one station attempts to use the medium at the same time they will interfere with each other, causing a collision between the signals of the two stations. A station can experience a collision during the initial part of its transmission (the collision window) before its transmitted signal has had time to propagate to all stations on the medium. Once the window has passed, the station has the medium and collisions will be prevented by other stations deferring to it.

When a collision is detected the MAC enforces the collision by transmitting a bit sequence called a *jam*. The duration of the jam is sufficient to ensure that all stations will detect the collision. After the jam has been sent transmission is stopped. Transmission is rescheduled to occur after a random period of time, called the back-off. If on retransmission a further collision is detected, the jam and back-off sequence is repeated with an increase in the back-off. This is continued until a maximum number of retries have been exceeded, in which case a transmission error is signalled to the LLC layer.

7.6.3 Network management and initialization

The CSMA/CD MAC scheme has no separate network management functions to provide detection, control and recovery from abnormal network conditions. This is because these functions are essential for the operation of the CSMA/CD control and are incorporated within it. In addition, as the network is a passive medium with fully distributed control, no medium or common network control initialization is required.

7.7 IEEE 802.4 token passing bus

The IEEE standard 802.4 describes a LAN based on a broadcast bus medium with a token passing access control scheme. The physical layer specifications cover a number of coaxial cabled systems with hardware derived from cable television (CATV) systems. The number of stations connected and the total length of the network depend on the physical layer scheme used.

7.7.1 Physical layer

Three different physical layer options are described in the standard. These are:

- An omnidirectional bus using phase continuous frequency shift keying at 1 Mbit/s.
- An omnidirectional bus using phase continuous frequency shift keying at 5 or 10 Mbit/s.
- A directional bus with active head end repeater using multilevel duobinary amplitude modulation/phase shift keying at 1, 5 or 10 Mbit/s.

(Details of the modulation techniques used, bus structure, etc., are described in standard texts on communication systems, e.g. Carlson, 1986.)

Each physical layer option provides a generic interface to the medium access control and to the station management. The MAC interface provides four primitives to control the operation of the physical layer. These are transmission initiation, reception notification, transmission mode control and end of reception (of any frame) indication. The

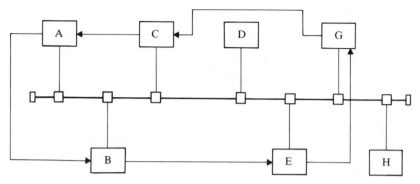

Figure 7.5   Logical token passing ring on bus medium

station management interface provides primitives for the configuration and control of the physical layer.

7.7.2 Medium access control

The medium access control scheme used is a token-based system where the token passes sequentially around a logical ring formed between stations on the network. The MAC provides the standard functions of data buffering, station address recognition, frame encapsulation, FCS generation and valid token recognition. The MAC also provides facilities for the maintenance of the logical ring. These are lost token recovery, token holding timeouts, distributed initialization, node failure error recovery and logical ring member addition.

The passage of the token between stations on the logical ring has little relationship to the physical connectivity of the stations, as can be seen in the example in Figure 7.5. The logical ring consists of stations A → B → E → G → C → A; the token is passed between these stations sequentially. Stations D and H are connected to the physical layer and can therefore receive messages but cannot initiate transmission. However, they can transmit a response when a station which is a member of the logical ring issues a *request_with_response*.

Frame formats

The format for a frame is shown in Figure 7.6. The start and end delimiters are unique octets that contain a special symbol which will never be present in the data. This *non-data* symbol is to aid in the identification of the delimiters. The frame control field is used to distinguish between the different frame types, including tokens, frames containing LLC data and special purpose frames which are discussed in the following sections.

PREAMBLE	SD	FC	DA	SA	DATA_UNIT	FCS	ED

PREAMBLE	Pattern for clock sync
SD	Start delimiter
FC	Frame control
DA	Destination address
SA	Source address
DATA_UNIT	Information, 0 or more octets
FCS	Frame check sequence
ED	End delimiter

Figure 7.6 Token bus frame format

MAC operation

The operations performed by the MAC layer involve the management of ordered access to the medium, providing a means for the addition and subtraction of stations (changes to the logical ring) and fault handling. The operation of the MAC is based on the following observations about a station's operation:

1 When a station transmits to the medium its signal can be received by all stations; no station can predictably alter the transmission but a transmission can be corrupted.

2 The transmitting station can assume that all active stations received some transmission but cannot assume that its message was received correctly by any station.

3 When a station successfully receives a frame it can assume that all other stations were aware of the transmission.

4 It is not necessary for all stations to be involved in the transmitted sequence, i.e. be members of the logical ring.

5 Each station must be able to handle multiple or lost tokens; there is no common monitor to perform this function.

MAC management functions

The normal operation of the network, when the logical ring has been established and is not changing, simply involves the passage of the token between a station and its logical successor after it has completed its transmissions.

A station holding the token may temporarily delegate its right to transmit to any station, including those not in the logical ring, by sending a *request_with_response* data frame. When a station receives a *request_with_response* it can respond with a *response* data frame; after this the right to transmit returns to the original station.

After a station passes the token it listens to ensure that its successor sends a valid frame indicating that the token was accepted successfully. If no valid frame is heard the sender attempts to determine the reason why the token was not transferred.

If the sender hears noise or an invalid frame it must attempt to recover the token. If it was just to repeat the token, it could create two tokens in the network. If the station hears a noise burst it continues to listen for four more *slot times* (maximum delay for a MAC action). If nothing more is heard it assumes that its own token was corrupted and, as in the case of no activity, repeats the token. The second token is monitored as before; if there is still no activity the sender assumes that its successor has failed. It then attempts to find out its own place in the logical ring by transmitting a *Who_follows* frame with its previous successor's address. All stations check the address and the station whose predecessor was the successor of the current station responds with a *Set_successor* frame. If there is still no response the token holder has to find another station on the logical ring by sending a *Solicit_successor* frame containing its own address. Any station wishing to be part of the logical ring responds using a *response window*. A response window is a controlled interval of time after the transmission of a MAC control frame during which the sender waits for a reply to be started. If the attempts to solicit a successor fail, either all stations have left the logical ring or an error condition exists and the logical ring needs to be reestablished.

Stations are added to the logical ring by active stations sending a *solicit_successor* frame. This specifies a range of addresses between the sender and its successor. Stations in this range respond during the response window.

It is possible that more than one station will wish to respond. If contention occurs the sending station starts an arbitration procedure by sending a *resolve_contention* frame. A station that responds chooses a delay (based on the station's address) of 0–3 slot times before transmitting. If during the delay another transmission is heard the station removes itself from the contention resolution. If nothing is heard it replies, this initiates further *resolve_contention* requests. Finally only one station will successfully respond, resolve the contention and become the successor.

The initialization of the logical ring is a special instance of adding stations. This action is started when the bus has been idle for too long. When this occurs, a station transmits a *claim_token* frame and contention is resolved in a similar manner as before. The station that wins the contention holds the token; it can solicit a successor and build the logical ring.

7.8 IEEE 802.5 token ring

The IEEE standard 802.5 describes a ring LAN using a token passing access control scheme for the connection of up to 250 stations with multiple levels of message priority.

Information is transmitted sequentially from one station to the next, each active station acting as a repeater for the information. The destination station copies the information from the ring. The information is removed from the ring by the source station. Error detection and recovery facilities and control of the number of bits in the ring are provided by one of the stations acting as a network monitor. This is protected by backup capability in all other ring stations.

7.8.1 Physical layer

The IEEE token ring uses shielded twin twisted pair cable with data transmission at either 1 or 4 Mbit/s. The frequency is fixed for a particular implementation of the system. Each station is connected into the ring via a trunk coupling unit (TCU) that enables an active station to be inserted into the ring or by-passed from the ring when inactive or failed. Symbols are transmitted to the medium using differential Manchester-type coding which is characterized by having two signal elements per symbol. Two non-data symbols are used within the frame delimiter octets.

The active monitor

The physical layer requires that all stations have the capability to become the network monitor. Only one station acts as the active monitor and signals its presence by regularly broadcasting active monitor present (AMP) MAC frames. The first station to receive the AMP frame resets a timer and sends a standby monitor present (SMP) frame. If the timer expires, indicating the failure of the active monitor, the standby monitor starts the procedure of becoming the active monitor. The SMP frame is passed between all other stations; these stations use a similar procedure to back-up the first standby monitor.

The active monitor checks the ring for correct transmission and will attempt to restore correct operation when an error is detected. It also checks for AMP frames that are not its own, indicating that two active monitors are in operation. If this occurs it returns to the standby state.

The active monitor station provides a variable-length delay buffer, the *latency buffer*, that maintains the minimum length of the ring, specified

		Operation
Identity	Length (octets)	
SD	1	Starting delimiter
AC	1	Access control
ED	1	End delimeter

(a) Token format

SD	AC	FC	DA	SA	INFO	FCS	ED	FS

Identity	Length (octets)	Operation		Identity	Length (octets)	Operation
SD	1	Starting delimiter		INFO	≥ 0	Information
AC	1	Access control		FCS	1	Frame check sequence
FC	1	Frame control		ED	1	End delimiter
DA	6 or 2	Destination address		FS	1	Frame status
SA	6 or 2	Source address				

(b) Frame format

Figure 7.7 Token ring

in bit times. This must be greater than the number of bits in the token sequence, 24 bits. The active monitor also acts as the master oscillator for the network, providing the clock to which all other stations synchronize. Although this provides the frequency to which all segments of the ring should be synchronized, it is possible for segments of the ring to be operating at slightly different frequencies and thereby change the total ring latency. These errors can be up to ± 3 bits, so to maintain a constant ring latency of ≥ 24 bits, a variable length buffer, of up to 6 bits, is added to the buffer.

7.8.2 Format of transmitted information

Two formats of information are transmitted around the ring, tokens and frames. The token (Figure 7.7(a)) represents the right for a station to transmit. The frame format is used for the transmission of medium access control (MAC) and logical link control (LLC) messages between stations (see Figure 7.7(b)). To meet the requirements of the LLC each

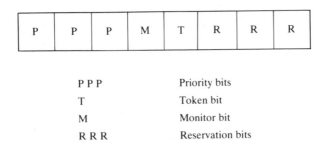

P	P	P	M	T	R	R	R

P P P	Priority bits
T	Token bit
M	Monitor bit
R R R	Reservation bits

Figure 7.8 Token ring access control field format

station should be capable of receiving frames up to and including 133 octets.

The *end delimiter* contains an intermediate frame bit, I, and an error detected bit, E. The I bit is set to a 1 to signify that another frame from the same transmitter will follow, and is set to 0 for the last or only frame transmitted. The E bit is set to 0 by the source transmitter. Each station in the ring checks tokens and frames for errors as they are repeated. If an error is detected the E bit is set to 1 in the frame in error; otherwise the E bit is repeated as received.

The *access control field* (Figure 7.8) contains bits for the identification and control of tokens, for message priority and monitor control. The token bit is set to 0 in a token and 1 in a frame. The priority bits specify the priority of a token allowing a station to determine if it can use the token. There are eight levels of priority with level 0 being the lowest. A station that has a protocol data unit (PDU) awaiting transmission with equal or higher priority than the token's priority may change the token to a start-of-frame sequence and commence transmission of the PDU.

The *monitor bit* is set by the transmitter to 0 in all frames and tokens. When a monitor bit of 0 is detected by the active monitor it is converted to a 1 to mark its first passage around the ring. If the active monitor detects an access control field with the monitor bit set to 1 and a priority greater than 0 then the frame or token is aborted; this is because a frame or token should have been removed by its transmitting station. The exception is a priority 0 token which will circulate around the ring when the network is idle.

The *frame status* field is used to enable the transmitting station to receive information about the passage of a frame around the ring. Two bits, the address recognized, A, and frame copied, C, are sent as 0 by the transmitting station. If a station recognizes the destination address of the frame as its own it sets the A bit to 1. If the station also copies the frame into its buffer, it sets the C bit to 1. From the possible combinations of these bits the originating station can determine:

1 If the destination station is inactive on this ring.

2 If the station exists but did not copy the frame (busy).

3 That the frame was copied by the destination station.

7.8.3. Medium access control

The access to the ring, or medium, is controlled by the token. MAC rules specify the usage of the token, the control of the priority of messages and management of the network in the case of errors or failures.

Frame transmission

For transmission, the MAC formats the frame by adding the address and control fields and then waits for the reception of a token with a priority equal to or less than that of the awaiting frame.

When a usable token is received it is converted into a start-of-frame sequence and the frame transmission is started. The frame check sequence is calculated and appended to the information field. Frame transmission will continue until either there are no more queued PDUs with the same or higher priority than the current service priority or the token holding time (THT) would be exceeded by transmitting the PDU. The THT ensures that one station does not hog the network and allows fair network usage.

When a station has a PDU of higher priority than the current service priority it requests a priority token. This is done by raising the value of the reservation bits of a repeated token or frame using the following algorithm:

> *if (PDU priority > received priority) set reservation bits*
> *else leave reservation bits unchanged*

Token transmission

A transmitting station checks that it receives its own frames by looking at the source address of the incoming frames and also checks the response bits to determine if each frame was successfully received. When it has completed its transmissions, it sends a token and continues to receive until all frames that it transmitted have been cleared from the ring. It then repeats the incoming signal.

When a token is inserted in the ring it may be necessary to increase its priority either because the received reservation value or the priority value of a frame still queued is higher than the priority of the received token. The station transmits a token of raised priority with the reservation bits set to zero. A station that raises the priority level has the responsibility for subsequently resetting it. To do this the station stacks separately both old and new priority values and then claims every token of the new priority to check the reservation bits as follows:

if (received reservation value > old priority)
  [increase priority of token
  replace value at top of new priority stack
  claim tokens at new priority]
else
  [transmit token at old value
  pop both stacks
  if stacks are empty stop claiming tokens]

The stacks are required because a station may need to raise the priority of the token more than once before returning it to its original value.

Frame reception

All stations that are repeating the incoming signal check the incoming data for frames that require some action. If the frame is a MAC frame its bits are acted upon. If the destination address matches the station's own, or is a group or broadcast address, the frame is copied into the receive buffer and the event is signalled to the appropriate sublayer.

7.9 The Cambridge ring

Although there have not been as many implementations of slotted rings as of other LAN types, the most recent approaches to multiservice networks (discussed in Chapters 14 and 15) are based on multiple access to a slotted network structure. For this reason it is valuable to discuss the Cambridge ring as an example of a slotted ring LAN.

The Cambridge ring was developed, from 1975, by Wilkes, Wheeler (Wilkes and Wheeler, 1979) and Hopper (1980) at the University of Cambridge Computer Laboratory. It has been a popular choice in the UK academic community where it has formed the basis for research into such areas as distributed systems and network interconnection. Up to 254 stations can be connected to a single ring, the total ring length being limited only by the total ring delay. Devices are connected to the ring by *repeaters* which pass on the circulating bit stream. Various technologies can be used in the ring transmission system to cater for differing interrepeater distances.

Information is written into a circulating slots in the form of a *minipacket*. The slot structure is established by a special station: the monitor. In addition to initial start-up the monitor provides the facility for the control of lost minipackets.

Bit	1	2	3	4–11	12–19	20–35	36–37	38–39	40
Field	L	FE	MB	DEST	SRCE	DATA	TYPE	RESP	PAR

L	Leader. Slot framing bit; always 1
FE	Full/empty. Set to 1 to mark slot as full
MB	Monitor pass. Set to 1 by transmitter. Set to 0 by monitor
DEST	Destination address
SRCE	Source address
DATA	16-bit data minipacket
TYPE	Type bits A and B
RESP	Response bits
PAR	Even parity

(Bit 1 transmitted to ring first)

Figure 7.9   Cambridge ring minipacket format

7.9.1 Physical layer

The basic transmission scheme uses a twin twisted pair cable providing reliable communication at 10 Mbit/s. The system uses a modulation scheme that enables a 10 Mbit/s data and clock to be transmitted while keeping the bandwidth used to 10 MHz. The scheme uses the two pairs, changing the state of both pairs to indicate a 1, and alternately changing pairs to indicate a 0. This scheme will suffer from any differential delays between the two pairs; this is a significant factor in limiting the interrepeater distance over copper cables to 200 metres. For greater distances fibre optic links (Hunkin and Litchfield, 1983; Roworth and Cole, 1983) can be used to extend the network over a wide physical area.

7.9.2 Ring access scheme

The latest Cambridge rings use a 40-bit minipacket (see Figure 7.9) separated by a gap of at least 4 bits. The type bits are used to distinguish between different sequences of minipackets. The data is transmitted at approximately 10 Mbit/s; the frequency is adjusted to ensure that the ring delay is equal to a whole number of bit times. In small rings, where the ring length is less than 44 bits, the ring is padded by a delay in the monitor so that a whole minipacket and a 4-bit interpacket gap exist.

An empty minipacket is filled by the transmitting station which sets the full bit and fills in the address and data fields as required. When the minipacket is received by the destination it copies the address and data fields into its internal buffer and sets the response bits to indicate that it *accepted* the data. If the destination station's register is full it sets the response bits to *busy*. Each station has a source select register which can be set to any station, a specific address or no station. This register is used to control reception so that minipackets from other stations do not interfere with a sequence of minipackets from one station. If the source of the minipacket is from a station the destination has not selected, the station does not copy the data, but sets the response bits to *unselected*. If the destination is inactive or not present on the ring the response bits return *unchanged*, to indicate that the minipacket was ignored by all stations on the ring. The transmitting station waits for the minipacket to return, sets the full bit to empty and stores the response bits for the controlling software.

The restriction of waiting for the minipacket to return limits the maximum data rate achievable in a ring with more than one minipacket. The system also prevents a station from reusing the same minipacket, forcing it to send it empty to the next station on the ring. This prevents one station from hogging the ring, resulting in load sharing among the ring stations.

7.9.3 Ring error control

The Cambridge ring uses the monitor station to maintain the ring size and detect circulating minipackets. The monitor station is also used to fill empty minipackets with random data which is checked for errors if the minipacket returns unused to the monitor, so that the error performance of the ring can be monitored.

With the Cambridge ring it is possible to determine if the device is inactive by examining the status of the response bits of the minipackets sent to it. If these return ignored then the device is off the ring. This method is used in the Cambridge distributed system (see Chapter 12) where a logging device sends minipackets at regular intervals to known devices on the ring and logs any change of status.

7.9.4 Cambridge ring standardization

The system has received considerable interest in the United Kingdom and is available commercially from a number of manufacturers. The standard consists of two documents, one defining the hardware and its interface (Sharpe and Cash, 1982) and the second defining the protocols for communication using the ring (Larmouth, 1982). These standards are known as the CR 82 standards.

7.10 The fibre distributed data interface (FDDI)

The fibre distributed data interface (FDDI) (Ross, 1986) is a token ring network that has many similarities to the IEEE 802.5 standard but is specifically intended for the faster speeds offered by optical fibres, and is designed to run at 100 Mbit/s. Optical fibres have the additional advantages of being lightweight and unaffected by electromagnetic interference. The FDDI was designed to be compatible with the IEEE 802 LAN standards and shares the common LLC level, thus making it particularly suitable as a backbone to interconnect IEEE 802 LANs. It was also envisaged as a general purpose interface channel between processors. The physical medium is multimode optical fibre using light emitting diodes (LEDs) transmitting at 1300 nms (Burr, 1986). The FDDI can support up to 500 stations with a maximum distance between stations of 2 km and a maximum ring circumference of 200 km. It uses a 4 out of 5 code (5 transmitted bits for each 4 bits of information sent) which implies signalling at 125 Mbit/s, and is intended to achieve error rates of less than 1 in 2.5×10^{10}. Clock synchronization is obtained not by transitions in every symbol as in the Manchester code, but by the receiver synchronizing to a stream of signals at the beginning of each frame. Of the 32 available bit patterns in the 5-bit code, 16 are used for data, 8 are used for delimiters, control and hardware signalling and the remaining 8 to ensure that signal transitions are sufficiently frequent. The maximum frame size is 4500 octets. FDDI products are now commercially available.

An FDDI network consists of two counterrotating rings. Stations are of two types: those of type A which require a high level of reliability are connected to both rings, stations of type B are connected to one of the rings. Reliability is provided by three techniques—station by-pass switches, ring connections that allow reconfiguration and the provision of wiring concentrators. The concentrators allow alternative paths to be formed and also allow type B stations which have a single connection to the concentrator to make use of its reconfiguring properties (see also Chapter 8).

FDDI tokens, which offer a priority mechanism similar to IEEE 802.5, are illustrated in Figure 7.10. Information is transmitted in a frame (also shown in Figure 7.10) when the token is held. Frames may carry either synchronous or asynchronous traffic depending on token priority. The frame status (FS) field is set by the destination and checked by the source station which removes its frame from the ring and generates another token. Unlike IEEE 802.5 it is possible for several frames to be carried on the ring at the same time, enabling the much larger potential size of an FDDI ring to be used efficiently. Thus a station (say station X) will transmit a token immediately after it has completed its transmission, even if the transmitted frame has not started

Frame

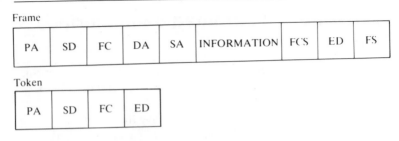

Token

PA Preamble (16 or more symbols)
SD Starting delimiter (2 symbols)
FC Frame control (2 symbols)
DA Destination address (4 or 12 symbols)
SA Source address (4 or 12 symbols)
FCS Frame check sequence (8 symbols)
ED Ending delimiter (1 or 2 symbols)
FS Frame status (3 symbols)

Figure 7.10 FDDI—frame and token formats

to arrive back. A station further round the ring is allowed to insert its own frame (subject to the normal priority rules) after the frame from station X and before regenerating the token.

To ensure a guaranteed bandwidth for synchronous traffic, the FDDI uses a timed token rotation protocol. Synchronous traffic using the token passing portion of the ring is given a guaranteed bandwidth by ensuring that the token rotation time does not exceed a preset value. If not required, this bandwidth is made available to asynchronous traffic. A target value of the token rotation time (TRT) is agreed when the ring is initialized. A station is allowed to capture an arriving token and transmit synchronous traffic provided it has sufficient synchronous allocation (agreed by the station management procedures). Asynchronous traffic may only capture a token when the time since it last received a token is less than the TRT. The MAC protocol ensures that on average the TRT is less than the target TRT and that the maximum TRT is not more than twice the target TRT.

A later version of the protocol called FDDI-2 has a more complex protocol and carries circuit switched traffic. It is compatible with FDDI-1 networks and reverts to this mode if no circuit switched traffic is being carried. For circuit switched traffic the capacity of the ring is divided into cycles by a master station, part of each cycle (a minimum of 1 Mbit/s still being available for use under the token passing scheme. Circuit switched traffic is allocated in increments of 6.144 Mbit/s to isochronous channels, in which information at a guaranteed bandwidth is carried from a fixed source to a fixed destination at an agreed fixed position in the channel. (Note that 6.144 Mbit/s divides neatly into three 2.048 Mbit/s (European) channels or four 1.536 Mbit/s (North American) channels.) The circuit switched option makes FDDI-2 particularly suitable for the connection of PABXs.

7.11 Personal computer networks

There have been a number of developments for personal computers (PCs) and their clones to support communications between hosts. There are a considerable number of proprietary implementations of networking systems by manufacturers, for example:

NETBIOS
Novell's Netware
3-Com Networks
IBM PS/2 LAN interfaces
SUN's NFS for PCs
The Macintosh network system Appletalk

For the IBM PC and its clones, the release of Microsoft's operating system MS-DOS 3.1 provided the necessary conventions for file management with file locking and multiuser functions, to enable networking software to be added. The features of the disk operating system (DOS) have been enhanced further in later releases.

7.11.1 NETBIOS

IBM's network basic input/output system (NETBIOS) was announced in 1984 and provided a basic building block for networking. IBM published the details of the NETBIOS interface to encourage other manufacturers to emulate their own NETBIOS products. IBM was committed to NETBIOS as the interface on future LANs and for the IBM PS/2 range of computer systems. NETBIOS provided the mechanism for PC-to-PC communication without the previously used approach of having an intermediate disk server to control the PC network. The aim of NETBIOS was to provide a fixed interface between the operating system and the LAN which was independent of the network hardware or software. It was effectively an extension to the standard PC basic input/output system (BIOS).

NETBIOS may be viewed as an interface between the presentation and session layers of the ISO reference model, thus offering peer-to-peer communication. It has support for both virtual circuits and datagram services. Implementation may be provided as a ROM (as is the case for the standard BIOS), which is often located on the network card itself. The IBM PC network and token ring use a card with its own microprocessor which relieves much of the network protocol load from the host PC. To communicate with NETBIOS, an application must give the system commands in the form of a data structure called the network control block, which is used to pass information in both directions across the NETBIOS interface. Up to 32 independent sessions may be supported simultaneously.

NETBIOS uses a distributed LAN management approach. If one PC wishes to access a named resource on another PC, it broadcasts its request and all PCs on the network must interpret the request; if the resource is found then a connection is made. When a new resource is named, the name is broadcast to ensure that it does not already exist. This broadcasting strategy represents a significant overhead on the PCs and some manufacturers have modified this approach to provide a centralized hierarchical method to naming.

The PC network program produced by IBM for their CSMA/CD-based PC network, and later for the token ring, provides simple techniques for file serving, printer spooling and simple PC-to-PC messaging. When the token ring product was released, IBM provided an extension to their existing systems network architecture (SNA) products (see Chapter 6). This was machine and operating system independent, and enabled personal computers with a 3270 connection card to emulate an IBM terminal or printer, to execute host application programs, or to transfer files and use host printers or disks.

7.12 Summary

This chapter has concentrated on the details of the lower levels of LAN systems and described, in detail, some example systems. For a LAN to be used in a real system we need to consider its performance and the operation of the software that controls it; these are described in the next chapter.

References

References to the following texts have been made in this chapter. Suggestions for further reading on LANs appear at the end of Chapter 8.

Bertine, H. V. (1980) 'Physical level protocols', *IEEE Trans. Commun.*, **COM-28** (4), April, 433–444.

Burr, W. E. (1986) 'The FDDI optical link', *IEEE Commun. Mag.*, **24** (5), May, 18–23.

Carlson, A. B. (1986) *Communications Systems*, McGraw-Hill, New York.

Clark, D. D., K. T. Pogran and D. P. Reed (1978) 'An introduction to local area networks', *Proc. IEEE*, **66** (11), November, 1497–1517.

Hafner, E. R., Z. Nenadal and M. Tschanz (1980) 'A digital communication system', *IEEE Trans. Commun.*, **COM-28** (6), June, 877–881.

Hopkins, G. T. (1979) 'Multimode communication on the Mitrenet', in *Proc. Local Area Communication Network Symp.*, Boston 1979, N. B. Meisner and R. Rosenthal (eds), Mitre and NBS, 169–177.

Hopper, A. (1980) 'Local area computer communication network', Technical Report 7, University of Cambridge Computer Laboratory.

Hunkin, D. J. and G. W. Litchfield (1983) 'An optical fibre section in the Cambridge digital ring', *IEE Proc. Computers and Digital Techniques*, **130**, Part E (5), September, 154–158.

IEEE 802.1 (1989) 'Relationship of the IEEE 802 standards with the ISO OSI model', American National Standards Institute.

IEEE 802.2 (1990) 'Logical link control specifications', American National Standards Institute.

IEEE 802.3 (1990) 'Carrier sense multiple access with collision detection (CSMA/CD) access method and physical layer specification', American National Standards Institute.

IEEE 802.4 (1990) 'Token-passing bus access method and physical layer specifications', American National Standards Institute.

IEEE 802.5 (1990) 'Token ring access method and physical layer specifications', American National Standards Institute.

Kleinrock L. and Y. Yemeni (1978) 'An optimal adaptive scheme for multiple access broadcast communication', *Proc. IEEE Int. Conf. on Communications*, Toronto, Canada, 1978, 7.2.1–7.2.5.

Larmouth, J. (1982) 'Cambridge ring 82 protocol specifications', Joint Network Team, Rutherford Appleton Laboratory, Chilton, Didcot, Oxfordshire.

Metcalfe, R. M. and D. R. Boggs (1976) 'Ethernet: distributed packet switching for local computer networks', *Commun. ACM*, **19** (7), July, 395–404.

Myers, W. (1980) 'Towards a local network standard', *IEEE Micro*, **2** (3), August, 28–45.

Pliner, M. S. and J. S. Hunter (1982) 'Operational experience with open broadcast local area networks', *Online Conferences on Local Area Networks and Distributed Office Systems*, 71–86.

Ross, F. E. (1986) 'FDDI—a tutorial', *IEEE Commun. Mag.*, **24** (5), May, 10–17.

Roworth, D. and M. Cole (1983) 'Fibre optic developments for the Cambridge ring', in *Local Net 83*, March, On-Line, London, 519–535.

Sharpe, W. P. and A. R. Cash (1982) 'Cambridge ring 82 interface specifications', Joint Network Team, Rutherford Appleton Laboratory, Chilton, Didcot, Oxfordshire.

Shoch, J. F., Y. K. Dalal, D. D. Redell and R. C. Crane (1982) 'Evolution of the Ethernet local computer network' *IEEE Computer*, **8**, August, 10–27.

Stallings, W. (1984) 'Local networks', *ACM Computing Surveys*, **16** (1), March, 3–41.

Tanenbaum, A. S. (1981) *Computer Networks*, Prentice-Hall International, Englewood Cliffs, New Jersey.

Wilkes, M. V. and D. J. Wheeler (1979) 'The Cambridge digital communication ring', in *Local Area Communications Network Symp.*, Boston, 1979, N. B. Meisner and R. Rosenthal (eds), Mitre and NBS, 47–60.

8 Local area networks: performance, reliability and protocol suites

SIMON JONES

This chapter discusses the factors affecting the operation of an LAN. These factors are the performance of the transmission medium and its medium access control, the reliability of the network and its interfaces, and the structure of the protocols used to pass data between the connected systems.

8.1 Local area network performance

A number of papers have been published on the performance of various network architectures and access schemes. These vary from theoretical studies of models of generalized network systems (Bux, 1981) to the direct measurement of existing systems (Shoch and Hupp, 1980). Comparisons between networks are affected by the nature of the system loading. Some performance studies have been reviewed by Stallings (1984). For a broader and more detailed coverage see Hammond and O'Reilly (1986). Detailed studies of the performance of computer communication networks generally make use of queueing theory, an extensive field that is beyond the scope of this book. Interested readers should refer to Schwartz (1987).

This section introduces the basic performance factors for an LAN. Further performance criteria and a detailed discussion of the performance of voice and data integration on an LAN are presented in Chapter 14.

8.1.1 Network dependent performance

For the lower levels of a local area network, the performance may be measured in terms of the utilization of the capacity of the network. For an individual node, performance is measured in terms of its mean data rate or the mean delays in transmission (Stuck, 1983).

Measurements of actual network installations show that most computer networks are rarely used to their full capacity, with typical usage figures of less than 10 per cent. Under these conditions most networks will appear to give the full capacity of the network to an individual node for the duration of the communication. Performance evaluations are, however, undertaken at a much higher network utilization when it is likely that a number of simultaneous transmission attempts will occur. This is valid, as studies have shown that most data communication is of a 'bursty' nature with periods of high data transfer followed by idle periods. If the bursts of a number of transmissions occur at the same time, the presented load may be higher than the maximum network capacity so delays are introduced. It is the performance evaluation of the different networks during these periods of high utilization that gives the most interesting results.

Slotted and register insertion rings

For these networks, as more devices attempt simultaneous transmission, the mean delay between packets transmitted from a station increases. This is because the number of free packets is restricted. For a *slotted ring* each device may not reuse a slot after a transmission, so this slot is made available to the next device. This means that the available network capacity is evenly shared between the competing devices. Also, the worst case delay, before a station can transmit again, is defined as the time for a minipacket to be inserted by all other stations, i.e. one slot delay per station. For a *register insertion ring* the delay increases as more registers are inserted into the network.

Multiple access bus with CSMA/CD

For a contention network, as the number of devices attempting to transmit increases, so does the probability of collisions. The utilization of the network therefore depends on the ability of the system to prevent and detect collisions. The effect of collisions and subsequent retries depends on the size of the data packets sent, because for smaller packets the proportion of time spent in resolving collisions to that spent transmitting data is higher. These effects mean that under high loads the system performance is probabilistic and hence that the delay in transmitting a packet is indeterminate. This can be a problem in certain applications, particularly where real time data transmission is required.

Token passing systems

With token passing systems there is the overhead of the transmission of the token, but, as the load is increased, the probability of the token being passed to a device that does not require to transmit is reduced. The throughput of the system will not be wasted in collisions, so high utilization can be achieved. The delay for the transmission of a packet depends on the action of the other devices; it also depends on the token holding time. If the packets have a maximum size then the maximum delay can be calculated.

8.1.2 Performance comparisons

Executing fair performance comparisons is difficult due to the differences between systems and the variance in the parameters for a single system. Performance comparisons have therefore concentrated on specific pairs or groups of LAN systems.

Comparing token passing ring and bus

The simple comparison presented here, which follows the technique discussed in Stuck (1983), shows the effect of token passing on the data rate achievable from one station. For the *token ring*, the point-to-point data rate between two stations, ppDR, can be calculated as follows. It is assumed that only one station requires to transmit data in fixed size frames, whose data portion is of length Framesize. In this case the token must pass through every station before another transmission occurs:

$$ppDR = \frac{Framesize}{T_{frame} + T_{prop} + N \times T_{interface}}$$

Here T_{frame} is the time taken to transmit a frame, T_{prop} is the total propagation delay around the network and $T_{interface}$ is the delay through each station. N is the number of stations in the ring. If all N stations are active the token need only be passed on to the next station, and the maximum point-to-point data rate becomes

$$ppDR = \frac{Framesize}{N(T_{frame} + T_{prop}/N + T_{interface})}$$

As an example for a 10-station ring operating at 1 Mbit/s, with frames of 1000 bits, a propagation time of 50 μs and a delay of 20 μs through each station, the following could be achieved:

For one active station: ppDR = 800 kbit/s
For each station when all 10 stations are active: ppDR = 97.5 kbit/s

For the *token bus*, similar formulae can be derived. In this case each

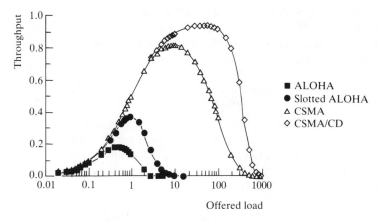

Figure 8.1 Throughput versus offered load for a variety of contention protocols

time the token is passed a propagation delay is incurred. For one active station:

$$ppDR = \frac{Framesize}{T_{frame} + N(T_{prop} + T_{interface})}$$

When all stations are active the token passing time is reduced, giving

$$ppDR = \frac{Framesize}{N(T_{frame} + T_{prop} + T_{interface})}$$

Using the same values as were used for the token ring calculation, the token bus achieves:

For one active station: $ppDR = 588.2$ kbit/s
For each station when all 10 stations are active: $ppDR = 93.4$ kbit/s

The examples show that token passing has a much greater effect on a lightly used bus network than on a ring network, and also shows how efficient the technique can be in a heavily loaded network. Note that we have assumed that the token holding time only allows one frame to be sent per station; a longer token holding time would give slightly higher data rates but data would be subject to longer delays.

CSMA/CD compared with earlier contention systems

In Chapter 2, the ALOHA and slotted ALOHA protocols were introduced and an analysis given of the throughput against offered load. Figure 8.1 shows throughput S (average number of successful transmissions) against offered load G (average number of arrivals per packet time) for ALOHA, slotted ALOHA, non-persistent CSMA and non-persistent CSMA/CD, all assuming a Poisson arrival distribution.

The curves for ALOHA and slotted ALOHA use equations (2.4) and (2.5) respectively. The two CSMA protocols are plotted according to the following formulae derived in Hammond and O'Reilly (1986). For non-persistent CSMA

$$S = \frac{Ge^{-aG}}{G(1 + 2a) + e^{-aG}}$$

Here a is the normalized propagation delay, i.e. the ratio of the propagation delay to the time taken to transmit a packet. In Figure 8.1, $a = 0.01$, typical of a local area network. Note that this is non-slotted CSMA. Slotted CSMA offers a slightly better performance. For non-persistent CSMA/CD we have

$$S = \frac{Ge^{-aG}}{Ge^{-aG} + 3aG(1 - e^{-aG}) + (2 - e^{-aG})}$$

where a is defined as above and $a = 0.01$ for Figure 8.1. It is also assumed that the jam time is equal to the maximum propagation delay of the network. The Figure shows that the maximum throughput of CSMA/CD (about 94 per cent) is greater than all the other protocols illustrated under these conditions, and is maintained high over a wide range of offered loads. (Hammond and O'Reilly show that slotted 1-persistent CSMA/CD gives a slightly worse performance and slotted non-persistent CSMA/CD gives a slightly better performance than for the CSMA/CD protocol illustrated.)

Comparing token passing with CSMA/CD

A performance comparison between CSMA/CD, token ring and token bus systems is presented in Stuck (1983). Stuck derived performance equations for the three systems; the calculations were for the total network data rate, not that achievable between individual stations. If the load sharing is fair then each station will achieve an equal portion of the network data rate. The analysis of the different systems yielded the following conclusions:

1 For the parameters used, the smaller the mean packet length, the higher the maximum mean throughput rate for token passing compared with that for CSMA/CD. This reflects the fact that, for a given amount of data, smaller packets suffer more collisions under CSMA/CD.

2 The token ring is the least sensitive to applied load.

3 CSMA/CD offers the shortest delay under light load, whereas it is the worst under heavy load.

4 When a single station is transmitting, the token passing bus is the least efficient due to the high overhead of the token passing.

Cambridge ring performance

A detailed performance study of a slotted ring, the Cambridge ring, was made by Blair (1982). The Cambridge ring transmission access rules restrict the maximum slot utilization for one transmitter to be one slot in every $(n + 2)$ slots, where n is the number of slots in the ring. This is because the station must wait for a minipacket to return around the ring and pass the slot on empty, to enable sharing. It then has to examine the response bits at the end of the minipacket, to determine the next minipacket for transmission; while this happens the next slot is allowed to pass. A ring built to the CR 82 standard (Sharpe and Cash, 1982) has 40-bit packets and a 10 Mbit/s data rate. Only 16 of the 40 bits in the slot are used for data. Ignoring the size of the gap, the maximum point-to-point data rate (ppDR) is

$$ppDR(max) = \frac{16 \times 10}{40(n + 2)} \text{ Mbit/s}$$

$$ppDR(max) = 1.33 \text{ Mbit/s (1 slot ring)}$$

The actual figure may be lower as the gap can be between 2 and 41 bits depending on ring length. As the access scheme prevents the reuse of a packet, this point-to-point data rate can be obtained between two pairs of communicating nodes. The advantage of the scheme is that if all nodes wish to transmit simultaneously then each node can use one out of every m slots $(m > n + 2)$, where m is the number of stations. This gives a worst case bandwidth (again ignoring the size of the gap) of:

$$ppDR(min) = \frac{16 \times 10}{40 \times m} \text{ Mbit/s}$$

$$ppDR(min) = 400 \text{ kbit/s (for 1 slot and 10 active stations)}$$

From the above discussion it can been seen that the performance of the Cambridge ring is dependent on the number of slots and the delay around the ring. The effect of the size of the gap can be seen in Figure 8.2, which assumes one active station and plots the point-to-point data rate against the ring length in bits. As stations are added to the ring, the delay increases and the bandwidth decreases, until there is sufficient delay to incorporate another slot into the ring which restores some of the lost bandwidth.

Comparing CSMA/CD with slotted and token rings

The performance comparison between CSMA/CD and slotted rings is quite difficult due to the variation of the performance of rings when their parameters are changed. The significant parameters are:

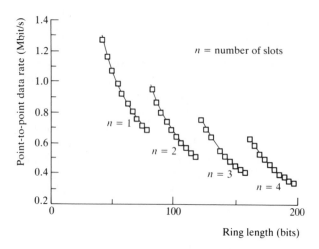

Figure 8.2 Cambridge ring performance (one active station)

- *Slotted ring.* Slot size, overhead bits per slot, ring length, number of slots.
- *CSMA/CD.* Slot size, network length, number of stations.

Figure 8.3 compares the delay/throughput performance of CSMA, slotted and token rings (Bux, 1981) operating under similar conditions at 10 Mbit/s. CSMA/CD shows the best performance at low loads, but for loads greater than 0.4 the delay increases rapidly, showing the degradation due to collisions. The token ring has much more stable delay characteristics. The slotted ring performs worse than the token ring, largely because of the minipacket overheads and the delay caused by the destination release of slots. (Chapter 14 returns to this comparison.) Blair and Shepherd (1982) comment that, although the delays for the Cambridge ring are greater than for Ethernet, the Cambridge ring does have the advantage of a lower variance of delay and the hardware construction of the network interface is simpler and therefore of lower cost and higher reliability.

8.2 Network reliability and fault detection

The usability of an LAN is strongly affected by the reliability of its components, the ease with which any faults can be detected and located and the measures taken to repair the faults.

8.2.1 Network reliability

The reliability of an individual device connected to a network is usually considered to be less important than the reliability of the total network.

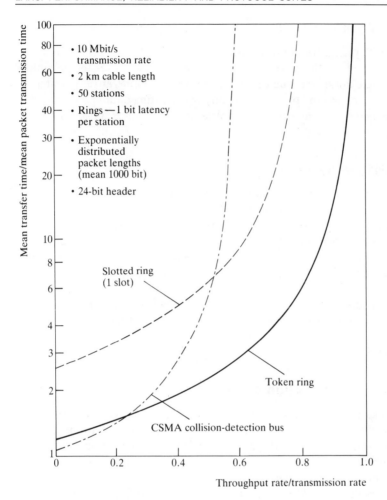

Figure 8.3 Transfer delay–throughput characteristics for CSMA, slotted ring and token ring (Bux, 1981; © 1981 IEEE)

If an individual device connection fails, this only affects the processes running on the device and other processes wishing to access the device. If the network fails then all communication processes are affected. Therefore networks are designed so that failures are unlikely to affect the network as a whole.

All systems are susceptible to the failure of the interconnecting cable. In a star topology, any failure will only cause the loss of the device connected by the faulty cable. In a ring or a bus a shorted cable will cause the failure of the network section. An open circuit cable will cause the complete failure of a ring system but will divide a bus into two sections which may continue to operate.

Reliability in bus networks

In a bus system, the control is decentralized and the connection is usually made to the cable using passive taps. This reduces the possibility of a single failure causing the system to become inoperative. Failures such as shorting the bus, permanently transmitting data and incorrect access behaviour can cause a complete or partial interruption to other devices on the bus.

Reliability in ring networks

In a ring system, the transmitted data is repeated at each device, requiring an active component to be included in the ring. A single failure can thus cause the system to become inoperative. In most systems the operation of the repeater is very simple, consisting of regeneration and retransmission of the incoming signal, the value of the signal only being changed when data is transmitted.

Such repeaters must always be active; they must be provided with a power supply even when the connected device is turned off or if the device is disconnected. This is arranged by distributing the power to the repeaters round the ring, often using the same cores in the cable as used by the signals.

8.2.2 Fault detection and location

The location of faults, particularly intermittent ones, can also be improved by the connection of devices that are dedicated to the monitoring of the network performance and error detection. The Cambridge distributed system incorporated an error logger (Needham and Herbert, 1982) which produced a hard copy of the source and type of errors detected. The printout produced enabled failing devices to be identified and, due to the error reporting of the Cambridge ring hardware, ring breaks or repeater failures to be detected. In all ring systems individual stations can detect errors in the structure of the incoming signal stream.

On CSMA/CD networks two methods of fault detection are possible. Firstly, periodic enquiries can be made to determine which stations are active. This can be combined with a knowledge of the network configuration to locate faults in sections of the system. Secondly, special monitor stations can be provided which act in *promiscuous mode* (i.e. listen to all transmissions on the network). Abnormal behaviour of individual stations can then be detected.

8.2.3 Improving reliability

A possible protection against failure is to provide an alternative communication path. For a bus system the communication path is

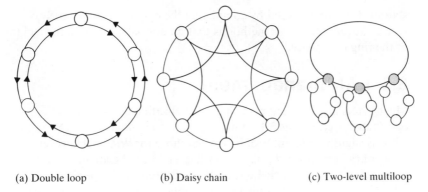

(a) Double loop   (b) Daisy chain   (c) Two-level multiloop

Figure 8.4   Fault-tolerant ring configurations

duplicated to provide an alternative path for use when the main path
fails. For ring systems, extra links between repeaters can be added.
(Figure 8.4 shows some possible alternative arrangements.) Then it
becomes possible either to form an alternative route for the ring using
the extra data paths or to by-pass nodes completely. Further discussion
of providing alternative communication paths to overcome failures is
provided by Zafiropulo (1974).

Wire centres for ring networks

Location and circumvention of faults can be difficult in a ring network
which snakes in and out of many offices. One solution employed, for
example, in some IEEE 802.5 ring installations is to wire stations in a
star arrangement, bringing the wires together at *wire centres* where the
ring is completed. An example with two wire centres is shown in Figure
8.5, with the solid lines indicating the normal ring configuration. It is
then a simple matter to remove one of the stations from the ring using
by-pass relays (indicated as dotted lines on the diagram). For example,

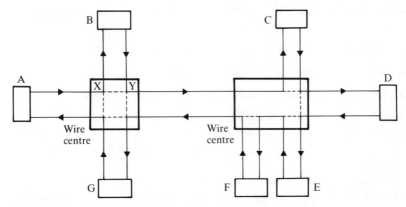

Figure 8.5   Example of wire centres for ring networks

station B can be excluded by completing the by-pass relay from X to Y. One disadvantage of this method is that, in normal operation, the length of the ring is increased.

8.3 LAN management

Most LANs are operated by a single organization and employ a distributed access protocol, so management techniques for LANs must address slightly different problems from those for WANs. However, many of the principles discussed in Chapter 13 can be applied. Fault management can be carried out as discussed above. The performance of a LAN can be assessed using a promiscuous station to determine, for example, network utilization and to give an early indication of possible overload conditions. This technique can also be used for accounting management. Access control mechanisms are also needed and these are discussed in Chapter 11.

8.4 Protocols above the MAC layer

The discussions above and in the preceding chapter have been concerned with the operation of the lower levels of a LAN system. To consider the operation of a full system, the higher levels have to be included. These higher levels consist of *protocol suites* or families.

Different protocol types have developed to meet system requirements and provide various levels of service to the user processes. In general, the higher the level of service and the more features provided, such as guaranteed delivery, the greater the complexity of the protocol and hence the overhead of operation. Three classes of protocols can be defined: connectionless, transaction and connection-oriented protocols. In general, transaction protocols are considered as special options of connectionless protocols.

Each of the different protocol types may support three modes of addressing that control the number of stations to which a message is sent:

1 Individual addressing. Sent to one station only.

2 Multicast addressing. Sent to a group of stations.

3 Broadcast addressing. Sent to all stations on a network.

Addressing types 2 and 3 are more suited to connectionless protocols. In principle, once the details of the network are hidden, i.e. once we get above the transport layer, any accepted protocol can be used (see Chapter 5).

Protocol suites have been developed for each major LAN type. In addition, the IEEE 802.2 LLC offers a standard to complete the data link layer for both connectionless and connection-oriented protocols.

DSAP address 8 bits	SSAP address 8 bits	Control 8/16 bits	Information

DSAP address	Destination service access point address
SSAP address	Source service access point address
Control	16 bits with sequence numbers, 8 bits otherwise
Information	0 or more data octets

Figure 8.6   IEEE 802.2 LLC protocol data unit

8.4.1 IEEE 802.2 logical link control

Logical link control provides that portion of the data link layer which is common to every IEEE 802 access technique. The specification (IEEE 802.2) details the service interfaces with the adjacent levels (i.e. the network layer above and the MAC layer below) and the protocol with peer LLC entities across the network.

Two classes of procedure may be supported. Class I is just the connectionless service. Class II provides both connectionless and connection-oriented services.

The fundamental unit of information that is passed between two LLC endpoints is the LLC protocol data unit (PDU), whose structure is shown in Figure 8.6.

The control field is used by the LLC to indicate which type of data transfer is being carried out—datagram or connection-oriented. In addition, this field contains further control parameters associated with the data being transferred, such as sequence numbers used for flow control. Finally, the information field is made available to the higher level software. This field contains an integral number of octets and can be zero. The maximum size of the information field depends on the MAC layer that the LLC is using for the data transfer. The protocol suffers from only having 8 bits for the service point address and has to incorporate facilities for the control of duplicate addressing within a connected system. If more bits were used the chance of duplicate addressing would be considerably reduced and is generally avoided by other protocol suites.

8.4.2 Cambridge ring protocols

The protocols for the Cambridge ring (Larmouth, 1982) have been developed to provide connectionless, transaction and connection-oriented communications. Connectionless communication is provided by the basic block protocol (BBP). All the other protocols rely

Header minipacket	Size minipacket	Route minipacket	Data minipackets	Checksum minipacket

Figure 8.7   Cambridge ring basic block format

on the BBP to provide a packet or frame communication similar to that supported by the MAC levels in the IEEE 802 systems.

Basic block protocol

Because the Cambridge ring delivers only 16 bits of data in every minipacket, some method had to be devised to send longer packets of information between two stations on the network. A sequence of such minipackets sent between two stations is called a *basic block* and is illustrated in Figure 8.7.

The first minipacket of the block contains a header that indicates that this is a basic block. The next minipacket is optional and contains the count of octets in the basic block, enabling both stations to know when the block has been completely transmitted and received. The route minipacket includes a 12-bit field called a *port* which is used to address a specific application process in the destination system.

Single shot protocol

The Cambridge ring protocols provide a transaction protocol called the *single shot protocol* (SSP) for simple request/reply transactions. The request contains the port to which the reply should be returned; the station addresses will be found in the ring minipacket and the destination port in the BBP header of the basic block in which the request is sent. Although all this information is in different places, the request includes the full identification of both the destination and the address for the reply.

Byte stream protocols

In order to provide a reliable stream of bytes in both directions between two processes and to ensure that no data is lost, corrupted or duplicated, the *byte stream protocol* (BSP) (Larmouth, 1982) was devised. The BSP protocol fragments the user process' byte stream into BSP packets which are sent using the basic block interface. Acknowledgements are used by the BSP protocol to ensure data integrity. However, as the ring error rate is low no negative acknowledgement is used for packets in error; instead a timeout is implemented at the transmitter to recover from unacknowledged packets and retransmit them.

During periods when there is no data to send or no available receive buffer, the transmitter or the receiver can stop the traffic and resume responsibility for restarting it. To enable a broken connection to be distinguished from a stopped connection an idle handshake is used in which the two parties infrequently exchange their last messages. This is particularly useful when operating through interconnected rings as it will prevent the path from timing-out and being cleared by the interconnecting systems (bridges).

The transport service byte stream protocol (TSBSP) is a superset of the BSP including primitives for the provision of a full transport level service in line with interim UK standards (Larmouth, 1982).

8.4.3 DARPA internet protocols

Many computer systems attached to LANs implement the Internet protocol (IP) and the transmission control protocol (TCP) at the higher levels. These protocols are collectively known as *TCP/IP*. This suite of protocols has much wider significance than its use for LANs, as it is the basis for an Internet—a system of interconnected networks of many types (Postel, 1980). In particular, most of the major research organizations and universities in the United States are connected by an Internet called the Defense Advanced Research Projects Agency (DARPA) Internet, and since 1983 all hosts connected to the DARPA Internet have been required to use TCP/IP. A full description of TCP/IP principles, protocols and architectures can be found in the book by Comer (1988). We shall also be returning to the DARPA Internet as an example of connectionless network interconnection in Chapter 9.

The hierarchy of the Internet protocols is shown in Figure 8.8. The IP is a connectionless network layer protocol; the Internet control message

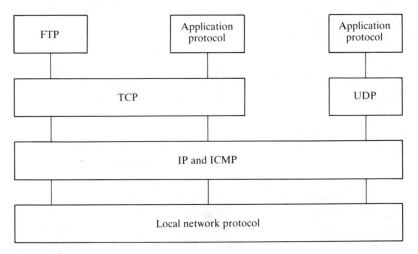

Figure 8.8   Internet protocol hierarchy

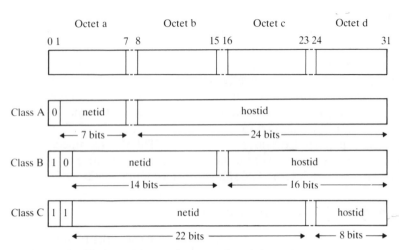

Figure 8.9 Three address classes of an Internet

protocol (ICMP) is used for network control messages. Above the IP, connection-oriented working is provided by the transport level TCP; for applications requiring only datagram reliability the connectionless user datagram protocol (UDP) is provided. Above the TCP and UDP, application protocols such as file transfer protocol (FTP) and remote log-in use the TCP or UDP as appropriate.

Internet addressing

In order to provide a user perspective of an Internet in which all hosts appear to be accessible, a standard form for network addressing is necessary. For every host on the DARPA Internet, a network connection is assigned a unique 32-bit *Internet address*. A host that exists on more than one network will have an address for each network on which it resides since the address defines a host's connection to a particular network and not an address for a specific host computer. Three classes of addressing A, B and C are possible, as shown in Figure 8.9. The three high-order bits are used to indicate which form of addressing is applicable. (It should be noted that a fourth class (class D) is reserved for multicasting.) The three main address formats are assigned different sized fields for network address (netid) and host address (hostid), appropriate to networks of large, intermediate and small sizes, respectively. To ensure that hosts of different architectures understand the Internet address, the Internet specifies that bytes should be sent in the big-endian format (most significant byte first). The 32-bit address is usually written in a form described as the *dotted notation*. Thus an Internet address may be written as four decimal integers separated by dots where each integer gives the value of one octet (e.g. 32.0.3.122).

Address resolution protocol (ARP)

The knowledge of a host's Internet address is not sufficient to allow communication with it; the physical hardware address is also required. In order that applications can be written which require only a knowledge of the Internet address, a mapping must occur at a lower level. This mapping is commonly known as *address resolution*. One mechanism for address resolution would be for hosts to keep tables consisting of pairs of Internet and physical addressees. Alternatively, the physical address could be mapped using some algorithm to the Internet address. Both methods are unattractive, particularly for Ethernet, which has a large physical address space of 48 bits assigned by the interface or chip manufacturer.

The Internet solution is both simple and elegant—a low level protocol provides dynamic binding of addresses. When a host wishes to find the physical address of another host on the Internet, it broadcasts a packet which contains the Internet address of the required host, requesting it to respond with its physical address. All hosts on the network receive the request, but only the host with the specified address should respond. The physical address in the response can then be used for direct communication with the specified host. To reduce the overhead of broadcast packets caused by the protocol, a table of ARP replies is held in a cache on each system for future reference. Also, the initiator of the ARP request sends its own physical address and Internet address in every ARP request, enabling recipients to update the binding information in their local ARP caches. The ARP message is treated as data in the Ethernet frame; details can be found in RFC 826.

The Internet protocol (IP)

The fundamental data delivery service of the TCP/IP suite is the IP which provides a connectionless datagram delivery service, i.e. an unreliable but best attempt packet delivery service (Postel *et al.*, 1981). The IP datagram is *encapsulated* (carried as data) in the MAC data frame. The IP datagram contains both header and data parts. Ideally the IP datagram should fit into one MAC frame, making the transmission fairly efficient. Ethernet allows a data unit of 1500 octets but some networks have much smaller data units available. To maintain efficiency on networks such as Ethernet and yet allow networks with a smaller frame size to carry IP datagrams, a strategy of splitting datagrams into fragments is used. These fragments, which are identified by bits in the header, must be reassembled at the destination.

Figure 8.10 shows the datagram header fields. The major items of interest concerning the header will be briefly presented. However, the detail of this packet header is not discussed here and the reader is referred to RFC 791 and RFC 894 for details of transmission of the IP

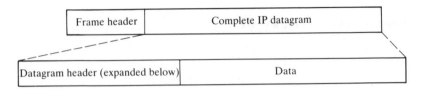

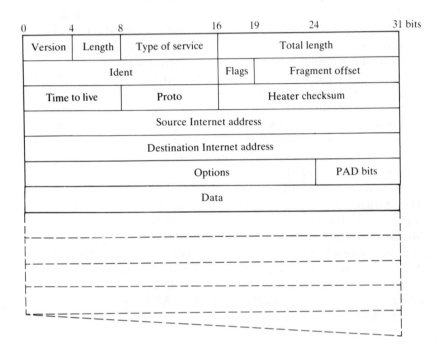

Figure 8.10 Internet protocol datagram format

across an Ethernet. The *version* field is used to check that all packet handlers are operating with the same version of software for IP data handling. The *length* field specifies the size of the header counted in 32-bit words (20 octets is the standard size and hence the length is equal to 5). The *total length* is the length of the whole datagram including the header and data and is counted in octets.

The *type of service* field has several parts. The *precedence* part is generally unused but would allow control messages to be given a higher priority over standard data messages, for example, to implement congestion control. Quality of service can be specified using the *D, T and R* bits: D requests low delay, T requests high throughput and R requests high reliability. The type of service field may be consulted at gateways and applied to the selection of a particular route from several alternatives.

The *ident, flags and fragment offset* fields control fragmentation and reassembly. The ident field (unique for each datagram) together with the source address allows the destination to associate the packet fragment with the correct datagram.

The *time to live* field specifies the time that the datagram should be allowed to remain in the Internet and hence ensures that datagrams do not travel around the Internet forever if routing difficulties occur. The field is decremented by one unit at each gateway on the assumption that each traverse between networks takes one unit of time; when it reaches zero, the datagram is discarded.

The *proto* field is used by higher level protocols to specify the protocol in use. The *header checksum* is used to ascertain the integrity of the header field data, the *options* field is largely used for testing and the remaining fields should be self-explanatory.

The Internet control message protocol (ICMP)

The ICMP protocol is used by gateways and hosts to send messages concerning error detection and network control. The ICMP messages are transmitted in the data part of an IP datagram and allow messages between the Internet software in two machines. An Internet host might, for example, send an echo request message; if a reply is received the destination is alive and hence reachable. Details of ICMP message types can be found in RFC 792.

The user datagram protocol (UDP)

The Internet user datagram protocol (UDP) provides a datagram service to allow user application programs to send messages to other programs with a minimum of protocol mechanism. The protocol does not guarantee delivery or provide protection. The UDP header and data is encapsulated within an IP packet, and the format is shown in Figure 8.11.

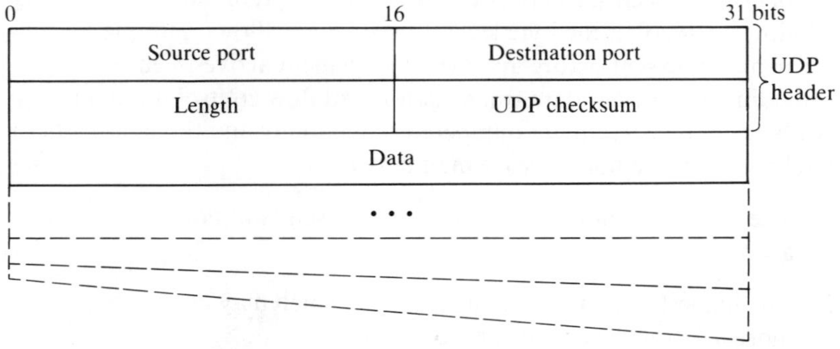

Figure 8.11 UDP message datagram format

The UDP is achieved by an abstraction for subaddressing called a *port*. The data is thought of as originating from or being directed to a host at a specific port number. The UDP protocol is normally implemented with a user interface which provides the following:

1 The ability to create new receive ports.

2 Receive operations on the receive ports which return the data octets and an indication of source port and source address.

3 An operation that allows a datagram to be sent, specifying the data, source address and port and destination address and port.

UDP port numbers may be permanently assigned or bound to an application when it is initiated. The UDP protocol is described fully in RFC 768.

The transmission control protocol (TCP)

The TCP makes few assumptions about the underlying network over which it runs. It provides a reliable stream transport mechanism for data and consequently its implementation is rather more complex than the previously discussed Internet protocols. The TCP is responsible for delivering error free data to the destination in the same order as it left the source. The TCP provides a virtual connection which will remain established while the stream is operating. TCP connections are established using a three-way handshake which enables both parties of the communication to be ready for transfer and allows them to agree on the initial sequence numbers. Control packets offer facilities for setting up calls, sending and detecting acknowledgements, flow control, multiplexing and security. Data transferred across the network is buffered to provide efficient transfer of data in reasonable sized datagrams, but this may be overridden by a *push* mechanism, forcing data to be transferred immediately.

The data byte stream is divided by the TCP into entities called *segments*, and each segment is transmitted in an IP datagram. A sliding window protocol at the byte level is used which allows multiple segments to be sent before an acknowledgement arrives, hence providing an efficient data transmission and flow control mechanism. All bytes in the stream are numbered sequentially and three pointers to the byte sequence numbers are maintained:

1 The sequence number reached for bytes sent and not yet acknowledged.

2 The highest sequenced numbered byte which may be sent before more acknowledgements are received.

3 The current byte count for transmission.

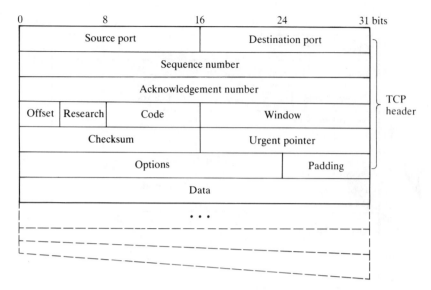

Figure 8.12 TCP segment showing header and data

The size of the window may be changed dynamically since the acknowledgement contains information on the number of additional bytes that the receiver is willing to accept. The TCP uses a go-back-N scheme and employs timeouts to ensure retransmission if acknowledgements are lost. The length of the timeout can be adapted to reflect the round-trip delay encountered on a particular virtual circuit. Details of mechanisms for window management may be found in RFC 813.

Figure 8.12 shows the format for a TCP segment with its header and data parts. The header contains a number of fields including the *source and destination ports* which are used as in the UDP for directing a stream to the correct application program. The *sequence number* identifies the position in the transmitted byte stream of this data segment. The *acknowledgement number* is used to indicate the sequence number of the highest byte received in the return data stream. Further detail on the fields in the header may be found in RFC 793.

8.4.4 Xerox network systems protocols

The Xerox network systems (XNS) protocols (Xerox, 1981) were principally developed for operation over Ethernets and the interconnection of Ethernet systems over public packet networks. All the protocols are built on the lowest level protocol, the Internetwork datagram protocol (IDP).

Octet
number

0	Checksum
2	Length
4	Transport control / Packet type
6	Network number ⎫
10	Host number ⎬ Destination address
16	Socket ⎭
18	Network number ⎫
22	Host number ⎬ Source address
28	Socket ⎭
30	Data (0–546 octets)

Figure 8.13 XNS Internet datagram format

Internetwork datagram protocol

This is the basic XNS protocol and has the format shown in Figure 8.13. Both addressing fields fully specify the location of the source and destination of the datagram. The packet type field (see Figure 8.14), identifies the function of the packet and its association, if any, with a higher level protocol. The checksum provides for additional error detection for the complete datagram, in addition to that provided by the MAC layer.

Number	Protocol
1	Routing information
2	Echo
3	Error
4	Packet exchange
5	Sequenced packet
16–31	Reserved

Figure 8.14 XNS Internet packet types

Octet
number

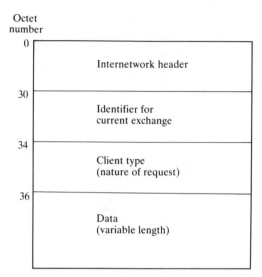

Figure 8.15   XNS PEP packet format

Packet exchange protocol

The XNS packet exchange protocol (PEP) is the transaction protocol of
the XNS protocol suite. PEP is implemented above the IDP and uses
type 4 packets. The format of a PEP packet is shown in Figure 8.15.

When a packet has been sent the sender will expect a reply. Timeouts
are used to retry the exchanges. The ID field is unique for each
exchange and is used to identify request and reply pairs.

Sequenced packet protocol

The sequenced packet protocol (SPP) is a connection-oriented protocol
which guarantees the integrity and sequence of the packet flow and
undertakes retransmission in the event of errors. The protocol offers
three services:

1   Message service, in which a message may comprise a number of
    packets.

2   Sequenced packet service, in which message boundaries are not
    distinguished.

3   Reliable packet mode, in which packets are presented in order of
    arrival without duplicates.

The data is carried by type 5 IDP packets. The format is shown in
Figure 8.16. Connection identifiers (IDs) are unique and identify the
call; the call establishment involves the process of agreeing IDs between

Octet
number

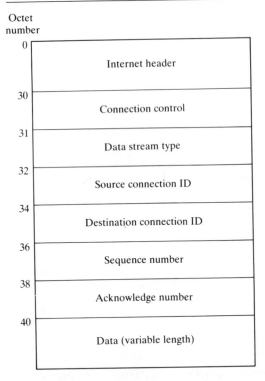

Figure 8.16 XNS SPP packet format

the parties. The caller has the responsibility for terminating a call, which is achieved by sending an 'end' packet.

8.5 Software interfaces for LAN protocols

The above discussion has concentrated on the requirements of network protocols and the facilities provided by implementations. For any protocol implementation to be useful, a software interface has to be provided for the applications programmer. This interface should ideally be uniform between different protocol families and implementations so the programmer can produce portable programs. Uniform interfaces are required for both connectionless and connection-orientated protocols.

As an example, the Berkeley UNIX interprocess communications environment offers an endpoint of communication, called a *socket*, for different protocol families (see also Chapter 12). Three separate domains are supported: the UNIX domain for processes on the same system, the Internet domain for processes communicating between machines using the DARPA Internet protocols described in Section 8.4.3 and the NS domain for processes communicating between machines using the XNS protocols described in Section 8.4.4.

8.6 LAN development directions

The development of local area networks has reached a state where
stable commercial products are established in the market place. This
does not mean that development and research have stopped but the
direction has been changing from new concepts to new applications.
The increase in interworking between computers enabled by local area
networks is creating an increasing need for interworking between
systems on different networks via gateways and bridges. As network
interfaces and protocols are standardized, interworking will be
simplified, enabling a wide variety of networks and systems to coexist
commercially.

As a wider choice becomes available the cost/performance of a
network will become the overriding factor in choosing a network for an
application. Terminals (personal workstations) will be interconnected
via low-cost networks offering local services and shared resources.
These networks will be interconnected by gateways to higher
performance networks connecting mini and mainframe systems offering
large-scale applications. Again these networks will be interconnected on
a metropolitan, regional or national level.

Two areas of development that will have a great influence on local
area networks are optical networks, offering higher performance
systems, and large-scale integration, which will enable the cost of a
network to be reduced due to the scale of manufacturing production.
These topics are discussed briefly in Chapter 15.

8.7 Summary

Local area computer networks provide a means of providing high-speed
data communications between processors. Since LANs cover a limited
distance they are able to provide high reliability and low error rates at a
reasonable cost.

Many LAN systems have been developed experimentally and some
have become commercial products. The choice of a LAN for an
application will be influenced by the type of communication required,
e.g. transactions, bulk data or voice. The choice of topology is only
important for highly reliable systems where error recovery is required.
For normal applications, the topology will be determined by the system
that meets the performance and cost requirements. In practice the
overriding factor in LAN selection is often the commercial availability
of a LAN compatible with the systems to be interconnected. Currently
the Ethernet has the greatest number of suppliers and interfaces for
different systems and is emerging as the LAN to which all others are
compared.

References and further reading

References

Blair, G. S. (1982) 'A performance study of the Cambridge ring', *Computer Networks*, **6** (1), 13–20.

Blair, Gordon S. and Doug Shepherd (1982) 'A performance comparison of Ethernet and the Cambridge ring', *Computer Networks*, **6** (10), 105–113

Bux, Werner (1981) 'Local area subnetworks: a performance comparison', *IEEE Trans. Commun.*, **COM-29** (10), 1465–1473.

Comer, D. (1988) *Internetworking with TCP/IP Principles, Protocols and Architecture*, Prentice-Hall, Englewood Cliffs, New Jersey.

Hammond, J. L. and P. J. O'Reilly (1986) *Performance Analysis of Local Computer Networks*, Addison-Wesley, Reading, Mass.

IEEE 802.2 (1985) 'Logical link control specifications', American National Standards Institute.

Larmouth, J. (1982) 'Cambridge ring 82 protocol specifications', Joint Network Team of the Computer Board for Universities and Research Councils, Rutherford Appleton Laboratory, Chilton, Oxfordshire.

Needham. R. M. and A. J. Herbert (1982) *The Cambridge Distributed Computing System*, Addison-Wesley, Wokingham.

Postel, J. B. (1980) 'Internet protocol approaches', *IEEE Trans. Commun.*, **COM-28** (4), 604–611.

Postel, J. B., C. A. Sunshine and D. Chen (1981) 'The ARPA Internet protocol', *Computer Networks*, **5**, 261 ff.

RFCs are 'requests for comments'. This is an extensive range of documents about DARPA Internet protocols and their ancestors and application protocols. Copies can be obtained from SRI International, Network Information Center, 333 Ravenswood Avenue, Menlo Park, California 94025, USA. Annotated lists and details of how to obtain RFCs by network mail can be found in Comer (1988).

RFC 768 User datagram protocol
RFC 791 Internet protocol (IP)
RFC 792 Internet control message protocol (ICMP)
RFC 793 Transmission control protocol
RFC 813 Window and acknowledgement strategy in TCP
RFC 826 Address resolution protocol
RFC 894 A standard for the transmission of IP datagrams over Ethernet networks

Schwartz, M. (1987) *Telecommunication Networks—Protocols, Modeling and Analysis*, Addison-Wesley, Reading, Mass.

Sharpe, W. P. and A. R. Cash (1982) 'Cambridge ring 82 interface specifications', Joint Network Team, Rutherford Appleton Laboratory, Chilton, Oxfordshire.

Shoch, John F. and Jon A. Hupp (1980) 'Measured performance of an Ethernet local network', *Commun. ACM*, **23** (12), 711–721.

Stallings, William (1984) 'Local network performance', *IEEE Commun. Mag.*, **22** (2), 27–36.

Stuck, Bart W. (1983) 'Calculating the maximun mean data rate in local area networks', *IEEE Computer*, **16** (5), 72–76.

Xerox Corporation (1981) 'Internet transport protocols', Xerox System Integration Standard 028112, December 1981.

Zafiropulo, P. (1974) 'Performance evaluation of reliability improvement techniques for single-loop communications systems', *IEEE Trans. Commun.*, **COM-22** (6), 742–751.

Further reading

The following is a list of a few of the many books that have been published on the subject of local area networks that are not directly referenced from within the text.

Beauchamp, K. G.: *Computer Communications*, Van Nostrand-Reinhold, London, 1987.

Dallas, I. N. and E. B. Spratt (eds): 'Ring technology local area networks', in *Proc. IFIP WG6.4 Workshop*, Canterbury, UK, North-Holland, Amsterdam, 1983.

Flint, D. C. *The Data Ring Main; An Introduction to Local Area Networks*, Wiley, Chichester, 1983.

Hopper, A., S. Temple and R. Williamson: *Local Area Network Design*, Addison-Wesley, Wokingham, 1986.

Hutchison, D.: *Local Area Network Architectures*, Addison-Wesley, Wokingham, 1988.

Hutchinson, D., J. Mariani and D. Shepherd (eds): 'Local area networks: an advanced course', Lecture Notes in Computer Science 184, Springer-Verlag, Heidelberg, 1985.

Meijer, A. and P. Peeters: *Computer Network Architectures*, Pitman, London, 1982.

9 Network interconnection

GILL WATERS

9.1 The need for interconnection

Networks must be interconnected to allow a user on one network to access services or information located on another network. To satisfy immediate needs it may be possible to connect the two networks together directly or in some ad hoc fashion, but in the more general case it will be necessary to interconnect a variety of networks to achieve communication between many different endpoints, and indeed to transport information between two networks that are situated a long distance apart.

As we have seen in previous chapters, a wide variety of networks already exists: proprietary networks such as SNA, heterogeneous WANs based on X.25, token passing and contention based LANs. Standardization should reduce this diversity, but even the standards offer choices (e.g. protocol subsets, the IEEE 802 options for LANs) and there is a transitional phase as standards are specified, refined and adopted. Networks are often chosen for their suitability for their main body of users, rather than for ease of connection. Also, new types of network are likely to be introduced implying a continuing need for interconnection strategies.

LANs have size limitations, so that interconnection with other networks is essential for wider geographical coverage. All networks must be maintained and managed; separate networks offer a well defined boundary for management control, decisions about the installation and security of a particular network being the responsibility of a single organization. Small networks offer better reliability and performance than large networks; by interconnecting a number of small networks in which the majority of communication takes place within individual networks these advantages can be preserved while wider interworking is made possible.

9.2 General problems of interconnection

In this chapter we give an introduction to the issues involved in network interconnection. A key discussion paper on interconnection issues is

Cerf and Kirstein (1978). Other general sources of information are *IEEE Computer* (1983), Currie (1988), Stallings (1988) and Tanenbaum (1988). The *IEEE Journal on Selected Areas in Communication* has devoted issues to LAN interconnection (*IEEE JSAC*, 1987) and to heterogeneous computer network interconnection (*IEEE JSAC*, 1990). Both of these issues include annotated bibliographies.

The nearer the architectures of the networks concerned, the easier it is to connect them. Linking a datagram network to a virtual circuit network presents obvious problems, as does connecting a high-speed local area network with simple access mechanisms to a slower more complicated wide area packet switched system. The following list shows some of the characteristics, which may differ between network architectures and which will therefore have to be considered:

- Addressing and routing
- Multiplexing
- Maximum packet size
- Directionality (e.g. full or half duplex working)
- Packet sequencing or resequencing
- Error control (e.g. error detection or error correction)
- Connection-oriented and connectionless working
- Flow control
- Expedited flow (priority packets)
- Classes of service

9.2.1 Terminology

A network interconnection system usually connects two networks, but some connect more than two. There are several terms in common use for such systems and examples are shown in Figure 9.1:

1 A *repeater* is often used to connect segments of a LAN and is an extension at the physical level, layer 1 of the OSI model.

2 A *bridge* connects LANs at the medium access control layer (layer 2); bridges may connect similar or dissimilar LANs.

3 A *gateway* is the term normally used when interconnecting two unlike networks, e.g. connecting a LAN to a WAN; the connection is made at layer 3. (In North America, the term *router* is also often used for layer 3 interconnection systems.)

4 A *protocol converter* is needed where protocols differ between interconnected networks. This could conceivably be at any level, but the term has been used to describe interconnection at a higher level, e.g. to convert between two message handling systems.

Bridges, gateways and protocol converters generally consist of both hardware and software. A gateway would, for example, have to be

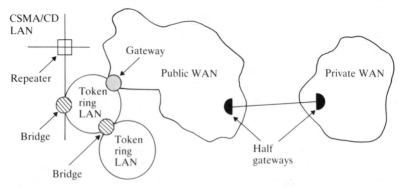

Figure 9.1 Examples of interconnection systems

capable of implementing protocols on both or all of its interconnected networks up to layer 3, and would also have to make routing, accounting and other decisions associated with information crossing between the networks. Because the actual implementation of such a system may well be at the boundary of two different organizations, or the boundary between a public and a private network provider, in practice a gateway often consists of two processing systems separated by a communications link. In this case the gateway functions are split and each system is called a *half gateway* (also shown in Figure 9.1). The interconnection problem is then reduced to agreeing a protocol along the communications link connecting the two half gateways.

A strategy for the interconnection of LANs on a single site is to connect each of the LANs to a *backbone network* (for example FDDI; see Section 9.7). Over a wider area LAN interconnection can be achieved though a metropolitan area network (see Chapter 15).

9.2.2 Internetworking in the OSI framework

In the OSI model, the decisions and actions involved in forwarding information between open systems are called *relaying*. The highest layer at which relaying can be done is the network layer. Thus the network service ensures that the transport and higher layers need not be concerned with routing and relaying. Indeed, in OSI terms open systems may communicate over a *network* consisting of a variety of interconnected *subnetworks*, each of which may, for example, be a LAN, a public WAN or a private data network. (For the remainder of the chapter the terms network and subnetwork will be used in this way.)

There should be no problems in interconnecting subnetworks that provide the full OSI network service. However, where networks that do not conform are to be connected together the network layer can be divided into sublayers as shown in Figure 9.2. An individual

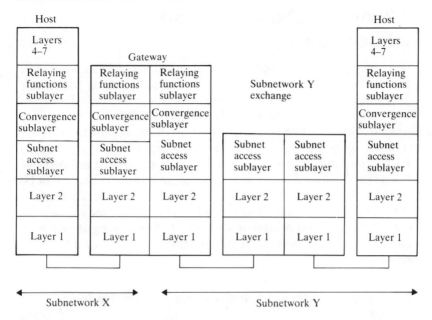

Figure 9.2 Convergence provided for interconnection at the network layer

subnetwork will need to use its own lower level services and must be modified (generally enhanced) to converge with standard relaying functions in order to comply with other subnetworks.

9.2.3 How should the gateway look to the rest of the subnetwork?

Should a gateway look like a host or a packet switching exchange to the rest of the subnetwork? If the two subnetworks are very similar, the switching exchange solution may seem easier as the gateway can perform similar functions such as flow control and routing. However, there are problems of addressing. Each exchange will have to know about the address space of the other subnetwork in order to perform routing. If the interconnected subnetworks have totally different strategies for routing and flow control, the problem becomes very complex and a common access interface is needed. If the gateway looks like a host, the operation of the exchanges of the subnetwork is not affected. This technique is suited to the transmission of datagrams, where there is no requirement for packet sequencing or guaranteed delivery. A datagram typically will be wrapped up in the protocol of a specific subnetwork for its passage through the subnetwork, and then unwrapped at the destination host or ongoing gateway.

9.3 Approaches to network interconnection

Because requirements have predated the availability of standards, there have been and continue to be a number of ad hoc solutions to network interconnection. However, there are three main areas where standards work has been most evident. The first is the CCITT X.75 recommendation which is intended for the interconnection of connection-oriented public packet switched networks. The second are *Internet* protocols which have been used for some time on the DARPA Internet and offer a connectionless service, and form the basis of an ISO standard. Thirdly, for the interconnection of local area networks, the IEEE project 802 has been looking at LAN bridges for the various IEEE medium access control options. We shall be looking at each of these three areas in more detail, but first discuss more general approaches.

9.3.1 Connection-oriented versus connectionless working

The arguments as to whether interconnection should be based on a connection-oriented or connectionless approach are very similar to those between the two techniques in a single subnetwork. With a connection-oriented approach it will be necessary to enhance connectionlesss subnetworks by providing virtual circuit facilities, and this may be very difficult. However, the advantages of shorter headers, packets being kept in sequence and quick recovery from errors are maintained. Where reliability and predictable response time are of prime importance the connection-oriented approach is preferred (Clyne, 1988/9). With the connectionless approach it is much easier to provide connectionless facilities on a connection-oriented subnetwork, and packets can be routed individually thus offering robustness in the event of gateway failure. However, any error correction or flow control must take place end to end at a higher layer. The connectionless approach may also be suitable for services such as packetized speech which do not require every packet to be delivered reliably (but the problems of variable delays caused by alternative routing must also be considered).

9.3.2 A general approach to protocol conversion

The following general technique for protocol conversion between different subnetwork architectures which is based on the OSI model was put forward by Green (1986).

We can consider conversion as a *mapping* between the protocols of one subnetwork architecture to those of another subnetwork architecture. The first step is to concentrate on the differences between

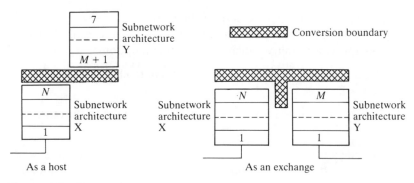

Figure 9.3 Protocol conversion boundaries (after Green, 1986)

the architectures. In particular, we must find an appropriate conversion level in the hierarchy of each network. In general it will be necessary to map the top of the Nth layer of architecture X onto the top of the Mth layer of architecture Y, as illustrated in Figure 9.3. The level of interconnection is often dictated by existing system structure; for example specialist hardware may be available to implement the protocols up to a certain level in one subnetwork. The technique can be applied to both host-based and exchange-based gateways, and to a number of conversion methods such as protocol substitution or enveloping one protocol inside another.

Having decided to connect subnetwork X at level N to subnetwork Y at level M, the next step is to examine the services provided by the N-layer and M-layer protocols and look for differences. The protocols of layers M and N (and also any control or management protocols that use these layers) must also be taken into account. Since each protocol may have a variety of transient or steady states it is also necessary to examine differences over a certain length of time.

Having done this comparison we are left with a well-defined protocol *mismatch*, which must be corrected. This may be very simple (e.g. protocols may be subsets or variants of each other). It may prove impossible (e.g. architecture A is allowed to drop packets, but this is not allowed by architecture B). Green calls this a *hard mismatch*. On the other hand, it may be possible to perhaps impose some degradation of performance such as extra delay (a *soft mismatch*). The gateway must then overcome the mismatch; in the case of a hard mismatch this may be achievable only by inserting an extra complementary layer to enhance the existing architecture(s). This technique has been shown in the extra sublayers of Figure 9.2.

Service versus protocol data unit mapping

In a more recent paper, Bochmann and Mondain-Monval (1990) distinguish two approaches to the mapping between subnetworks—a vertical approach (mapping service primitives) and a horizontal approach (mapping protocol data units between peer levels). When mapping service primitives there are three main options:

1 Reduce the service available to a minimum level provided by both networks.

2 Enhance the services in one network to correspond to the more powerful services of the other.

3 Provide a common set of extra services on top of existing ones which can be used to enhance both networks to an agreed common level.

The advantages of the service approach are that it is conceptually simple and can use existing protocol implementations which provide the existing services of the interconnected subnetworks.

In the protocol conversion approach, the protocol data units (PDUs) from one subnetwork are converted to PDUs of the other subnetwork, thus conforming to the appropriate protocol in each network. This approach is more complex and more difficult to implement than the service mapping approach, but is easier to optimize.

9.3.3 Addressing

The problems of addressing are similar to those within a single subnetwork as discussed in Chapter 3. The options are:

1 *Unique global addresses* This would be the simplest scheme to implement but very hard to manage, and it is already too late to include all subnetworks. The IEEE 802 standard LANs support global addressing with 1 bit in the 48-bit address field being used to indicate whether addresses are administered universally or locally.

2 *Address mapping* Addresses from outside a subnetwork are mapped onto unused addresses within the subnetwork. By performing the mapping only while a call is in progress, a small range of addresses is needed, though care has to be taken not to reuse addresses too soon. This technique was used, for example, in the Universe project, an experiment in high-speed LAN/WAN interconnection (Burren and Cooper, 1989).

3 *Hierarchical schemes* The address contains a field specifying the subnetwork agreed globally and a field for the address within that subnetwork allocated by the managing body for each subnetwork. It is probably the simplest and most satisfactory solution when different organizations are involved. Variations of this method are used in X.75 and for the Internet protocols as discussed later.

9.3.4 Routing

Gateways and bridges must keep routing tables which may be fixed or may change. Connectionless schemes offer more flexibility than connection-oriented ones, allowing each datagram to be routed independently and quickly adapting to fluctuations in subnetwork availability.

Another desirable feature is *source routing* in which the user specifies the route to be taken. For example for security purposes the user may wish to avoid traversing a subnetwork operated by a rival organization or a user may wish to ensure that a single satellite subnetwork is used rather than a concatenation of several interconnected terrestrial subnetworks.

9.3.5 Packet size and fragmentation

All packet networks have a maximum packet size. When a packet enters a subnetwork whose maximum packet size is less than the length of the packet, it must be fragmented by the gateway. Each fragment must be appropriately addressed and labelled. Fragmentation is relatively easy, but at some point the fragments must be reassembled. If the packet is destined for a host on a subnetwork of small maximum packet size, that host can be expected to cope with reassembly. If the packet is simply passing through such a subnetwork there are two options:

1 Reassemble the packet at the outgoing gateway (all fragments must be sent via the same gateway). Packets travelling through a succession of subnetworks may be subjected to this procedure several times. The gateways must provide buffer space and avoid reassembly lockup.

2 Reassemble only when the packet reaches the destination host. In this case the host must be aware of the fragmentation procedure, but it will save effort and memory in the intermediate gateways.

It may be necessary to indicate within fragments whether the receiving host is capable of reassembly; if not then reassembly must be done by the gateways.

9.4 Connection-oriented approaches

In this model the transport layer asks the network layer to open a connection to a specific destination, and that connection will consist of a sequence of virtual circuits created across each subnetwork between source and destination. Each gateway must provide resources, remember all the paths which are currently active, and provide appropriate conversion of packet formats. The result is a transport layer

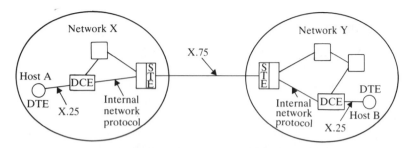

Figure 9.4 Public networks interconnected using X.75

end-to-end connection. Obviously all packets on a single call must pass through the same sequence of gateways. (Packets that have crossed a connectionless subnetwork may arrive at the outgoing gateway out of order and must be resequenced by the gateway.)

9.4.1 CCITT recommendation X.75

The CCITT recommendation X.75 is principally aimed at international data communications between public switched data networks. It is based on the virtual circuit model and defines the interface between signal terminating equipments (STEs) in two packet switched networks. The STEs must be implemented in the exchanges of each network. Figure 9.4 shows a typical sequence of virtual circuits and the procedures followed for communication from host A on network X to host B on network Y.

X.75 is very similar to X.25 on which it is based. In X.75, the call request packet must contain the full CCITT X.121 address. (X.121 defines the international addressing conventions applied to DTEs connected to public data networks. The address consists of 14 decimal digits. It should be noted that the range of subnetwork addresses within each country is very small because the recommendation is aimed at public rather than private networks.) The X.75 call request packet has a field for utilities specific to network interconnection such as transit network identification and throughput class. The user facilities of the X.25 call which is being carried are conveyed transparently in the X.75 call request packet. Additional diagnostic codes are defined in the clear request packets which will be conveyed across each of the virtual circuits in turn to the source or destination host.

9.5 Connectionless approaches

The ISO *Internetwork protocol (IP)* standard (ISO 8473) is based on an earlier US Department of Defense (DOD) standard. It offers a connectionless service, with the ability to send and receive datagrams, and the protocol is effectively a common datagram format. Since the standards are based on the DARPA Internet protocol we now describe this.

9.5.1 The DARPA Internet protocols

One of the best examples of practical network interconnection is the Defense Advanced Research Projects Agency (DARPA) Internet (Postel, 1980; Hinden *et al.* 1983; Comer, 1988). The Internet consists of over 150 subnetworks of several types including packet radio networks, LANs, commercial networks, satellite networks and the ARPANET. It is in everyday use as a vehicle of communication between researchers in the United States and Europe.

The Internet protocol (IP) is implemented by all the hosts attached to the Internet and is based on the connectionless approach discussed in Section 9.3.1. Reliability and sequenced delivery are provided if necessary by the transmission control protocol (TCP), a higher level protocol implemented in the hosts and offering end-to-end flow control. (The TCP/IP suite of protocols is described in Chapter 8.)

The gateways of the Internet have several functions including store and forward for the datagrams, providing interfaces to the appropriate networks, routing, monitoring and control facilities. The Internet address (shown in Figure 8.9) uses a common format, and is therefore easily interpreted by each gateway. IP datagrams (whose format is shown in Figure 8.10) are enclosed within the local subnetwork header for transmission across each subnetwork, as shown in Figure 9.5. The

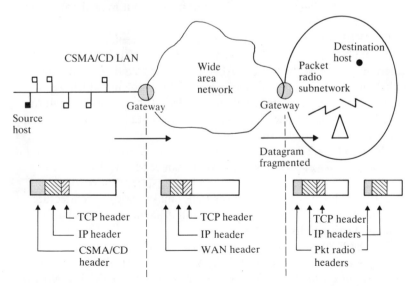

Figure 9.5 Local subnetwork protocol wrapping and unwrapping by gateways

local header is stripped off when a datagram arrives at the gateway. A datagram may be destined for the gateway itself (e.g. for routing information), for a host on the connected network (in which case it is wrapped in that subnetwork's local header for transmission) or for a host on another network. In the last case the gateway chooses the most appropriate neighbouring gateway for the datagram's onward route, wraps it in the subnetwork's local header and sends it to that gateway. Datagrams are fragmented by the gateway if necessary and reassembled at the destination host.

Gateways communicate with each other by a special protocol which allows them to send out echo and probe packets to determine the status of their neighbours. The information obtained is then sent to other gateways as routing updates so that a gateway's view of the status of the network is obtained by dynamic learning. To avoid datagrams circulating endlessly within the Internet each datagram has a *time to live* field which is decremented on its passage through a gateway. Monitoring systems poll the gateways to collect information on traffic carried.

9.6 Bridges

A bridge connecting two identical high-speed LANs is much simpler than a gateway. The interface on each side is identical and some buffering is needed in case the receiving network is busy, but the main objective is to forward information as quickly as possible. Bridges normally make 'best efforts' to forward packets but may discard them if congestion occurs.

The IEEE 802 standardization procedure has resulted in three MAC options: CSMA/CD (802.3), token bus (802.4) and token ring (802.5). Interconnection of the three MAC types is not trivial. Tanenbaum (1988) gives a detailed picture of some of the problems which we summarize here. Each standard has a different frame format requiring translation in the bridge. They have different data rates making congestion more likely when forwarding, for example, from 802.3 (10 Mbit/s) to 802.5 (4 Mbit/s). They all have different maximum frame lengths. This is very serious as no MAC can cope with fragmented frames, so frames that are too long to be transported have to be discarded. The result of these mismatches makes bridge architecture complicated and in general a degradation of facilities occurs when interconnecting unlike LANs.

The IEEE 802 committee have also produced two options for the routing strategies for bridges (*IEEE Network*, 1988). The first, produced as an extension to IEEE 802.1 and applicable to all IEEE 802 networks, is the *transparent bridge*, which can simply be plugged in and will learn from its environment. Initially all frames are flooded onto all possible connected networks. As each frame arrives its source address indicates where a particular host is situated, so that the bridge learns which way to forward future frames to that address. To prevent looping, the bridges form a *spanning tree*, a tree structure that has one bridge as a root and in which there is exactly one route between any two bridges. The second bridge type, developed for IEEE 802.5 token rings, is the *source routing bridge*, which requires users to specify the route to be followed. A user can discover a route by sending a discovery frame which proliferates through the network using all possible paths to the destination; each frame gradually collects bridge addresses as it goes. The destination responds to each frame and the source chooses an appropriate route from these responses, for example the route with the minimum number of hops.

Thus shorter paths may be found for source routing than for users in an unfortunate position relative to the root of the spanning tree in transparent bridging. However, the proliferation of discovery frames can exert a serious extra load on the network. A comparison of the two techniques is made by Soha and Perlman (1988).

9.7 FDDI/LAN interconnection

As an example of practical interconnection problems, we consider the use of FDDI as a backbone to link together several lower-speed LANs (see Figure 9.6). This raises a number of issues for bridge design. The issues can be resolved reasonably easily if all the LANs are of the same type (e.g. IEEE 802.3), but more complexity is introduced if it is required to have more than one bridge to a single LAN (e.g. B1 and B2 in Figure 9.6), to interconnect LANs of different types or to allow access to systems directly connected to the FDDI backbone network (e.g. H in Figure. 9.6).

FDDI bridges will adopt one of the LAN bridge routing techniques discussed in Section 9.6—transparent bridging or source routing. We will briefly discuss some other interconnection issues: frame size, frame check computation and MAC address interpretation. It should be noted that there are further issues such as response bit setting for token rings. The maximum size of a frame on a CSMA/CD network is 1518 bytes, on a token ring it can be more than 2048 bytes and on FDDI the frame size could reach 4500 bytes. This can pose problems for frames that are too large to be forwarded onto a network of small maximum size. The easiest solution to this problem is for bridges to discard such frames. Extensions to the LLC layer which allowed fragmentation and reassembly might be possible, but seem less practical.

The second problem is what to do about frame check sequences—should these be passed on unchanged or should they be recomputed by the bridges? Different frame check algorithms are used

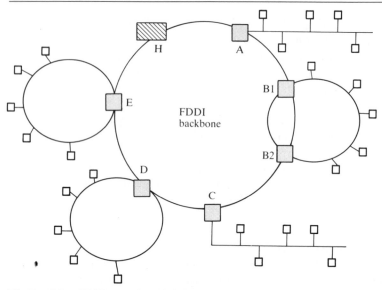

Figure 9.6   FDDI as a backbone to interconnect LANs

for each LAN type so recomputation is essential when connecting
LANs of different types. For bridging networks of the same type, it
would seem sensible to leave the check field unchanged, but as the
checksum is often provided by the LAN interface hardware
recomputation may be easier. Any doubt about the validity of a
checksum at this level may be removed by providing higher layer
reliability when it is needed, e.g. at the transport layer.

The third problem, which is considered to be the most serious, is the
interpretation of the MAC address. CSMA/CD networks transmit
MAC addresses starting with the least significant bit of each byte; token
rings start with the most significant bit of each byte. This is not a
problem for the bridge as the appropriate bit order can be provided for
each LAN, but it is a problem when the MAC address is used as data
by a higher layer (e.g. by the TCP/IP address resolution protocol
described in Chapter 8). A late decision by the IEEE 802.1 committee is
that the least significant bit should be sent first; this implies that all
token ring interfaces should be changed, a daunting task. A similar
problem could apply if CSMA/CD stations wish to communicate
directly with FDDI stations because FDDI is based on token ring
conventions.

Commercial Ethernet to FDDI bridges are available, most of which
employ *encapsulation*, that is the whole of each Ethernet frame is carried
unchanged within an FDDI frame across the FDDI network, and the
receiving bridge simply extracts the Ethernet frame for transmission
onto the destination Ethernet. This technique, although simple, is not
easily extended to enable other networks to be attached or to allow
communication directly with FDDI stations.

9.8 Summary

Bridges offer better performance than gateways because they are simpler; also they do not require all user systems to implement the same network layer protocol. Gateways, on the other hand, can resolve the more major differences between networks such as addressing strategies, packet length, bandwidth and error rate (see also Perlman *et al.*, 1988). Repeaters, which we have not discussed in detail, simply repeat all signals at the physical layer and generally consist exclusively of hardware. They are found in many LAN installations enabling small sections of the LAN to be handled separately for ease of installation and modification. Bridges, although more complex than repeaters have the advantages of filtering frames and overcoming the distance limitations of LANs.

We have discussed the architectural solutions to interconnection. There are also practical implementation problems requiring efficient hardware and software solutions. Implementation options are described in Benhamou and Estrin (1983). A number of case studies are described in *IEEE JSAC* (1990), including a discussion of possible solutions to the problems of interconnection of networks based on TCP/IP and those based on OSI (Rose, 1990). Both service mapping and PDU mapping approaches are considered. This is an important topic now that OSI is becoming more widely adopted, especially for the transitional period when investment in TCP/IP implementations needs to be protected and a graceful transition to OSI is needed.

The experiences of the DARPA Internet community and the wide availability of simple Ethernet bridges and repeaters are two examples of how successful network interconnection can be. Although some of the interconnection systems discussed in this chapter are not yet available commercially, the architectural and implementation solutions are becoming clearer.

References

Benhamou, E. and J. Estrin (1983) 'Multilevel internetworking gateways: architecture and applications', *IEEE Computer*, **18** (9), September, 27–34.

Bochmann, G. V. and P. Mondain-Monval (1990) 'Design principles for communication gateways', *IEEE J. Selected Areas in Commun.*, **8** (1), January, 12–21.

Burren, J. W. and C. S. Cooper (1989) *Project Universe: An Experiment in High-speed Computer Networking*, Clarendon Press, Oxford.

CCITT recommendation X.75 (1988) International Telecommunication Union, Geneva.

Cerf, V. G. and P. T. Kirstein (1978) 'Issues in packet network interconnection', *Proc. IEEE*, **66** (11), 1386–1408.

Clyne, L. (1988/9) 'LAN/WAN interworking', *Computer Networks and ISDN Systems*, **16**, 34–39.

Comer, D. E. (1988) *Interworking with TCP/IP: Principles, Protocols and Architecture*, Prentice-Hall, Englewood Cliffs, New Jersey.

Currie, W. S. 'Extended LANs and interworking', in *LANs Explained—A Guide to Local Area Networks*, Ellis Horwood, Chichester, Chapter 14.

Green, P. E. (1986)'Protocol conversion', *IEEE Trans. Commun.*, **COM-34**, (3), 257–268.

Hinden, R., J. Haverty, and A. Sheltzer (1983) 'The Darpa Internet: interconnecting heterogeneous computer networks with gateways', *IEEE Computer*, **16** (9), September, 38–48.

IEEE Computer (1983) Special issue on 'Network interconnection', **16** (9) September.

IEEE Network (1988) Special issue on 'Bridges and routers', **2** (1), January.

IEEE JSAC (1987) Special issue on 'Local area network interconnection', *IEEE J. Selected Areas in Commun.*, **SAC–5**, (9), December.

IEEE JSAC (1990) Special issue on 'Heterogeneous computer network interconnection', *IEEE J. Selected Areas in Commun.*, **8**, (1), January.

ISO 8473: (1986) 'Protocol for providing connectionless mode network service' (Draft).

Perlman, R., A. Harvey, and G. Varghese (1988) 'Choosing the appropriate ISO layer for LAN interconnection', *IEEE Network*, **2**, (1), January, 81–86.

Postel, J. (1980) 'Internetwork protocol approaches', *IEEE Trans. Commun.*, **COM–28**, (4), 604–611.

Rose, M. T. (1990) 'Transition and coexistence strategies for TCP/IP to OSI', *IEEE J. Selected Areas in Commun.*, **8**, (1), January, 57–66.

Soha, M. and R. Perlman (1988) 'Comparison of two LAN bridge approaches', *IEEE Network*, **2**, (1), January, 37–43.

Stallings W. (1988) 'Internetworking', in *Data and Computer Communications*, Macmillan Publishing, New York, Chapter 14.

Tanenbaum, A. S. (1988) 'The network layer', in *Computer Networks*, 2nd edition, Prentice-Hall, Englewood Cliffs, New Jersey, Chapter 5.

10 Coding for data security

CHRIS SMYTH

10.1 Introduction

In this chapter and the next we survey the important subject of data security. In Chapter 11 we consider the authentication and access control aspects of security. Here we are concerned primarily with the use of secret codes (or *ciphers*) to encrypt data. The data should then be incomprehensible if, during storage or transmission, it is tapped by an eavesdropper. Computer networks can be designed with the possibility of data encryption built in. A suitable position in the OSI model for an encryption service is the presentation layer, although it may be placed in other layers (see Chapters 4 and 5).

Of course the design of ciphers is very different from that of codes used for other purposes (data-compression codes, error-control codes, line codes and so on). Clearly, the primary requirements of a good cipher system is that encryption and decryption should be easy, fast and reliable for authorized users, but essentially impossible for unauthorized users. (However, in public key systems this applies only to decryption—anyone can encrypt—see Section 10.4.) These requirements are not easy to meet. It is particularly difficult even for experts to assess reliably how strong a cryptosystem is. Historically, their inventors, purveyors and users have had a very inflated opinion of the level of security they provide. Working for the French government and army in the nineteenth century, for instance, *Etienne Bazieres* had ruined numerous cryptographic inventions by showing how test enciphered texts could be deciphered. Later he invented a cylinder cipher of his own, and declared in 1901 'Je suis indéchiffrable'. One of his victims, *De Viaris*, took his revenge and cracked the method.

Indeed, such unjustified claims persist to this day. Recent evidence (see Kochanski, 1987) shows that much commercially available security software for microcomputers is cryptographically very weak. Although they scramble files and prevent immediate reading of the contents, they fall readily to serious cryptological analysis.

10.1.1 Early ciphers

We now look briefly at two early methods of enciphering. Firstly, we mention the ancient Greek *Scytale*. This simple device consisted of a strip of parchment or hide wound helically around a cylindrical or conical wooden baton. The message was written *across* the coiled strip, so that when it was unwound the message was jumbled. It could only be deciphered by someone again winding the strip around a similar baton.

The Caesar cipher, attributed to Julius Caesar (first century BC), shifts the alphabet by (say) four places: A maps to E, B to F, ... Z to D. So the message

$$m \; = \; \text{HELLO THERE}$$

is encrypted to

$$e \; = \; \text{LIPPS XLIVI}.$$

To decode e, the alphabet shift is reversed four places. For any cipher involving a simple shift of this kind, the whole cipher is completely determined by what A is mapped to (E here). We call E the *key*.

10.1.2 General cipher structure

Although very simple, the Caesar cipher illustrates nicely the structure of a general cipher system. We have the original data or message m, called the *plaintext*, written in some alphabet (which could be just the binary alphabet 0,1). Then we have a set of possible keys K, called the *key space*. This must be very large for a secure cipher; the Caesar cipher has only 26 keys. Each key K determines a function f_K which maps m to a coded message $e = f_K(m)$, which is called the *ciphertext*. Further, for decoding, the inverse function f_K^{-1} computes m by $m = f_K^{-1}(e)$. For instance, in the Caesar cipher, f_K is the four-place alphabet shift and f_K^{-1} the corresponding reverse shift.

10.1.3 What may an eavesdropper know?

It is important to be clear what our assumptions should be concerning how much information an eavesdropper may be able to find out about our cipher system. Clearly we must assume that some ciphertext might be accessible (otherwise there would be no need for a cipher at all!). However, it is also easy to envisage situations whereby an eavesdropper might gain access even to chosen plaintext and its corresponding ciphertext.

What, moreover, should we assume an eavesdropper knows about K, and about f_K? We must, unfortunately, assume that our adversary knows the general class of method we are using. This is because our cipher philosophy evolves gradually over a long period: new methods cannot be introduced without extensive work and it is perceived as too risky to introduce radical new ideas suddenly. We therefore cannot expect that the general method is secret. Hence, we must concentrate our efforts on keeping the key K secret. Knowing the general method, the eavesdropper can be assumed to know the key space. So, knowing m and e, in principle K could be found by searching through the key space to find a key K such that $e = f_K(m)$. This implies that the key space must be large. The converse need not hold, however—one cannot conclude that the cipher is secure merely because the key space is large. So we see that these assumptions imply that our cipher must be such that, even if the eavesdropper knows m and $e = f_K(m)$ for possibly many different messages m, all this information does not make it possible to deduce the key K.

It may seem as though we are being very generous in our assumptions of what our adversary knows, and that almost certainly he or she will know far less than this. While this *may* well be true, there are two reasons for this apparent generosity. First of all, bitter experience (most dramatically with the Allied breaking of the ENIGMA cipher in World War II; see Kahn, 1968) shows that it is fatal to underestimate the enemy. Secondly, these assumptions clarify for us precisely where we believe the strength of our cipher lies, namely, as above, that K cannot be found even from knowing m and $f_K(m)$.

In the following sections we look at particular classes of ciphers. In Section 10.2 we consider block ciphers, where the data is enciphered in blocks of fixed length, concentrating on the data encryption standard. In Section 10.3 we look at stream ciphers, where the data is encrypted symbol by symbol. In Section 10.4 we discuss public key ciphers, a recently developed kind of cipher where the encryption key need not be kept secret.

10.2 Block ciphers and the data encryption standard

We assume for this section that both plaintext and ciphertext are binary files. An alphanumeric file could of course be translated into binary using a (non-secret!) code such as ASCII or EBCDIC.

A *block cipher* enciphers and deciphers in blocks of a fixed number (L, say) of bits at a time. So if the message is

$$m = m_1 m_2 \cdots m_N \qquad \text{(each } m_i \text{ } L \text{ bits)}$$

then the ciphertext is

$$e = f_K(m) = f_K(m_1) f_K(m_2) \cdots f_K(m_N)$$

(i.e. the blocks are encoded separately and then strung together). The Caesar cipher could be regarded as a (single symbol per block) block code, as

$$f(\text{CAT}) = f(\text{C})f(\text{A})f(\text{T}).$$

Note that in a true block code, any block m_i is always encoded as the same ciphertext block. Thus, if the same block occurs twice in the plaintext, a corresponding block will be repeated in the ciphertext.

Many cryptologically strong block ciphers work on the following basic principle. They use two kinds of transformations of the data:

- *transpositions*, which merely reorder the symbols without changing them. The Scytale was a simple transposition cipher.
- *substitutions*, which replace symbols (or groups of symbols) by other symbols or groups, without any reordering. The Caesar cipher was a simple substitution cipher.

It is believed that while ciphers which employ only one of these two principles are invariably weak, those which employ *both* transposition and substitution, usually over and over again, can be very strong ciphers.

The data encryption standard (DES) employs this idea. Since in detail it is rather complicated, we look first at a toy version of DES which we shall call MINIDES. Although cryptologically very weak (if only because the size of its key space is very small, $2^{12} = 4096$ keys), it does illustrate most of the features of DES, while, unlike DES, being simple enough to encipher by hand.

10.2.1 MINIDES block cipher

The MINIDES cipher uses a 12-bit key, applied to 8-bit blocks, which we shall call *words*. As with DES, frequent use is made of the XOR (exclusive OR) operation on bit strings. We denote it by $\oplus$, so that e.g. $1101 \oplus 1011 = 0110$. Enciphering consists of three 'rounds', each containing the following four steps.

Step 1

Split word into left half L (4 bits) and right half R and apply 'wire cross' to R:

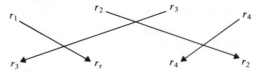

Step 2

Replace R by $R \oplus$ (4 bits of key). (For each round use next 4 bits of key.)

Step 3

For each 4 bits of R, use outside two bits to determine row and the inside two bits to determine column, and use the following fixed look-up table (*S-box*, S standing for substitution):

	00	01	10	11
00	A	0	C	8
01	5	B	3	2
10	1	9	E	4
11	7	6	F	D

Interpret table as a hex number, to be written as 4 bits.

Let us call $f(R,K)$ the result of steps 1, 2, 3.

Step 4

Make new 8-bit block by

$$\text{new } L = \text{old } R, \qquad \text{new } R = \text{old } L \oplus f(R, K)$$

After repeating the four steps three times, flip the final L and R at the very end. Each round takes the 8 bits $L\ R$ to $R\ L \oplus f(R, K)$.

Deciphering MINIDES is almost identical to enciphering. The only difference is that the three 4-bit blocks of the key are used in reverse order (see Exercise 2 below). The following example illustrates how it works.

Example (MINIDES)

Key 1101 0110 0111

ENCIPHER

Message block		1001	0001
Round 1	Wire cross		0010
	First block of key		1101
	⊕		1111
	S-box look-up		1101
	L		1001
	⊕		0100
New block		0001	0100
Round 2	Wire cross		0001
	Second block of key		0110
	⊕		0111
	S-box		0010
	L		0001
	⊕		0011
New block		0100	0011
Round 3	Wire cross		1010
	Third block of key		0111
	⊕		1101
	S-box		1111
	L		0100
	⊕		1011
New block	0011 1011		
Switch	1011 0011 = ciphertext		

DECIPHER

Ciphertext block		1011	0011
Round 1	Wire cross		1010
	Third block of key		0111
	⊕		1101
	S-box		1111
	L		1011
	⊕		0100
New block		0011	0100
Round 2	Wire cross		0001
	Second block of key		0110
	⊕		0111
	S-box		0010
	L		0011
	⊕		0001
New block		0100	0001
Round 3	Wire cross		0010
	First block of key		1101
	⊕		1111
	S-box		1101
	L		0100
	⊕		1001
New block	0001 1001		
Switch	1001 0001 = message		

The reader might like to work through the following two straightforward exercises, to get a feel for how MINIDES works.

Exercise 1

Show that, if $\overline{m}$ denotes the ones complement of m (i.e. m with ones replaced by zeros and zeros replaced by ones) then

$$\text{MINIDES}_K(\overline{m}) = \overline{\text{MINIDES}_K(m)}$$

To show this, just check that

$$f(\overline{R}, \overline{K}) = f(R, K)$$

so that using $\overline{K}$ in each round takes

$$\overline{L}\,\overline{R} \text{ to } \overline{R}\,\overline{L \oplus f(R, K)}$$

Thus encrypting using $\overline{K}$ takes $\overline{m}$ to

$$\overline{\text{MINIDES}_K(m)}.$$

Exercise 2

Prove that encoding $e = \text{MINIDES}_K(m)$ with $K^* = (K$ with blocks in reverse order) decodes e to m, i.e. that $\text{MINIDES}_{K*}(e) = m$.

10.2.2 The data encryption standard (DES)

In May 1973, the US National Bureau of Standards (NBS) asked for 'cryptographic algorithms for the protection of computer data during transmission and dormant storage'. Among other things, NBS imposed the following requirements on these algorithms:

1   They must be completely specified and unambiguous.

2   They must provide a known level of protection.

In 1977 they announced that IBM's submission, a development of their LUCIFER cipher, was to become the data encryption standard as published in US Federal Information Processing Standards Publication No. 46 (DES, 1977). Later in 1981, it was adopted as an ANSI standard. It is used extensively throughout the world for electronic funds transfer. It is also used on computers running UNIX for password encryption.

The DES encrypts 64-bit blocks to 64-bit blocks, using a 56–bit key (key originally 64 bits, but 8 bits not used for encryption).

General description of DES

An initial permutation is performed on the message block. It is then subjected to 16 'rounds' similar to the 3 in MINIDES. For each round, 16 different 48-bit 'keys' are used, obtained from the original key. Finally, the inverse of the initial permutation is performed to obtain the ciphertext block.

Detailed description of DES

The detailed description of DES given in this section is complete except that only one of eight S-boxes is reproduced. The deciphering algorithm DES^{-1} is identical to DES, except that, as with MINIDES, the 'keys' $K_1, K_2, \ldots, K_{16}$ are used in the reverse order $K_{16}, K_{15}, \ldots, K_2, K_1$. Figure 10.1 illustrates the basic structure of DES.

The initial permutation (IP) is defined as:

$$IP = \begin{array}{cccccccc}
58 & 50 & 42 & 34 & 26 & 18 & 10 & 2 \\
60 & 52 & 44 & 36 & 28 & 20 & 12 & 4 \\
62 & 54 & 46 & 38 & 30 & 22 & 14 & 6 \\
64 & 56 & 48 & 40 & 32 & 24 & 16 & 8 \\
57 & 49 & 41 & 33 & 25 & 17 & 9 & 1 \\
59 & 51 & 43 & 35 & 27 & 19 & 11 & 3 \\
61 & 53 & 45 & 37 & 29 & 21 & 13 & 5 \\
63 & 55 & 47 & 39 & 31 & 23 & 15 & 7
\end{array}$$

i.e. The first step of DES takes the bitstring $m_1 \ldots m_{64}$ to $m_{58}\, m_{50} \ldots$ $m_{15}\, m_7$. (It seems to achieve little cryptological purpose.) It remains only to define the function f and then to show how the expanded keys $K_1, K_2,$ $\ldots K_{16}$ are generated from the original key.

To work out $f(R, K)$, where

$$R = r_1 \ldots r_{32}$$
$$K = k_1 \ldots k_{48}$$

proceed as follows.

Step 1

Write R as

$$\begin{array}{cccc}
r_1 & r_2 & r_3 & r_4 \\
r_5 & r_6 & r_7 & r_8 \\
\cdot & & & \cdot \\
\cdot & & & \cdot \\
\cdot & & & \cdot \\
r_{29} & r_{30} & r_{31} & r_{32}
\end{array}$$

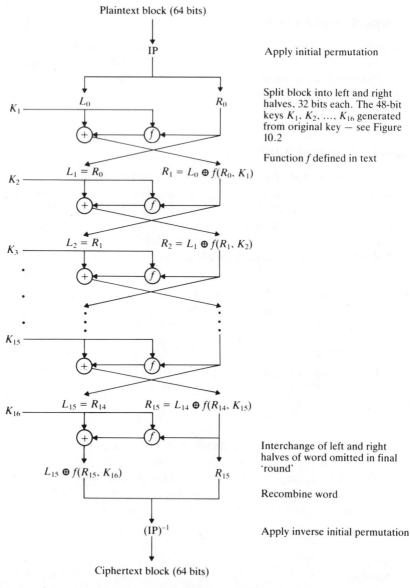

Figure 10.1 The basic structure of DES (after Denning, 1982)

and then add 'next' bits on the left and right to obtain eight 6-bit words (total 48 bits):

$$
\begin{array}{cccccc}
r_{32} & r_1 & r_2 & r_3 & r_4 & r_5 \\
r_4 & r_5 & r_6 & r_7 & r_8 & r_9 \\
& & \cdot & & & \\
& & \cdot & & & \cdot \\
& & \cdot & & & \\
r_{28} & r_{29} & r_{30} & r_{31} & r_{32} & r_1
\end{array}
$$

Step 2

Add K to this, mod 2:

$$r_{32}\oplus k_1 \quad r_1\oplus k_2 \quad r_2\oplus k_3 \quad r_3\oplus k_4 \quad r_4\oplus k_5 \quad r_5\oplus k_6 \quad (1)$$
$$\cdot \qquad\qquad\qquad\qquad\qquad\qquad\qquad\qquad \cdot \qquad (2)$$
$$\cdot \qquad\qquad\qquad\qquad\qquad\qquad\qquad\qquad\qquad\qquad \cdot \quad \cdot$$
$$\cdot \qquad\qquad\qquad\qquad\qquad\qquad\qquad\qquad\qquad\qquad\qquad \cdot$$
$$r_{28}\oplus k_{43} \qquad\quad \cdot \qquad\qquad \cdot \qquad\qquad\quad \cdot \qquad\quad r_1\oplus k_{48} \quad (8)$$

Step 3

Process the 6-bit words (1), (2), ..., (8) as follows: interpret the outer two bits as row number (0 1 2 3) and the inner four bits as column number (0 1 ... 9 A B C D E F) and substitute according to value in respective S(ubstitution)-box number, e.g. use box (1) for word (1), etc. Output is 4 bits (the hex value in the S-box). The total output from boxes (1),...,(8) is 32 bits. S-box (1) is as follows:

	0	1	2	3	4	5	6	7	8	9	A	B	C	D	E	F
0	E	4	D	1	2	F	B	8	3	A	6	C	5	9	0	7
1	0	F	7	4	E	2	D	1	A	6	C	B	9	5	3	8
2	4	1	E	8	D	6	2	B	F	C	9	7	3	A	5	0
3	F	C	8	2	4	9	1	7	5	B	3	E	A	0	6	D

S-boxes (2), (3), ..., (8) all similar, but different (see Mayer and Matyas, 1982, p. 667).

Step 4

Apply the 32-bit permutation P where

$$P = \begin{array}{cccc} 16 & 7 & 20 & 21 \\ 29 & 12 & 28 & 17 \\ 1 & 15 & 23 & 26 \\ 5 & 18 & 31 & 10 \\ 2 & 8 & 24 & 14 \\ 32 & 27 & 3 & 9 \\ 19 & 13 & 30 & 6 \\ 22 & 11 & 4 & 25 \end{array}$$

which takes the bitstring $b_1\, b_2 \ldots b_{32}$ to $b_{16}\, b_7\, b_{20}\, b_{29} \ldots b_4\, b_{25}$.

The expanded keys $K_1, K_2, \ldots, K_{16}$ are generated as in Figure 10.2. The number of left shifts at the ith iteration is given by the table:

Number of left shifts	1	1	2	2	2	2	2	2	1	2	2	2	2	2	2	1
Iteration number i	1	2	3	4	5	6	7	8	9	10	11	12	13	14	15	16

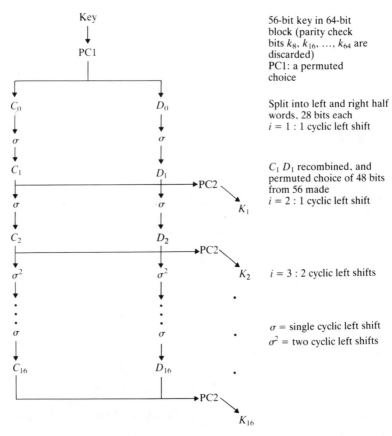

Key

PC1

56-bit key in 64-bit block (parity check bits k_8, k_{16}, ..., k_{64} are discarded) PC1: a permuted choice

C_0 D_0

Split into left and right half words, 28 bits each $i = 1 : 1$ cyclic left shift

σ σ

C_1 D_1 → PC2

$C_1 D_1$ recombined, and permuted choice of 48 bits from 56 made $i = 2 : 1$ cyclic left shift

σ σ K_1

C_2 D_2 → PC2

σ^2 σ^2 K_2 $i = 3 : 2$ cyclic left shifts

σ σ

σ = single cyclic left shift
σ^2 = two cyclic left shifts

C_{16} D_{16} → PC2

K_{16}

Figure 10.2 Generating the extended keys K_1, K_2, ..., K_{16} from the original DES key (after Denning, 1982)

The permuted choices PC1 and PC2 are given as

PC1
57	49	41	33	25	17	9
1	58	50	42	34	26	18
10	2	59	51	43	35	27
19	11	3	60	52	44	36
63	55	47	39	31	23	15
7	62	54	46	38	30	22
14	6	61	53	45	37	29
21	13	5	28	20	12	4

PC2
14	17	11	24	1	5
3	28	15	6	21	10
23	19	12	4	26	8
16	7	27	20	13	2
41	52	31	37	47	55
30	40	51	45	33	48
44	49	39	56	34	53
46	42	50	36	29	32

(bits 8,16, 24, 32, . . ., 64 discarded) (chooses 48 from 56).

Strengths and weakness of DES

1   A change of one bit of (a) plaintext or (b) key produces totally different ciphertext. So $DES_K(m)$ is a highly 'non-continuous' function of K and m, a highly desirable feature of any cipher.

2   Extensive statistical tests can find no correlation between plaintext and ciphertext.

3   Extensive efforts by skilled cryptanalysts have not been able to find (or at least have not published!) any significant weaknesses in DES. It is *not*, however, *provably* secure. DES has the property

$$DES_{\bar{K}}\ (\overline{m}) = \overline{DES_K(m)}\ (\overline{\ } = \text{ones complement})$$

(as has MINIDES). It also has some 'weak keys' (for instance K all zeros) which should not be used, as the resulting keys $K_1, \ldots, K_{16}$ are too regular.

4   The 56-bit key of DES was, in 1977, thought long enough to guard against exhaustive key search for the foreseeable future. With the great technological improvements in computer power of the past decade or so, it is no longer clear that this will continue to be the case for many more years.

Thus while DES is a cipher of the present rather than of the future, it is at least a strong cipher whose detailed structure is openly published, and so can be assessed.

In the United Kingdom, British Telecom has a block cipher, called BCRYPT, thought to be similar in general structure (though not in detail) to DES. They, however, have no plans to publish details of the BCRYPT enciphering algorithm.

For some time the International Standards Organization (ISO) had planned to make DES one of *their* standards (as well as a US Federal and ANSI standard). However, in 1987, they decided to opt out of establishing standard ciphers, perhaps out of fear of encouraging overdependence on a single coding system. Instead, they are instituting a *register* of encrypting algorithms, giving the relevant parameters (blocklength, keysize, etc.) of each but probably without giving any indication of the supposed strength of the algorithm (Price, 1988).

Modes of operation of DES

Being a block cipher, DES suffers from the disadvantage that a repeated block (for instance a sequence of blanks) is always encrypted to the same ciphertext block. This may make it possible to deduce the structure of the message. Therefore, in practice, straight DES is used only for very short messages—encrypting keys, passwords, initialization vectors, etc. For longer messages, blocks must be linked in some way to earlier message or ciphertext blocks, so that repeated blocks are encrypted differently. Cipher block chaining (see below) illustrates a way of doing this. Another way is cipher feedback (Denning, 1982). Further, in another mode, output feedback, DES can be used as part of a stream cipher (see Section 10.3). When contrasted with these methods, straight DES is called the electronic code book (ECB) mode.

Cipher block chaining (CBC)

Suppose that the message consists of N 64-bit blocks $m_1 m_2 \cdots m_N$. Then, by using an initialization vector IV we encrypt by

$$e_1 = \text{DES}_K (\text{IV} \oplus m_1)$$
$$e_2 = \text{DES}_K (e_1 \oplus m_2)$$

and generally

$$e_i = \text{DES}_K (e_{i-1} \oplus m_i)$$

To decrypt,

$$m_1 = \text{IV} \oplus \text{DES}_K^{-1} (e_1)$$
$$m_2 = e_1 \oplus \text{DES}_K^{-1} (e_2)$$

and generally

$$m_i = e_{i-1} \oplus \text{DES}_K^{-1} (e_i)$$

The IV must be known to both the sender and receiver.

DES chips have been produced for encryption in all modes of DES, some at very high bit rates. The AMP9518 and Z8068, for instance, have a speed in ECB mode of 13 Mbit/s, (Abbruscato, 1984).

In both ECB and CBC modes, a single bit error in the received block e_i will cause the decrypted block m_i to be totally wrong. Thus these modes should only be used in a noise-free environment or in conjunction with error-control coding.

10.3 Stream ciphers

A stream cipher is one where the plaintext is enciphered one symbol at a time. The Caesar cipher is an example. However, stream ciphers would be very weak if they always enciphered to the same symbol, and in general they do not. Slightly stronger than the Caesar cipher (but still weak) is the enère cipher (see Konheim 1981, p. 137).

Suppose the key is FORTY. Then the first, sixth, eleventh, . . . symbols are encoded with the Caesar cipher having F as key (i.e. A → F), the second, seventh, . . . with O as key, and so on. Therefore,

TOMORROWNEVERCOMES

encodes, using

$$\text{FORTYFORTYFORTYFOR}$$

to

$$\text{YCDHP \quad WCNGC \quad ASIVM \quad RSJ}$$

In *The Gold Bug*, Edgar Allen Poe, who was an amateur cryptanalyst, claimed

> Yet it may be roundly asserted that human ingenuity cannot concoct a cipher which human ingenuity cannot resolve.

That Poe was wrong is shown by our next example of a stream cipher, the *one-time pad* (due to Major J. O. Mauborgne, US Army Signal Corps). It is one of the few ciphers which is *provably* secure, assuming that the key is kept secret.

10.3.1 The one-time pad

Given a message

$$m = \text{THEOWLANDTHEPUSSYCAT} \ldots$$

one uses a key K:

$$K = \text{AXTYWUVLLPR} \ldots$$

and enciphers each letter as in the Caesar cipher, except changing the key for each letter, so that the nth letter of m is enciphered using the nth letter of K as key. So the above enciphers to

$$e = \text{TEXMSFVYOIY} \ldots$$

Why is the one-time pad unbreakable? Simply because we can find a key K to decipher the ciphertext e to any message m' we like. For instance, to decipher $e = \text{TEXM} \ldots$ above to, say, $m = \text{TWINKLETWINKLELITTLESTAR} \ldots$ we simply find K' such that m' encodes to e. We easily find $K' = \text{AIPZ} \ldots$.

Thus, not knowing the key, e could be deciphered to *any* message. However, the one-time pad has one great disadvantage, which prevents its general use: the fact that the key has to be as long as the message! Therefore, as the receiver of the message must have the key before he can decipher the message, and the key must be sent to him secretly, why not just send the message by the method (whatever that is) by which the key is sent? Of course, the key may be sent at leisure, but the message may need to be sent quickly. It is in such a situation where the one-time pad can be useful. It is, for instance, used for the Washington–Moscow hot line, with keys exchanged monthly through embassies. It is not practical in situations where long keys cannot be readily exchanged or kept secure.

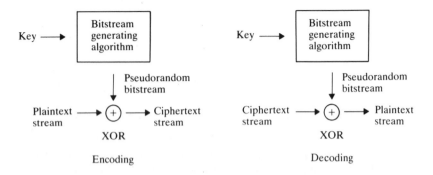

Figure 10.3 Vernam's stream cipher: encoding and decoding

When the alphabet is the binary one, the resulting cipher, the *binary one-time pad*, takes a particularly simple form. For a message bitstring m and a key bitstring K of the same length, the ciphertext is $e = m \oplus K$. For deciphering, m is simply given by $m = e \oplus K$, since

$$e \oplus K = (m \oplus K) \oplus K = m \oplus (K \oplus K) = m \oplus (0 \cdots 0) = m$$

10.3.2 The Vernam cipher

The idea of the Vernam cipher (Gilbert Vernam, AT&T, 1917; see Kahn, 1968) is to imitate the binary one-time pad by using a short key K to generate a long 'cryptobitstream' s (Figure 10.3). This would take the place of the message-length key in the binary one-time pad. Thus encoding is $e = m \oplus s$, and decoding is $m = e \oplus s$. There is a wide range of options for the actual algorithm to generate s from K.

The Vernam cipher is insensitive to errors in transmission, a single received bit error producing only a single bit error after decoding. Thus, it can safely be used in noisy environments. Note, however, that it is of course highly vulnerable to loss of synchronization between the cipher and the pseudorandom bitstream.

We now describe two possible ways to use the key to generate the cryptobitstream s. Of course, if the key were as long as the message we could just take s to be the key itself, producing the binary one-time pad. The simplest general method is to simply repeat the (binary) key over and over again. This, however, produces a binary version of the Vigenère cipher. What we want to do is to use our key to generate a cryptobitstream s which is as random as possible so that the cipher is similar to the binary one-time pad. Such an s is called *pseudorandom*.

Use of linear feedback shift registers

One method of producing a pseudorandom sequence is by use of a binary-linear feedback shift register (Figure 10.4). Let our key be N bits

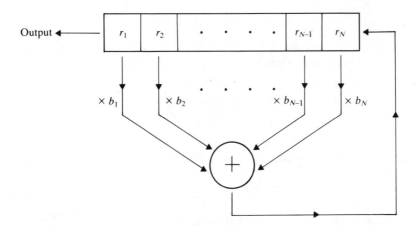

Figure 10.4 Generating a pseudorandom bitstream using a linear feedback shift register (after Denning, 1982)

$b_1, b_2, \ldots, b_N$ and the N initial values $r_1, \ldots, r_N$ of the shift register be fixed. At the kth clock pulse, the current r_1 is output as the kth crypto bit s_k, a new $r_N = b_1 r_1 \oplus b_2 r_2 \oplus \cdots \oplus b_N r_N$ is defined, and the old $r_2,$ $\ldots, r_N$ are shifted one place to the right to become the new $r_1, r_2, \ldots, r_{N-1}$. Since the current $r_1, r_2, \ldots, r_N$ are the next N bits to be output,

$$
\begin{array}{ll}
\text{New} & \text{Old} \\
\begin{pmatrix} r_1 \\ r_2 \\ \cdot \\ \cdot \\ \cdot \\ r_N \end{pmatrix} =
\begin{pmatrix}
0 & 1 & 0 & 0 & \cdots & 0 & 0 \\
0 & 0 & 1 & 0 & \cdots & 0 & 0 \\
\cdot & & & & & & \\
\cdot & & & & \cdot & & \\
\cdot & & & & & \cdot & \\
0 & 0 & 0 & 0 & \cdots & 0 & 1 \\
b_1 & b_2 & b_3 & b_4 & \cdots & b_{N-1} & b_N
\end{pmatrix}
\begin{pmatrix} r_1 \\ r_2 \\ \cdot \\ \cdot \\ \cdot \\ r_N \end{pmatrix} &
\end{array}
$$

and so

$$
\begin{pmatrix} s_{k+1} \\ s_{k+2} \\ \cdot \\ \cdot \\ \cdot \\ s_{k+N} \end{pmatrix} =
\begin{pmatrix}
0 & 1 & 0 & 0 & \cdots & 0 & 0 \\
0 & 0 & 1 & 0 & \cdots & 0 & 0 \\
\cdot & & & & & & \\
\cdot & & & & \cdot & & \\
\cdot & & & & & \cdot & \\
0 & 0 & 0 & 0 & \cdots & 0 & 1 \\
b_1 & b_2 & b_3 & b_4 & \cdots & b_{N-1} & b_N
\end{pmatrix}
\begin{pmatrix} s_k \\ s_{k+1} \\ \cdot \\ \cdot \\ \cdot \\ s_{k+N-1} \end{pmatrix} \tag{10.1}
$$

Here we are using $\oplus$ (XOR) in place of ordinary addition.

In particular

$$
s_{k+N} = b_1 s_k \oplus b_2 s_{k+1} \oplus \cdots \oplus b_N s_{k+N-1} \tag{10.2}
$$

Example

$N = 4, b_1 = b_2 = 1, b_3 = b_4 = 0$. Suppose initially $r_1 = r_2 = r_3 = r_4 = 1$. The sequence s_k starts at 1111, then using (10.2)

$$s_5 = s_2 \oplus s_1 = 0$$
$$s_6 = s_3 \oplus s_2 = 0$$
$$s_7 = s_4 \oplus s_3 = 0$$
$$s_8 = s_5 \oplus s_4 = 1$$

and so on. So the sequence is $s = 111100010011010101111\ldots$, repeating, of period $15(=2^4 - 1)$.

We now show, however, that using sequences s produced in this way gives cryptologically weak sequences. Those not interested in the mathematics that follows should skip these details.

The reason for this weakness is that if we know any $2N$ consecutive values $s_k,\ldots,s_{k+2-1}$ we can determine $b_1, \ldots, b_N$ explicitly, as follows. Writing equation (10.1) for $k, k + 1, \ldots, k + N - 1$ we obtain

$$
\begin{pmatrix} s_{k+1} & \cdots & s_{k+N} \\ \cdot & & \cdot \\ \cdot & & \cdot \\ \cdot & & \cdot \\ s_{k+N} & \cdots & s_{k+2N-1} \end{pmatrix} = \begin{pmatrix} 0 & 1 & 0 & 0 & \cdots & 0 & 0 \\ 0 & 0 & 1 & 0 & \cdots & 0 & 0 \\ \cdot & & & & & & \\ \cdot & & & & & & \\ 0 & 0 & 0 & 0 & \cdots & 0 & 1 \\ b_1 & b_2 & b_3 & b_4 & \cdots & b_{N-1} & b_N \end{pmatrix} \begin{pmatrix} s_k & s_{k+1} & \cdots & s_{k+N-} \\ \cdot & & & \cdot \\ \cdot & & & \cdot \\ s_{k+N-1} & \cdots & & s_{k+2N} \end{pmatrix}
$$

or $\qquad X_{k+1} = BX_k \qquad$ say

Hence $B = X_{k+1} X_k^{-1}$, which gives $b_1, \ldots, b_N$. (*Note*: X_k can always be assumed to be invertible, for otherwise the columns of X_k would be linearly dependent. This would mean that the sequence $\{s_k\}$ could be produced by a shift register with a smaller value of N.) Thus, for our example above, taking the first eight elements 11110001 of the sequence s and guessing that $N = 4$, we would have

$$X_1 = \begin{pmatrix} 1 & 1 & 1 & 1 \\ 1 & 1 & 1 & 0 \\ 1 & 1 & 0 & 0 \\ 1 & 0 & 0 & 0 \end{pmatrix} \qquad X_1^{-1} = \begin{pmatrix} 0 & 0 & 0 & 1 \\ 0 & 0 & 1 & 1 \\ 0 & 1 & 1 & 0 \\ 1 & 1 & 0 & 0 \end{pmatrix}$$

$$X_2 = \begin{pmatrix} 1 & 1 & 1 & 0 \\ 1 & 1 & 0 & 0 \\ 1 & 0 & 0 & 0 \\ 0 & 0 & 0 & 1 \end{pmatrix} \qquad X_2 X_1^{-1} = \begin{pmatrix} 0 & 1 & 0 & 0 \\ 0 & 0 & 1 & 0 \\ 0 & 0 & 0 & 1 \\ 1 & 1 & 0 & 0 \end{pmatrix} = \begin{pmatrix} 0 & 1 & 0 & 0 \\ 0 & 0 & 1 & 0 \\ 0 & 0 & 0 & 1 \\ b_1 & b_2 & b_3 & b_4 \end{pmatrix}$$

Thus $b_1 = b_2 = 1, b_3 = b_4 = 0$, as we know. (There is a more efficient method of finding $b_1, b_2, \ldots, b_N$, called the Berlekamp–Massey algorithm.)

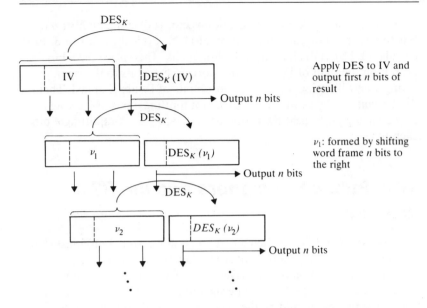

Figure 10.5 Generating a pseudorandom bitstream using the output

Incidentally, an N-bit register must output a *periodic* repeating sequence and the period cannot be longer than 2^N. Why? Because the shift register has only 2^N *states* corresponding to the 2^N possible values of $r_1, r_2, \ldots, r_N$. Therefore, after at most 2^N outputs, it must return to a previous state. Since the output is completely determined by the state, as soon as it repeats a state, it repeats the output too. In fact the period can be no longer than $2^N - 1$, for if the initial state is $r_1 = r_2 = \ldots = r_N = 0$ (the *zero* state) then the sequence $000 \cdots$ is output. Further, if the initial state is not the zero state, then the zero state cannot be reached. Sequences of period $2^N - 1$ are called maximal (linear) shift register sequences, or M-sequences. They contain all possible n-tuples except $00 \cdots 0$. They have been used as a cryptobitstream. Since the cryptobitstream can be determined from known plaintext and corresponding ciphertext (simply using $s_i = m_i \oplus e_i$) and only $2N$ consecutive s_i are needed to determine the whole cryptobitstream s_k, we see that this is not a secure cipher by any means.

Use of DES OFB mode in Vernam cipher

DES can also be used to generate a pseudorandom bitstream, which could be used in ciphers such as the Vernam cipher. The output feedback (OFB) mode of DES generates, from an initial 64-bit vector, a pseudorandom bitstream s. Statistically, it appears to behave as a random sequence, although this cannot be proved. The method is illustrated in Figure 10.5.

We first fix a key K, an initialization vector (IV) and a number n of bits to be output on each application of DES. Typically n is 1 or 8. First calculate DES_K (IV) and output its first n bits. Then place the n bits output on the right of IV, move the word frame n bits to the right, giving a new bitstring v_1. The leftmost n bits of IV are dropped. Then do the same for v_1 as was done for IV, outputting n bits and giving a new bitstring v_2. Repeat the procedure for v_2, v_3, ..., outputting n bits each time.

10.4 Public key ciphers and the RSA scheme

In classical cryptology, the sender (Alice) and receiver (Bob) must share a secret key, causing problems of secure key transport and necessitating key-encrypting keys, and so on (Denning, 1982). However, suppose there were a cryptographic scheme such that knowing the encrypting key K was no help in finding the decrypting key K'. Bob could then make K public, but keep K' secret. Then anyone could send him secret messages. He would be the only one who could decrypt the messages, since no-one else would know K'.

The concept of such schemes was proposed in 1976 by Diffie and Hellman. An actual practical scheme (*the RSA scheme*) was published in 1978 by Rivest, Shamir and Adleman.

To describe these ideas, we need some definitions and a little basic number theory.

10.4.1 Modular arithmetic

Suppose f is a function which takes integers to integers, i.e. given an integer n, $f(n)$ is an integer. Suppose also that f has an inverse, so that given an integer m, there is an n such that $f(n) = m$, or $n = f^{-1}(m)$. Such a function is called a *one-way* function if

1 Given n, it is computationally easy to work out $m = f(n)$.

2 Given m, it is much harder to work out $n = f^{-1}(m)$.

A one-way function is a *trapdoor* one-way function, if, knowing certain extra information, f^{-1} becomes easy to calculate.

To give an example of such a function, we need to recall some properties of modular arithmetic. For integers a, b and n, we say that $a \equiv b(\text{mod } n)$ if a and b have the same remainder when divided by n, or, equivalently, if n divides $a - b$.

Examples

$13 \equiv 6(\text{mod } 7) \quad 14 \equiv 0(\text{mod } 7) \quad -1 \equiv 8(\text{mod } 9) \quad 2^4 \equiv 1(\text{mod } 5)$

We now list some facts we need about modular arithmetic.

1   If $a \equiv b(\text{mod } n)$, then $ac \equiv bc(\text{mod } n)$.

2   If n and a have no common factors, then a has an *inverse* a' such that $aa' \equiv 1(\text{mod } n)$. (For instance, $3 \times 5 \equiv 1(\text{mod } 7)$ so 5 is the inverse of $3(\text{mod } 7)$ and 3 is the inverse of $5(\text{mod } 7)$.)

3   (Fermat's theorem, 1640.) If p is prime and a is not divisible by p, then $a^{p-1} \equiv 1(\text{mod } p)$. (e.g. $2^{12} \equiv 1 \ (\text{mod } 13)$).

4   If p and q are primes and a is divisible neither by p nor q, then $a^l \equiv 1(\text{mod } pq)$, where l = least common multiple (l.c.m.) of $p-1$ and $q-1$.

5   Suppose p, q and l are as in fact 4. Then if m and pq have no common factors, and k and l have no common factors, and $e \equiv m^k \ (\text{mod } pq)$ then $e^{k'} \equiv m(\text{mod } pq)$, where $kk' \equiv 1(\text{mod } l)$.

Facts 1 to 4 can be found in Denning (1982) or in Knuth (1973). To prove 5, first note that by 2, k' exists, so $kk' = 1 +$ multiple of $l = 1 + ul$, say. Hence $e^{k'} \equiv (m^k)^{k'} \equiv m^{kk'} \equiv m^{1+ul} \equiv m^1 \times (m^l)^u \equiv m(\text{mod } pq)$ as $m^l \equiv 1(\text{mod } pq)$ by fact 4.

Calculating $m^k(\text{mod } n)$

We shall need to calculate m^k (mod n) in the following section. A neat algorithm for doing this is the following, called the SX method (nothing to do with the University of Essex!). Write k in binary (e.g. $k = 11011101$ for $k = 221$), cross off the leading one and replace each one by SX, and each zero by S (for this example get SXSSXSXSXSSX). Then apply this sequence of operators to m, interpreting S as 'square it' and X by 'multiply it by m'. After each such operation reduce the result mod n, to obtain a number between 0 and $n - 1$. Thus at no stage in the calculation does a number larger than n need to be stored. For our example,

$$\begin{array}{cccccccccccc} \text{S} & \text{X} & \text{S} & \text{S} & \text{X} & \text{S} & \text{X} & \text{S} & \text{X} & \text{S} & \text{S} & \text{X} \\ m & \to m^2 & \to m^3 & \to m^6 & \to m^{12} & \to m^{13} & \to m^{26} & \to m^{27} & \to m^{54} & \to m^{55} & \to m^{110} & \to m^{220} & \to m^{221} \end{array}$$

10.4.2 The RSA public key scheme

We can now describe the RSA public key algorithm. Suppose Alice wants to send a secret message to Bob. B(ob) chooses two large primes p and q, $p \neq q$, say > 100 decimal digits each, and an integer $k < pq$, and having no common factor with $p - 1$ or $q - 1$ or p or q. B makes k and

pq public, *but not p or q individually.* A(lice)'s message m (a string of binary bits less than pq in value) is encoded to the ciphertext $e = m^k$ (mod pq), where $e < pq$.

To decrypt e, B uses fact 5 above. Since he knows p and q, he can calculate l, the l.c.m. of $p - 1$ and $q - 1$. Then he can calculate k' such that $kk' \equiv 1(\text{mod } l)$. This is done in practice by an algorithm called the Euclidean algorithm (Denning, 1982, 43–44). For small toy examples, however, one can look at $1 + l, 1 + 2l, 1 + 3l, \ldots$ until one of these is divisible by k. Suppose $1 + 6l$ is. Then $k' = (1 + 6l)/k$ as $k(1 + 6l)/k = 1 + 6l \equiv 1(\text{mod } l)$. (Of course k' has only to be calculated once.) Then from fact 5, $m \equiv e^{k'} (\text{mod } pq)$.

Small example

(p and q totally unrealistically small!)

$pq = 15, k = 7$ (the public information). Suppose $m = 2 (= 10$ binary). Then $e = m^k = 2^7 \equiv 8(\text{mod } 15)$. So $e = 8$. To decode, $l = $ l.c.m. $(3 - 1, 5 - 1) = 4$ and $1 + 5 \times 4$ is divisible by 7. Hence $k' = (1 + 5 \times 4)/7 = 3$, and $m \equiv e^{k'} \equiv 8^3 \equiv 2(\text{mod } 15)$.

The RSA algorithm depends for its effectiveness on the 'fact' (conjectured but not proved) that the function which takes m to m^k (mod pq) is a trapdoor one-way function. The truth of this hinges crucially on the experimental evidence, which indicates that given a large (> 200 decimal digit) number N which you know to be the product of two primes p and q, the factors p and q are impossibly time-consuming to discover.

Bob, of course, knows the factorization of $N = pq$ since he has *constructed* N from two large primes p and q. He then can easily calculate $l = $ l.c.m. $(p - 1, q - 1)$ and hence k' as described above.

Finding large primes

From fact 3 (Fermat's theorem) above, we know that if p is prime, then $2^{p-1} \equiv 1(\text{mod } p)$, etc. So if we want to find a prime number p in the range say $10^{150} < p < 10^{150} + 1000$, we first filter out those numbers divisible by, say, the first 50 primes. We then test the first survivor p to see whether $2^{p-1} \equiv 1(\text{mod } p)$, $3^{p-1} \equiv 1(\text{mod } p)$, etc. If these congruences are satisfied, then p is called a *pseudoprime*, and is almost certainly prime. If any are false, then p is not prime, and we try the next survivor, and so on. To actually prove p to be prime is more difficult, and probably unnecessary for practical purposes.

The RSA algorithm has been available as a chip for several years. Because of the exponentiation involved in the algorithm, performance is significantly slower than that of DES chips. Its use to date has therefore been mostly for short-message applications such as key encryption.

10.5 The Diffie–Hellman key exchange method

Diffie and Hellman (1976) have proposed a key exchange method that does not involve any initial physical exchange of keys. It is based on the experimentally likely assumption that the function $f(x) \equiv \alpha^x \pmod{p}$ is a one-way function. Here p is a large (> 120 decimal digits) prime and α is a number (a 'primitive root') such that $\alpha, \alpha^2, \alpha^3, \ldots, \alpha^{p-1} \equiv 1$ are all distinct mod p.

The procedure is as follows: each user i has a secret value X_i and stores the value $\alpha^{X_i} \pmod{p}$. When user i wants to communicate with user j, i sends α^{X_i} to j and j sends α^{X_j} to i. Then, knowing X_i, i works out $(\alpha^{X_j})^{X_i} \pmod{p}$ and j works out $(\alpha^{X_i})^{X_j} \pmod{p}$. Since $\alpha^{X_i X_j} \equiv \alpha^{X_j X_i}$, i and j both know this value, and it can be used as a shared key for communication. Even if an eavesdropper obtains the values α^{X_i} and α^{X_j} the one-way nature of the function $\alpha^x \pmod{p}$ means that these values are of no help in finding X_i or X_j or $\alpha^{X_i X_j}$.

Here is a toy example of how the method works. Take $p = 11$, $\alpha = 2$, $X_1 = 3$, $X_2 = 2$, $\alpha^{X_1} = 2^3 \equiv 8 \pmod{11}$, $\alpha^{X_2} = 2^2 \equiv 4 \pmod{11}$. Then the key is $(\alpha^{X_1})^{X_2} = 8^2 \equiv 9 \pmod{11}$ or $(\alpha^{X_2})^{X_1} = 4^3 \equiv 9 \pmod{11}$.

10.6 Summary

We have surveyed the main types of ciphers in use today: block, stream and public key ciphers. We have concentrated on the most important examples of each type: the DES block cipher, the Vernam stream cipher and the RSA public cipher. More detailed information on these and other ciphers can be found in Denning (1982), Welsh (1988), Mayer and Matyas (1982) or Konheim (1981). For an interesting account of the long and fascinating history of the subject, see Kahn (1968).

References

Abbruscato, C. R. (1984) 'Data encryption equipment', *IEEE Commun. Mag.*, **22** (9), September, 15–21.

Denning, D. E. R. (1982) *Cryptography and Data Security*, Addison-Wesley, Reading, Mass.

DES (1977) 'Data encryption standard', US Federal Information Processing Standards Publication 46, National Bureau of Standards, Washington, DC, January 1977.

Diffie, W. and M. E. Hellman (1976) 'New directions in cryptography', *IEEE Trans. Inf. Theory*, **22**, 644–654.

Kahn, D. (1968) *The Codebreakers: The Story of Secret Writing*, Weidenfeld and Nicolson, London.

Konheim, A. G. (1981) *Cryptography: A Primer*, Wiley, New York.

Knuth, D. E. (1973) *The Art of Computer Programming*, Vol. 1, 2nd edition, Addison-Wesley, Reading, Mass.

Kochanski, M. (1987) 'A survey of data insecurity packages', *Cryptologia*, **11** (1), January, 1–15.

Mayer, C. and S. Matyas (1982) *Cryptography*, Wiley, New York.

Price, W. (1988) 'Standards for data security—a change in direction', in *Proc. CRYPTO '87*. Lecture Notes in Computer Science 293, Springer, Berlin.

Rivest, R. L., A. Shamir and L. Adleman (1978) 'A method for obtaining digital signatures and public-key cryptosytems', *Commun. ACM*, **21** (2), 120–126.

Welsh, D. (1988) *Codes and Cryptography*, Oxford University Press, Oxford.

11 Security, authentication and access control

SIMON JONES ✓ – useful

11.1 Introduction

Security is required in any environment where information or items are not intended to be freely available to all. In computer networks it is the security of the information contained within and passed between the interconnected systems that is of prime concern. The physical security of the hardware can be protected by conventional methods, such as secure rooms and card keys, and is beyond the scope of this discussion.

In situations where it is necessary to restrict access to the information within a network, security measures have to be introduced. These methods are determined by two factors: firstly, security requirement, which is the nature of the restrictions required, and, secondly, security risks, which are the types of attacks.

The aim of providing security mechanisms is to reduce the risk of exploitation of the vulnerabilities of a system. If the system is vulnerable to attack in many ways, only some of these will be exploitable. This is because the opportunity may not exist or the results obtained may not justify the effort and risk of detection.

In general, systems cannot be made fully secure without making them totally unusable. The types of security measures used have to reflect a difficult balance between cost, ease of use and risk of attack. Further reading on this subject can be found in the references (Davies and Price, 1984; Denning, 1982; Nessitt, 1987; Jamieson and Low, 1989; Grimm, 1989).

This chapter presents a framework for the analysis of the security of a computer system or network. It highlights the need for reliable identification and protection. The chapter then presents some mechanisms for protection and identification based on encryption and an access control scheme that uses these mechanisms. The final section is a description of the Internet worm, an actual network security attack.

11.2 Security analysis

When the security of a system is being specified a number of considerations have to be evaluated. The activities required are listed below, they are based on a set of requirements specified by the advanced networked systems architecture (ANSA) project which was part of the United Kingdom Alvey research programme. The aim of the ANSA programme was to 'provide a coherent set of system components for open distributed processing' (ANSA, 1987). Part of the programme defined the following sequence of steps to be performed for the security analysis of a system:

- *Vulnerability analysis and threat assessment*   Determine the vulnerable parts of the system and assess the threats that may exploit them.
- *Risk analysis*   Analyse the risk of each threat. Some of the threats may have a low risk associated with them and these may be discarded.
- *Result analysis*   Assess the potential benefit to an attacker of realizing each threat. Threats that have low benefits may usually be discarded.
- *Cost analysis*   Assess the cost to the attacker of realizing each threat. When this cost is higher than the benefit, the threat may be discarded.
- *Analysis of protective measures*   For each of the remaining threats identify the security mechanisms that may be used to counter them.
- *Justification of mechanisms*   For each threat assess the cost and impact on system performance of introducing each security mechanism.
- *Security requirement assessment*   The extent of security required has to be determined by the nature of the application using the network. Any application using a computer network would have to ensure that the facilities for security provided by the hosts involved and the interconnections between them were able to ensure sufficient security for its tasks. Assessment is a major task and is beyond the scope of this discussion which is concerned with the mechanisms for providing various levels of security.

Sections 11.3 and 11.4 define a framework that can be used for vulnerability analysis, threat assessment and risk analysis. Result and cost analysis are very system dependent, being closely related to the nature of the systems tasks; they are therefore not covered in detail. Sections 11.5 to 11.8 cover the provision and analysis of protective mechanisms. Section 11.9 covers security auditing which is important for maintaining a secure system.

In subsequent sections the definitions and symbols given in Figure 11.1 are used.

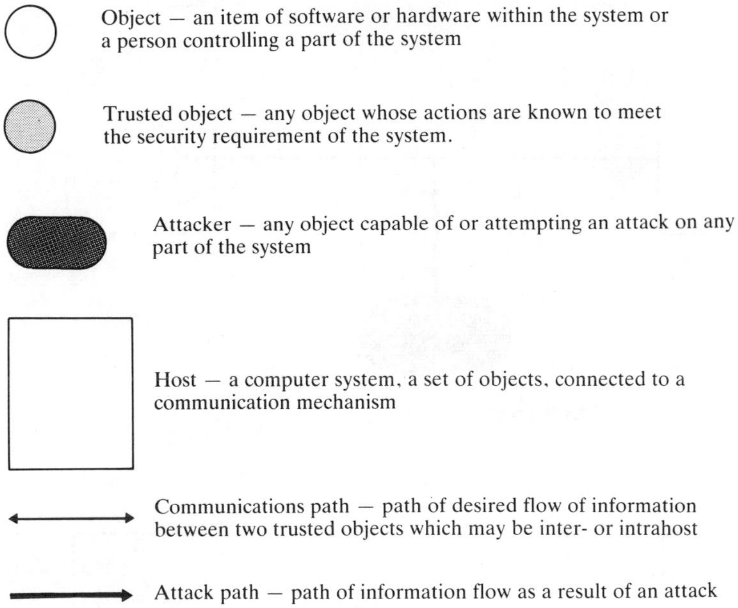

Object — an item of software or hardware within the system or a person controlling a part of the system

Trusted object — any object whose actions are known to meet the security requirement of the system.

Attacker — any object capable of or attempting an attack on any part of the system

Host — a computer system, a set of objects, connected to a communication mechanism

Communications path — path of desired flow of information between two trusted objects which may be inter- or intrahost

Attack path — path of information flow as a result of an attack

Figure 11.1   Definition of symbols

11.3 Classification of threats to the security system   ↗ Open

A threat is a potential exploitation of a vulnerability within a system. Threats can be classified according to type and objective. There are two types of threats—passive and active—and three threat objectives—disclosure, integrity violation and denial of service. The type of threat affects the ease with which an attacker may achieve the attack objective and the ease of detection and prevention of attacks.

A *passive threat* is a potential exploitation of a system where no modification is made to the system, its operation or its messages. This type of threat covers the unauthorized observation of the behaviour of the system and its information (see Figure 11.2).

An *active threat* is a potential exploitation of a system where alterations of the system, its operation or its messages could be made. This type of threat requires the attacker to influence the behaviour of the system to achieve the desired objective (see Figure 11.3).

The three objectives of threats are disclosure, integrity violation and denial of service. In a *disclosure attack* the attacker attempts to obtain information from an object with the objective of using or disclosing the information obtained. An example would be obtaining and disclosing personal information for a person's health records (see Figure 11.4).

When the objective of an attack is *integrity violation* the attacker

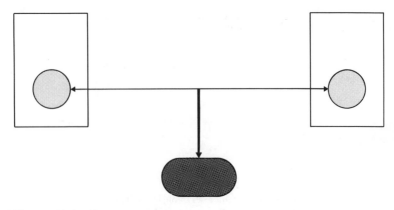

√ **Figure 11.2** Passive threat

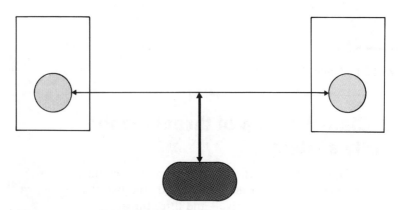

√ **Figure 11.3** Active threat

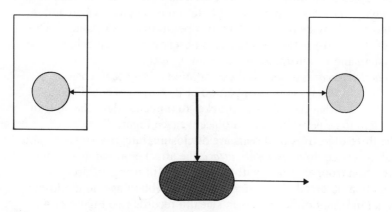

√**Figure 11.4** Threat objective: disclosure

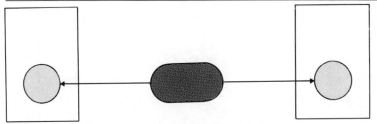

✓**Figure 11.5**   Threat objective: integrity violation

attempts to alter the operation of an object or the interaction between objects with the objective of altering the integrity of operation of the system. An example would be changing the information received when accessing a data base, e.g. improving a person's qualifications from a company's records (see Figure 11.5).

If the objective of an attack is to achieve *denial of service* the attacker alters the operation of an object, the interaction between objects or the communication between the objects to disrupt the service provided by the objects. An example would be preventing a person from accessing a data base (see Figure 11.6).

11.4 Methods of attack

An attack is the realization or activation of a threat; this is the activity that compromises the security of a system. Attacks in general can be classified into attacks on interacting objects, the communication between objects and on isolated objects.

With an *attack on interacting objects* the attacker attempts to observe or modify the operation of a selected object to achieve one of the three objectives described above (see Figure 11.7).

An *attack on communication* aims to observe or modify the operation

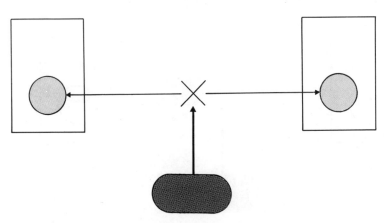

Figure 11.6   Threat objective: denial of service

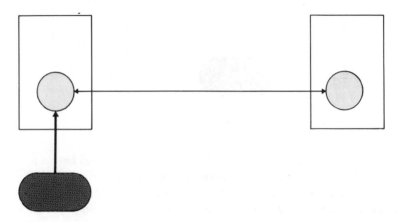

Figure 11.7 Attack on interacting objects

of the communication between objects to achieve one of the three objectives (see Figure 11.8).

In the case of an *attack on an isolated object* the aim of the attack is to establish a relationship with an isolated object in order to meet the attack objective (see Figure 11.9).

11.4.1. Classification of attacks on interacting objects

There are three terms commonly used to classify attacks on an object within a system that interacts with, and is trusted by, another object to perform its task (as shown in Figure 11.7). The names have a historical basis relating to breaches of physical security by persons; however, they have strong analogies within computer systems.

A *Trojan horse* attack is where an object is introduced into the system that has an unauthorized behaviour in addition to its authorized and visible behaviour. An example is to replace a log-in process with one

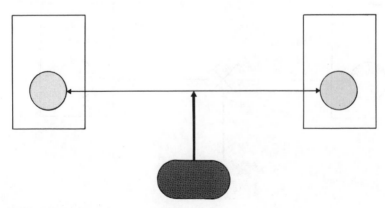

Figure 11.8 Attack on communication

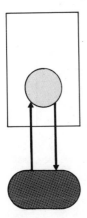

Figure 11.9 Attack on an isolated object

that, in addition to its required operations, records all user name–password pairs.

An *insider* attack is where an authorized person behaves in an unauthorized manner. This is similar to the Trojan horse attack except that it refers to a person. An example is an authorized user allowing an unauthorized user to use a restricted facility.

A *trapdoor* attack is when an object has been altered to produce an unauthorized effect when requested. This differs from a Trojan horse attack in that it requires activation; an example would be a log-in process that recognized a specific user and gave him or her special privileges.

/ 'ɪvz, drɑpəl / 偷聽者

✓11.4.2. Classification of attacks on the communication

When two objects require to interact a communications mechanism is required. Depending on the location of the two objects this mechanism can be internal to a host or external between two different hosts.

A passive attack where the attacker monitors the messages passing between two objects is called an *eavesdropper* attack. The only objective that can be achieved is disclosure of the information monitored. An example of this is the monitoring of a network connection between machines to obtain the data passed between them. Eavesdropping is often used for purposes other than realizing an attack, e.g. monitoring message traffic on a network to study its performance (see Figure 11.10).

In a *replay* attack the attacker records a sequence of messages passing between two valid objects. Later the attacker replays the messages to one of the objects in an attempt to appear to be the second object and perform the same operation. An example of this is to record a log-in sequence from a terminal to a host. When replayed the attacker aims to

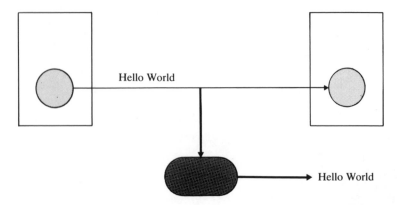

Figure 11.10 Eavesdropper

appear as the valid terminal user and effect a successful log-in (see Figure 11.11).

√ To achieve a *modification of messages* attack the attacker alters the operation of the communication channel between two objects so that

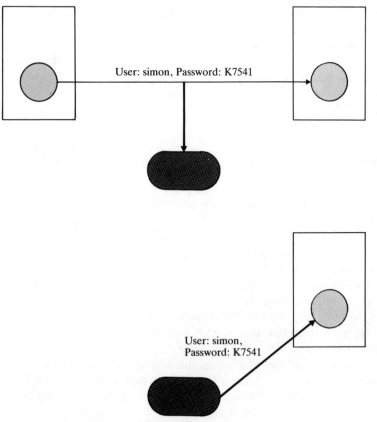

Figure 11.11 Replay of messages

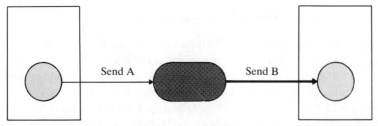

✓**Figure 11.12**   Modification of messages

the meaning of the messages at the receiver is altered, but messages still appear valid. An example of this would be modifying the contents of an electronics funds transfer message to increase the value of an attacker's bank balance (see Figure 11.12). ✓

An attacker can *masquerade* as a valid object and thereby be able to perform operations to achieve its objective. This type of attack often uses other forms of attack, such as replay and modification of messages. An example is the capturing and replaying of a valid authentication sequence that took place between valid objects. An attacker may masquerade as a log-in service to obtain names and passwords which it subsequently uses to achieve its objectives (see Figure 11.13).

Figure 11.13   Masquerade

✓The objective of a *disruption* attack is the denial of service to communicating objects by the disruption of the communication medium. An example of this would be an attack on a broadcast network by continually broadcasting messages and congesting the medium (see Figure 11.14). ✓

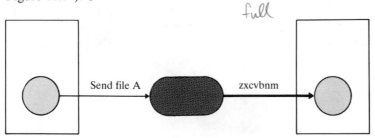

✓**Figure 11.14**   Disruption of communication

11.4.3 Example of attack on isolated object

The above examples have covered attacks on the interaction between two or more objects. There are cases where an object is not communicating but is still vulnerable to attack. An attack on an isolated object requires the establishment of a special or *covert channel* over which the information, obtained from the object, can be passed. A covert channel requires a Trojan horse or trapdoor attack to establish the attacker inside the system that generates the information.

A covert channel does not use a conventional, and thereby controlled, means of communication. Often some aspect of a system is changed to allow the information to be passed between the objects. This could be the number of messages sent during a particular time interval or the time taken for a process to start operation (see Figure 11.15).

Covert channels are hard to detect due to the multitude of system parameters that could be modified. However, data rates for covert channels are generally many orders of magnitude less than that of official communication channels. The mechanism relies on the fact that both objects have access to some shared object whose behaviour can be manipulated by one object and observed by the other.

An example of a covert channel is where two objects (processes) use processor memory as a common resource. The two processes may have no direct channel of communication but if they are running at the same time on the same processor they can request memory from the processor. If the sender uses memory in a controlled way, i.e. requests large blocks and then frees large blocks, a receiver can test this by making memory allocation requests of its own. The success of the receiver's requests will depend on the action of the sender so communication can be established. To send a '1' the sender would request memory to be allocated to it; for a '0' any memory allocated to the sender would be freed. The receiver would then make a request for memory to be allocated to it. If this succeeds it reads a '0' and if it fails

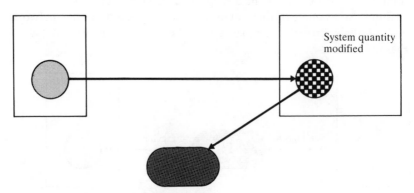

System quantity modified

Figure 11.15   Covert channel attack

it reads a '1'. Any memory allocated to the receiver would have to be freed before the next 'bit' could be sent.

The problem with this method is the synchronization required to ensure that the receiver only attempts one read for each of the sender's writes. This can be overcome by the use of a second channel to pass synchronization information between the processes. However, this increases the complexity of channel establishment requiring agreement between the Trojan process and the attacker. Directory access and processes creation/deletion can also be used as a means to provide covert channels.

11.5 Security measures

This section outlines the measures that may be taken to prevent attacks from occurring or to make them so expensive as to be impractical. All the methods are based on implementing barriers between an object and any potential attackers. The barriers divide the system into separate sections; communication between the sections can therefore be defined and hence controlled.

11.5.1 Separation

If the attacker and the target object are separated then they will not be able to interact. Full separation would prevent most attacks, including covert channels. However, ensuring full separation would involve separating the target object from all other objects, including valid ones, and hence would be impractical. Separation is therefore carried out to ensure that the separated objects can only interact by controlled, and visible, communication mechanisms. Separation can be performed at different levels (ANSA, 1987).

In *physical separation* objects are separated from potential attackers by placing them in different locations (see Figure 11.16).

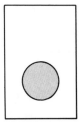 

Figure 11.16   Physical separation

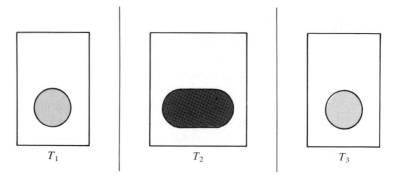

Figure 11.17 Temporal separation

With *temporal separation* objects are separated in time by activating them at different times (see Figure 11.17).

Logical separation is where objects are separated using both physical and temporal separation in order to cause the interaction between them to occur via defined and controlled channels.

With *encryption separation* objects are separated by the use of encryption, discussed in Section 11.6 and Chapter 10. Here it is accepted that attackers may be able to monitor communication; however, encryption ensures that they will not be able to interpret the information received (see Figure 11.18).

Effectiveness of separation

Separation is essential to the production of a secure system. Separation is based on the creation of boundaries between the objects that need to be protected from each other. These boundaries permit the control of interaction through them. Separation in itself does not counter any

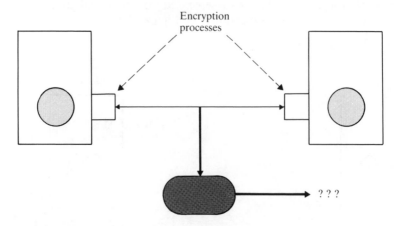

Figure 11.18 Cryptographic separation

threats; it can, however, serve as a principle for the specification, design and implementation of secure systems.

In the following section the term *object* is used to refer to the items that need protection; the term *subject* is used to refer to the items that cannot be trusted, i.e. persons or processes that may be potential attackers.

In a secure system there is a need for rules that specify which object should be separated from other objects or subjects. These rules are known as the *security policy*. The security policy classifies an item as an object or as a subject. The distinction between objects and subjects will vary within a computer network depending on the aspect of the system under consideration. An item that is considered an object for access to a network may be considered a subject by a host connected to a network. So the security policy for a computer network has to make provisions for the individual security policies of all applications using the network as well as specifying the security requirements for the network as a whole.

This further set of rules for each host is called *multilevel security*. These rules specify the separation that is required within the sets of objects and subjects. The set of objects may be divided into classes of objects according to, for example, the value, to an attacker, of the information they contain. The set of subjects may be similarly divided into subject classes, according to their security clearance.

A common aim of a computer network system is the sharing of resources and information between the connected systems. The security policy needs to specify in what way and to what extent the separation between classes of subjects and objects may be relaxed. The policy specifies precisely which classes of objects may interact with which classes of subject.

For any form of separation to be achieved, appropriate boundaries must be established and maintained; the area within a security boundary is called a *security domain*.

Implication of the separation methods

The implications of the different separation methods are considered below.

To physically separate objects they must be located on separate physical machines. The control of the physically separate objects then depends on the control of the communication between them. A computer network system accommodates the physical separation of objects by containing objects within different hosts on the network. This relies on each host being in full control of the flow of information through its network connection.

The temporal separation of objects that use a common resource, e.g. processor, can be achieved by time-sharing the resource between the objects. A particular host in the network could perform low security tasks when connected to the network, but disconnect when performing critical tasks. When switching between objects, it is important to ensure that any information relating to an object is removed from the common resource. This is to prevent the use of covert channels.

The cryptographic separation of objects may be achieved by using a unique cryptographic system, or unique cryptographic keys. This uniqueness is essential for the successful operation of cryptographic separation. Encryption creates separate logical channels through the communications network.

11.5.2 Communication security techniques

Within the communications mechanisms used for a computer network, encryption techniques can be used within any layer of the OSI model. However, as encryption is a data translation process it is normally seen as a presentation layer operation. In practice it is also common for encryption facilities to be provided within the data link layer of the system.

Encryption at the link layer can be easily implemented using encryption hardware within the network interface. The link layer provides for the encryption of data before it is passed into the communication medium so that all information passing through the physical connection is encrypted (see Figure 11.19). Providing link layer encryption makes connection to systems without link level encryption

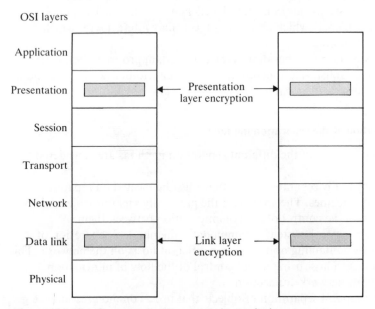

Figure 11.19 Communication security techniques

impossible, without introducing the facility to by-pass the encryption and hence weaken the system.

In presentation level encryption, encryption is provided as one of the presentation layer data translation facilities. Presentation layer encryption can provide encryption for a specific session. The encryption can be performed in different ways during a session, e.g. for passwords, for confidential data, for public-access data, etc. The encryption is provided between the two communication processes; it is also termed end-to-end encryption (see Figure 11.19).

These two methods differ in several respects. When link level encryption is used, the encryption method is chosen by the system administration rather that by the user. With end-to-end encryption users can select an encryption method to meet their requirements. The choice of method will depend on the security of the connected system. Link level encryption assumes that the security of the system is sufficient and that there is no danger of insider attacks, whereas end-to-end encryption can keep data in an encrypted form until application level processing.

11.6 Cryptography

The history of cryptography (the technique of making ciphers) and details of cryptography suitable for use in computer networks are given in Chapter 10.

Messages for encryption, known as the *plaintext*, are transformed by a function that is parametrized by a key. The output of the encryption process, known as the *ciphertext* or *cryptogram*, is transmitted. The attacker is assumed to be able to listen to or copy the message or the ciphertext. However, unlike the intended recipient, the attacker does not know what the key is and so cannot decrypt the ciphertext easily.

It is normally assumed that the attacker knows or can determine the method of encryption used. Therefore the method in itself does not provide the protection. The protection is provided by the use of the key. A key usually consists of a short string of characters, or bits, that selects one of many potential versions of the ciphertext. In contrast to changing the general method, which is usually complicated and expensive, the key can be changed as often as required.

From the attacker's point of view there are three possible situations to be conquered. Firstly, the attacker may have access only to the ciphertext and not the plaintext. Secondly, the attacker may have some matched plaintext and ciphertext. Finally, the attacker may have the ability to encrypt selected plaintext, known as *chosen* plaintext. A cryptographic security system has to be able to withstand attacks of all the above forms.

11.6.1 Encryption techniques for authentication and access control

Traditionally, encryption was based on the use of simple algorithms and relied on the use of long keys to provide security. Modern techniques, however, reverse the emphasis; the encryption algorithms are made extremely complex so that breaking them is prohibitively expensive. There are currently two encryption techniques that are commonly used for authentication and access control in computer networks, the data encryption standard (DES) and public keys systems. The basis for these techniques is described in Chapter 10.

11.7 Authentication and signature

If object B receives a message that is supposed be from object A, how can B confirm this? B needs to know if the messages is intact, i.e. it has not been altered, and that it is from A. This requires the *authentication* of the message.

The integrity of a message can be determined by including additional fields within the message that are functions of the body of the message, such as CRC fields within the data link layer. If these fields are correct then any modification to the message entails recalculation of these fields. If these fields are included within the encrypted section of the message, modification is extremely difficult to perform without breaking the encryption method or key. Alternatively, the method of calculation of the check field can be a common secret between A and B.

If the check bits are correct and the message decrypts correctly then the message must have come from A and so the message is authenticated by its coding and contents.

There is often a requirement for an outside object C to be able to prove B's statement that a particular message came from A. For example, if the message is a request for payment to B from A, B may have to prove that the request did come from A and was not created by itself. This requires that the message contains information that only A could create; this is called a *signature*, as it is equivalent to a conventional signature.

This message signature must be a function of the whole message in such a way that no part of the message can be changed without invalidating the signature. Also the signature must depend on information that only A knows, to prevent B from forging it. The final requirement is that there must be a mechanism by which B can validate the signature. These requirements can be met by using the public key cryptosystem such as the RSA cipher (Rivest *et al.*, 1978), which is described in Chapter 10.

In a public key signature system each object has two keys: a secret key, A_s, that only it knows and a public key, A_p, known to all. For

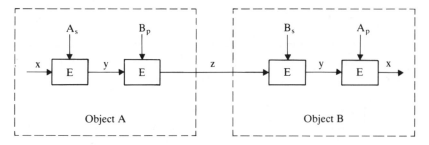

Figure 11.20   Public key signature

the method to work it also requires the ability to successfully decrypt, using an encryption key, a message encrypted with the corresponding decryption key, this is $A_s A_p = A_p A_s$. The method is shown in Figure 11.20. E is an encryption process using the indicated key, x is the plaintext, y the signed message where $y = A_s(x)$ and z a protected version of signed message where $z = B_p(y) = B_p(A_s(x))$.

To sign a message A will use the secret key A_s to encrypt the plaintext x into the message y. The message can be decrypted by any object knowing A's public key A_p. This means that the message is insecure but must have come from A. To ensure secure transmission A then encrypts the message y with B's public key B_p to form message z. This message can only be successfully decrypted by object B using its secret key B_s to form message y again. B finally decrypts again using A's public key A_p to yield the plain text x. This sequence not only provides authentication of the message's sender but also proves it by the presence of the signature.

11.8 Access control lists and capabilities

The control of access to objects such as data and programs can be achieved in a number of ways. A common factor is the idea of *identity*, both of the owner of the information and of the accessor. For successful access control a mechanism must exist to compare the identity of a user attempting to access an object with the list of valid users specified by the owner of the object.

The correlation of users and access rights are combined into an *access control matrix* (Davies, 1981) (see Figure 11.21). The objects which may be accessed are represented by the horizontal index i and the users which access them are represented by the vertical index j. The entry A_{ij} of the matrix contains the access rights of the subject C_j to the object O_i.

When a user completes the initial authentication required by the system it establishes processes for the user. The system passes the user's access rights to the processes. These rights may be further controlled by the user, perhaps to limit the extent of their operation and potential

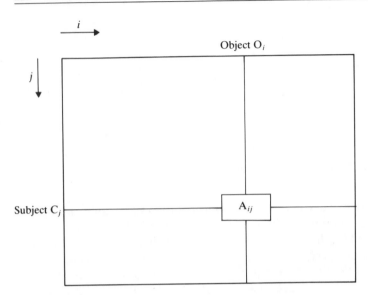

Figure 11.21   Access control matrix

damage. The entities requiring access can therefore be users or processes acting on their behalf. A group of these may share identical access rights, so it is common for them to share a single value for j. Objects sharing a common value of j can be considered to comprise a single protection class C_j.

The object to which access is controlled varies according to the context, the commonest examples being data and programs, but it is also possible for constructs such as queues to have access right attached to their functions, such as joining or being removed from a printer queue. Peripherals and communication paths can also appear among the objects O_i. The entry A_{ij} will define a form of access such as the right to read data, to write data, to execute programs, to append data to a file or to send data via a communications path.

Access control matrices are used for the specification of access rights within a system. In practice it will be necessary to consider the contents of a single row or column, i.e. a user's rights or a facility's users. For implementation, the matrix is too large and too sparsely filled to be practical. This stems from the observation that each of the N_i object's rights requires an entry in the matrix for each of the N_j potential users. This would require $N_i \times N_j$ entries. Most of the matrix would be empty due to the fact that many files or programs are for access by their owner only and therefore there would only be one entry in the object's column.

Therefore for practical systems it is required that this information is represented in a more compact form, and this can be achieved by access control lists.

Sparse access control matrix

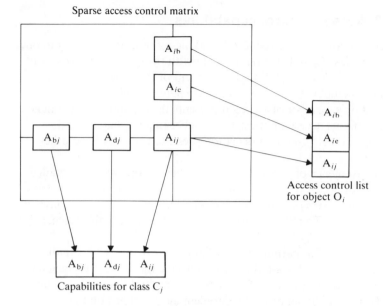

Access control list
for object O_i

Capabilities for class C_j

Figure 11.22 Access control list and capabilities

11.8.1 Access control lists

The access control list attempts to present the data contained within an access control matrix in a compact form.

The column of the access matrix under O_i when compacted so that it does not contain empty elements describes all protection classes having access to that object. This is the *access control list* for the object O_i. Each entry must contain a unique identifier for the appropriate protection class C_j. The access control list is located in association with the object and used to control access to O_i.

Ensuring a unique identifier for each protection class C_j is difficult to achieve in a network system. One approach is to give each system a subset of the protection classes. Users wishing to join a protection class first access their home system. The global protection class identifier is the concatenation of the local name and the name of the system that administers it. The identifier is used, in an encrypted form, as an ability to join a protection class. For additional security it is time-stamped and has a limited lifetime.

In networks that have a large number of objects in use at any one time access control of this kind is usually applied at relatively coarse levels such as to the whole of a file. Access control lists are shown in Figure 11.22.

11.8.2 Access control capabilities

The access matrix can be reduced by taking a row corresponding to one protection class C_j and storing this data with the class, rather than with the objects (see Figure 11.22). Each element of this new one-dimensional matrix is then called a *capability*. When access is requested by a member of a class it presents the capability, like a ticket, to verify that its access right exists. The checking mechanism is then very simple; it acts on the information presented and needs no other stored data.

The convenience of capabilities allows them to operate at a detailed level, e.g. for access to a remote file. Capabilities can be duplicated and given to other users or processes and, in addition, the copy can have its rights restricted. To prevent forgery, capabilities are usually protected by encryption.

With systems that permit the duplication of capabilities it rapidly becomes impossible for an object to determine the set of class members who have access rights to it. This is because multiple capabilities can exist. To reduce the scope of this problem most systems that permit copying of capabilities restrict this to copies of the originals, preventing the duplication of copies. This reduces the number of capability copies and can allow a trace of the capability origin.

11.8.3 Combining access control lists and capabilities

Access control lists provide strict control on the extent of access and the possibility of changing this extent. Capabilities provide a simple checking mechanism, but it is not easy to control the objects that have an access permission. Access control lists and capabilities can be combined by making the initial access to an object via the access control list and creating a new capability for any subsequent accesses.

√11.8.4 Capabilities within a computer network system

For capabilities to be securely passed between objects in a computer network it is necessary to protect them using encryption. The detail of operation will then depend on the method of encryption used—a multiple key or a public key system.

Multiple key systems

For a single computer system, capabilities are secured by a hardware mechanism such as a privileged instruction and can only be created by the operating system. For a computer network, capabilities can be protected by encipherment with a single-key cipher, like DES, using a different key per object–subject pair. This is shown in Figure 11.23.

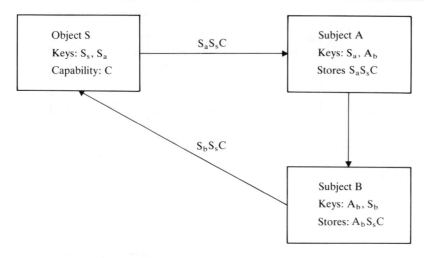

Figure 11.23   Encryption of capabilities

The capability C is created internally by an object S that is to be accessed. Externally only the enciphered form S_sC is used; S_s is the object's secret key. The coded form is known publicly and could be copied. To prevent the copying of S_sC during communication or storage it has to be additionally encrypted, as shown for the connection between S and A. The new form is S_aS_sC, which is the original encrypted capability encrypted by the key S_a which is known only to A and S. A could use the same encrypted form for storage. The capability can be passed from A to B as A_bS_sC and then used by B to obtain access to object S. This is possible as the object S will accept S_sC independently from any subject. Note that for a single key cipher, the key from S to B, S_b, is the same as the key from B to S, B_s.

Public key systems

Public key systems can also be used to encrypt capabilities, as shown in Figure 11.24. These overcome the need for many secret keys used between objects and subjects. The scheme discussed below is due to Fletcher and Donnelly and based on a description by Davies (1981). They used the term *server* to represent an object providing a service and *user* to represent a subject requiring use of an object. Server S has a secret key S_s and a public key S_p. $S_s(C)$ is the product of encrypting the capability C with the server's secret key S_s, and is a *signed* form of C. If this is signed again with S_s and then enciphered with A's public key A_p the result is $A_pS_s(S_sC)$, the signed capability protected by the other ciphers for its passage through the network.

The encryptions carried out on the capability are shown at the entry into the system boxes. The stored values are shown within the system

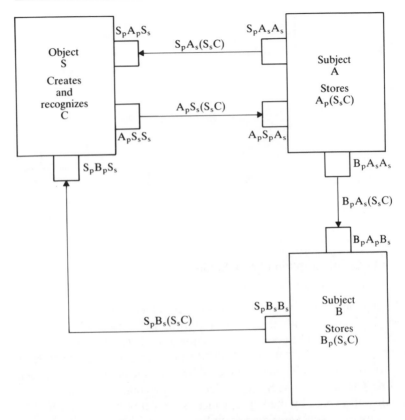

Figure 11.24 Public key encryption of capabilities

boxes. The capability is passed from S to A as $A_pS_s(S_sC)$. A performs A_s then S_p, then A_p, shown as $A_pS_pA_s$. This gives the result $A_pS_pA_s(A_pS_s(S_sC)) = A_pS_pC$ since $A_sA_p = 1$ and $S_pS_s = 1$. A uses A_p to decipher, S_p to check the signature and A_p to reencipher for storage. To allow checking of S_sC it should contain some well-known information. In practice C has a well-known form that can be examined by performing $S_p(S_sC) = C$. Each user treats S_sC in the same way, deciphering, checking the signature and reenciphering for storage. The capability is deciphered, signed and enciphered for transmission.

11.9 Security monitoring

The successful operation of a secure system depends not only on the security techniques used to build the system but also on the continued monitoring of these techniques to detect any attacks, successful or otherwise. Successful monitoring requires the ability to trace the operation or the behaviour of the objects within the system. This tracing provides for evasive or corrective action to be taken in the case of compromises to the system security.

To trace the security of a system, information on its status must be collected and processed by a security manager to form a security *audit trail*. The problem of audit trails is the volume of information that will accumulate over a period of time. The security object must either archive this information or summarize it. The problem of creating summaries is that detail will be lost. Also, if a breach of security is not detected before the information is summarized, the source of the attack may be lost. This implies that the degree and speed of the summary produced must depend on the level of security involved, which in turn depends on the significance of the information or objects being managed.

The audit information is a shared object and can itself become a vulnerability, providing a means by which a covert channel may be established. The degree and flow of audit information must therefore be controlled very carefully. The security object and the security manager must be trusted with the same degree of trust required by the most important object within the system.

11.10 An actual network security attack

The most widely published attack on the security of a computer network was the *Internet worm*, which occurred on 2 November 1988. A worm is a type of computer virus which is able to propagate itself through a computer system or network (Chess, 1989). The Internet worm was a program which, when started on one computer system, would attempt to exploit weaknesses in the system's security with the apparent aim of replicating itself on any other connected computer system.

The worm originated at Cornell University, though the first reported instance occurred on a machine at the Artificial Intelligence Laboratory at Massachusetts Institute of Technology (MIT) (Rochlis and Eichin, 1989). The attack rapidly spread throughout the Internet covering the whole of the United States and affected about 6000 computers. The worm affected Sun Microsystems' Sun 3 systems and VAX computers running particular versions of UNIX. Although substantial and successful measures were taken to counter the attack within 24 hours, the disruption continued long after the attack was identified and cured. This was because confidence in the Internet was damaged and many systems administrators delayed reconnection for fear that their systems would become reinfected. After three weeks some sites had still not been reconnected.

The worm caused significant disruption to the operation of individual computer systems. The computers would become heavily overloaded with extremely large numbers of processes running, thus denying or degrading services to authorized users. In some cases, computers would crash due to lack of resources needed to run these processes. This behaviour hampered initial attempts to identify the problem as it prevented the systems administrators from running the programs required to identify and remove the infecting processes. When the problem was first noticed at MIT and identified to be a virus or a worm, it was assumed that removing the offending processes would return the system to normal (Spafford, 1989). However, when the problem recurred, its origin was traced to the network and in particular to the mail system.

The mechanism of the worm was determined by a number of groups in the United States, in particular at MIT and the University of California at Berkeley. The process involved catching copies of the worm and then reverse engineering them to produce the C language program from which they were compiled. The task of capturing the worm was hampered as it made deliberate attempts to hide traces of its operation and to remove any files created.

The basic operation of the worm was to attempt to establish instances of itself on all computers connected to the computer on which it was running. The worm used three mechanisms to attempt connection, each exploiting some security weakness of the UNIX system software.

One attack attempted to exploit a bug in the 'fingerd' program, a utility for providing information on network users. The program runs as a background process listening for remote requests for information. A bug in the C language I/O library would allow an excessively large incoming request to overrun a buffer, and hence alter the operation of the program in such a way that termination of the request resulted in an attempt to break into the machine from which the request was initiated.

The second attack performed by the worm was to try to obtain user name and password pairs. In UNIX, it is possible to encrypt words and compare these with passwords held in a system file. The worm performed the attack by trying (among other strategies) permutations of the user's log-in identity and actual name, which were stored in the password file. Once the password was discovered, the worm tried to break into other systems on which the user had an account.

The third attack, which appears to have been the most successful, exploited a debugging facility coded into the UNIX mail receiving program 'sendmail'. This facility should have been turned off but in many cases had been enabled. This allowed a program to contact the mailer process, but instead of sending mail it entered a debug mode where commands could be executed. Where the debug facility was usable, the worm would send over a simple program that has been referred to as the *vector* or *grappling hook* program (Rochlis and Eichin, 1989). This small C language program was then compiled and executed on the new machine. It would attempt to establish a connection back to the first machine, and then copy over executable code versions of two programs—one to repeat the infections and one to attack the systems' passwords. (The executable code was suitable only for Sun 3 and VAX computers, which is why the attack was limited to these systems.) It would then attempt to run each program and then remove all traces of the copied programs in the filestore.

The person responsible for programming and entering the worm into the Internet was identified and prosecuted. The incident highlights the fact that weaknesses exist in all systems and are often relatively simple to exploit.

The attack confirms the need to protect computers and networks. However, just restricting network access is not a solution. The security weaknesses exploited by the worm were not in the network but in the systems software of the connected system. The connectivity of the Internet, which allowed the worm to spread so rapidly, allowed information about its detection and eradication to spread equally rapidly and allowed experts in different institutions to collaborate in solving the problem (Rochlis and Eichin, 1989).

11.11 Summary

This chapter has presented a framework by which the security of a computer system or network can be assessed and an example method by which access within a network can be made secure. However, this is only an outline of the problems associated in providing secure communication systems. Further information can be found in the books by Denning (1982) and Davies and Price (1984). A good source of references to more recent publications is a review paper by Nessitt (1987).

A journal covering the wider field of computer security which includes communication security is *Computers and Security*. The subject is of obvious commercial interest and there exist many products and services that attempt to control computer security problems; the *Information Security Yearbook 1988/89* is a useful source of information on these products and services.

References

The journal *Computers and Security* is published by Elsevier Science Publishers Ltd.

ANSA (1987) *Advanced Networked Systems Architecture*, ANSA Reference Manual, ANSA, Poseidon House, Castle Park, Cambridge, June.

Chess, D. M. (1989) 'Computer viruses and related threats to computer and network integrity', *Computer Networks and ISDN Systems*, **17** (2), July, 141–148.

Davies, D. and W. Price (1984) *Security for Computer Networks*, Wiley, Chichester.

Davies, D. W. (1981) 'Protection', in *Distributed Systems Architecture and Implementation*, B. W. Lampson, M. Paul and H. J. Seigert (eds), Springer-Verlag, Heidelberg, 211–264.

Denning, D. (1982) *Cryptography and Data Security*, Addison-Wesley, Reading, Mass.

Grimm, R. (1989) 'Security on networks: do we really need it?', *Computer Networks and ISDN Systems*, **17** (4/5), 315–323.

Information Security Yearbook 1988/89, IBC Technical Services Ltd, London.

Jamieson, R. and G. Low (1989) 'Security and control issues in local area network design', *Computers and Security*, **8** (4), 305–316.

Nessitt, D. M. (1987) 'Factors affecting distributed system security', *IEEE Trans. Software Eng.*, **SE 13** (2), February, 233–248.

Rivest, R. L., A. Shamir and L. Adleman (1978) 'A method for obtaining digital signatures and public key cryptosystems', *Commun. ACM*, **21** (2), February, 120–126.

Rochlis, J. A. and M. W. Eichin (1989) 'With microscope and tweezers: the worm from MIT's perspective', *Commun. ACM*, **32** (6), June, 689–698.

Spafford, E. H. (1989) 'The Internet worm: crisis and aftermath', *Commun. ACM*, **32** (6), June, 678–687.

12 Distributed computing systems

MARK CLARK

12.1 Introduction to distributed computing

A classical view of a distributed computing system requires the services to be distributed over a wide area and to be known by name and not location while the user should be unaware of the distributed nature of the system being used. All the functions at the user level should be similar to those that would be expected on a uniprocessor operating system. The distributed system may be nothing more than a computer network demonstrating a high degree of cohesiveness and transparency. However, a computer network does not automatically result in a distributed system since it depends upon the organization of the connected computer systems. In this chapter we will consider the more general case of computer systems connected by a network providing a distributed computing architecture and environment.

There are two approaches to distributed systems that may be taken. The first is to have an operating system onto which have been added networking capabilities to provide the feel of a distributed system, sometimes termed a *network operating system*, or NOS. This has the advantage that the user is free to use a familiar operating system in which there may have been a high degree of investment of both time and money. This is commonly called a *loosely coupled system* and is effective in exploiting the power of modern workstations and microcomputers, enabling them to access shared resources commonly called *servers* which are other processor systems offering a service to eligible users (see Figure 12.1). The second is to have a single homogeneous distributed operating system which has its parts distributed over the network making up the *distributed operating system*, or DOS, sometimes known as a global operating system, or GOS. These are often implemented as layers over a minimal, sometimes termed lightweight, operating system kernel upon which code can be loaded for execution (see Figure 12.2). The DOS environment is comprised of cooperating computers that strive to provide systems management decisions based on the entire system and its resources rather than a single computer system. The DOS must provide management facilities

Workstations/computers providing local processing
or acting as networked servers

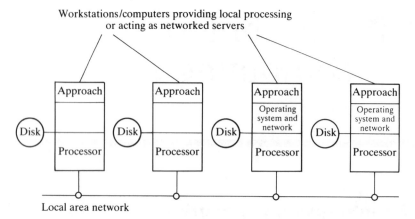

Local area network

The processors may be workstations or dedicated processors
providing a service accessible over the network

Figure 12.1   A loosely coupled distributed system

for processes or objects, memory, input and output, devices and also network access.

Should one consider the *tightly coupled* multiprocessor systems (see Figure 12.3), as providing a distributed system? This is of course a difficult question to answer and depends on the strictness of interpretation of the definition. A tightly coupled multiprocessor system does not usually meet the requirement of being spread over a wide area. For example, with the balance Sequent multiprocessor computer (Sequent Computer Systems, 1986) the user may access up to thirty-two processors sharing memory and address space. The contention problems in this computing system are significantly reduced by use of large cache memories for each processor. The user of the system, however, sees a standard UNIX processing environment and is unaware of the underlying system architecture.

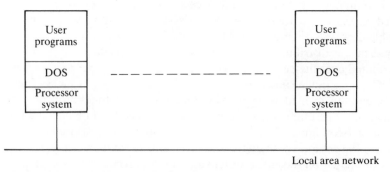

Local area network

Figure 12.2   Architecture of a DOS

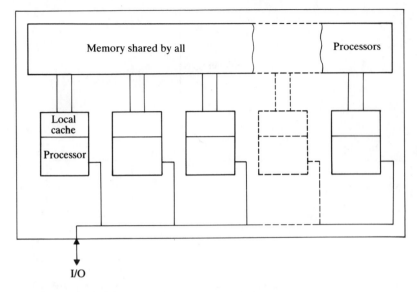

Figure 12.3 Tightly coupled computer system

We will look at the various architectures covering the spectrum of distributed systems. Further information can be found in journals and books such as Coulouris and Dollimore (1988) and Mullender (1989).

12.1.1 The hierarchical approach

One fairly common approach is to organize a system as a tree (Figure 12.4), with a powerful root processor system and less powerful machines at the branches and leaves. This architecture might be used for shop point-of-sale terminals handling both the financial details locally and the stock control issues centrally or even nationally for a multiple chain store.

In this approach each computer may handle tasks at different levels such that the level closest to the user may be providing no more than a tailored man–machine interface function.

12.1.2 The CPU cache approach

For some environments the hierarchical approach is not suitable since the workload exceeds the capacity of the central machine. An approach to this is the CPU cache (see Figure 12.5). Here part of the computation is done on one or more front end processors and some on the central system. If the central machine is busy it may be better to do more processing on the local processor and use the central facility when resources are available. Certain tasks would be better done on the local CPU, while array or processor intensive functions should be done centrally.

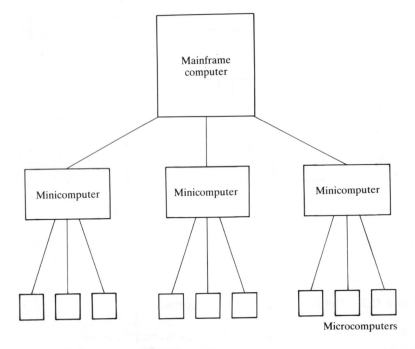

Figure 12.4 Hierarchical approach

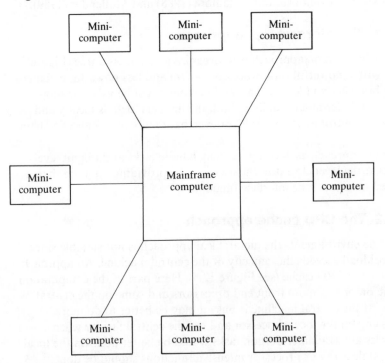

Figure 12.5 CPU cache approach

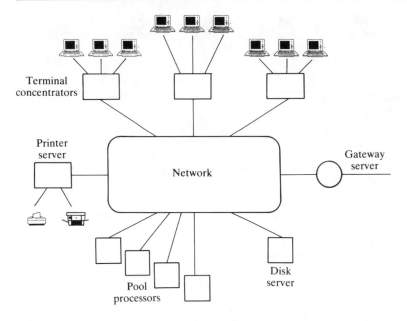

Figure 12.6 Processor pool approach

12.1.3 The processor pool approach

The user accesses the system through a terminal concentrator or personal computer on the network and is allocated a processor to perform the desired task from a pool of processors (see Figure 12.6). (The Cambridge distributed computing system as an example of this will be discussed in more detail later.) If the task requires more processing power then another free processor could be allocated. Some processors could perform a fixed task and require no loading of program code other than at the initial boot time (e.g. a multipass compiler could use several processors in parallel).

12.1.4 The user–server approach

The price–performance ratio for personal workstations (computers) has dramatically decreased. The user can have sufficient local processing power to meet most requirements, usually file editing and manipulation tasks, thus using the network to share access to expensive resources and peripherals such as array processors, laser printers or plotters. The networked resources may be either local or remote to the site of the user and also provide communication for mail, etc., and similar facilities. However, there may be savings in sharing disk resources together with the simplification of project management strategies. There has been an enormous economy of scale associated with large high-speed disk

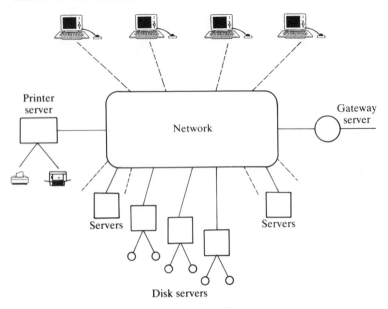

Figure 12.7 User–server approach

storage devices making file servers very economical in their operation. The idea of powerful diskless workstations has recently grown in favour with several commercial offerings. It may also be appropriate to provide access to mainframes for large numerical computations (see Figure 12.7).

12.2 Goals for distributed system design

12.2.1 Transparency and naming

One of the major objectives of the designer of any distributed system must be to try and construct the system such that the distributed nature of the system itself is largely transparent to the users. The users should be unaware of the location of the components (software and hardware) from which the system itself is constructed. This requires that the components of the system be known by name and not location, and mechanisms be provided to support the necessary transformations from names to addresses.

The servers must exist at network addresses which can be accessed at the network transport level. This requires that the clients can obtain these addresses which are combinations of network and port addresses (e.g. see the TCP/IP description in Chapter 8). The process of obtaining a transport address from a service name is termed *binding*. This process is performed in different ways for various systems. One static approach would be to configure the address of the remote service(s) at the compilation time of the application. However, this is not adequately dynamic for a distributed system where it is desirable for services to be able to migrate and perhaps be replicated. Another approach is to use a *nameserver* (also called a binder) which can be accessed by the client like any other server to obtain the mapping from service name to transport address. Servers are required to register with this special server as they are invoked, and the nameserver exists at a well-known address reserved for this function. The nameserver itself might be replicated, requiring the clients and servers to know about all possible nameserver addresses. Some systems require the host computers to obtain the address of the binder when booted (initially loaded), and thence verify its existence at regular intervals. A broadcast packet if supported might be employed to contain the binding request. The binder is expected to respond within a timeout period, after which a slave binder may respond and assume the role of master.

12.2.2 Security

The level of security and protection that should be applied to a distributed system is dependent upon the needs of the organization to which it is accessible, and consideration must also be given to what access is available to external systems and to what degree these may be trusted. Most operating systems provide simple password mechanisms to authenticate users to the system, and provide mechanisms to provide access control using a user's access privileges for resources such as files and devices. There are two forms of attack upon the system that must be considered. The first is that from the naive user who should be protected from his or her own misdoings perpetrated through ignorance. Secondly and more threatening are attacks from 'hackers', viruses and the like. In a distributed system, the component parts may not reside in a single administrative environment. In a large internetwork, the distributed system may encompass many administrative areas and support many separate distributed systems and support many applications. Smart card technology, a microprocessor system incorporated into a credit card package, offers an attractive mechanism to support authentication and access control even for overlapping security domains (see Jones and Clark, 1990).

The distributed nature of the system itself makes security more difficult, providing more nodes from which an attack may be originated and communication channels which are themselves vulnerable. Open system architectures increase the vulnerability to external attack since the interface is well known and available to those who wish to attempt to access. One commonly used mechanism to control access to objects is the *capability*, an identifier that grants access privileges for an object or resource. The capability cannot itself be accessed by the user processes but only through privileged calls associated with resource control for the object to be accessed. The storage of capaiblities for multiple overlapping domains of authority are being explored at Essex University in the context of the smart card technology. This technology permits users, who must authorize themselves both with their local host and the smart card, to carry with them their access privileges regardless of the node from which they access the system. It has the advantage that one host does not have to trust information obtained from another host with the possibilities for attack that this presents. Possession of the card itself is insufficient since the user must be authenticated against information stored in the card but not accessible to a card holder. Complicated mechanisms must be supported to allow the updating of capabilities when new objects are created. Previous single computer systems have made use of hardware supported capability mechanisms, but in a distributed system this would be impractical without technology similar to the smart card.

12.2.3 Statelessness

Servers may be designed to be stateful or stateless. A *stateful* server keeps track of the state of all client requests upon the server. In the case of a file server it would mean that files could be left open and the state of the file pointer within the file would be retained for subsequent calls upon the server. In a *stateless* server every request must be dealt with as a total entity. Hence for a file server dealing with repeated requests to perform read operations on a file, it would be necessary to specify for each read request the position within the file from which data must be read. The server would open the file for each request, seek to the required position and read the data to be returned to the client, and thence close the file.

A server providing access to files, a disk or file server is likely to be kept very busy and hence the extra overhead of tracking the state of all clients would be too great. The server is duty bound to provide all available resources to the task at hand, namely file access. Requiring the client to track the state of each of its requests does have some implications, however. For example, in the UNIX operating system the read/write pointer for an open file may be regarded as state information. If the server was to maintain this information for each of its clients, then in the event of a server crash and its following return to service, the clients would no longer have information concerning their state within the files. Furthermore, in the event of a client crash, following its return to service the information held by the server would be incorrect and not easily undone.

It should be clear that a stateless approach at the server will simplify the design and maximize the response of its operation. If the server crashes, the client can use a simple timeout mechanism to recognize that this situation has occurred. This is the approach used in the Sun network file system (NFS) discussed in more detail later.

12.2.4 Reliability

Distributed systems offer a higher degree of tolerance to failure as a function of their architecture. The loss of the network would be critical but the failure of a single node in a well-architectured system would present little impact on the system as a whole. The loss of a single server will not be crucial if its service is duplicated elsewhere. In the case of most servers, and particularly file servers, it is important that the data associated with the server (file information) be restored to a consistent state following a crash, and hence considerable research has been performed into atomic transactions (see Section 12.3.4).

12.3 Approaches for distributed system design

12.3.1 The client–server model

The client–server model is the most commonly used approach to the construction of distributed systems. This scheme requires the client application to make a request for services from a server process. Implied in this model is an asymmetry in the establishment of communication between client and server (see Figure 12.8). The client and server must use a set of well-known conventions for communication exchanges which may involve the use of acknowledgement messages where large volumes of data are to be transferred. The protocol of communication may be implemented by a message passing transport level protocol using a datagram service.

The server process may listen at a well-known address for a service request and then reply to the client process from whom the request was directed. This strategy is fine where only a small number of servers are existing on a system. However, if a large number of infrequently used servers are to reside in a single system it may well be better to use a superserver which will prevent clogging of system resources. The superserver may be used such that a request from the client for a particular service from the superserver results in a server being created to deal with that particular request and the connection between client and server being passed to the newly created server process. Thus a single contact point exists for all servers supported by the super server and no blocking will result from one client's request for a service upon another client. In the case of a frequently used server the extra overhead of the intermediate process may not be able to be tolerated and hence this service would not be supported by the superserver but would stand alone at a well-known address for that service.

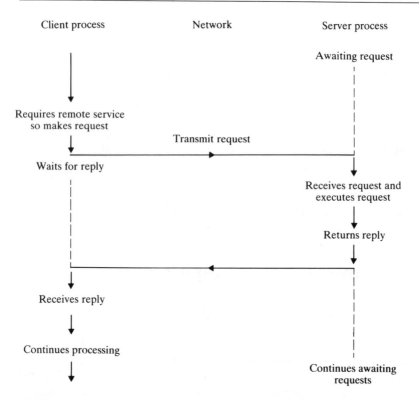

Figure 12.8   Client–server communication sequence

12.3.2 Remote procedure calls

The remote procedure call (RPC) is a mechanism for constructing distributed computing environments. The RPC is an extension of the familiar concept of a procedure or system call within the conventional programming environment where the procedure call is used in the decomposition of programs into modules for the purpose of program abstraction and data hiding. The distributed programming environment can use similar mechanisms since the RPC uses the same semantics at the program application level, but the procedure call itself is executed on a different processor to that from which it was called. In the distributed environment the remote procedure should be able to be called by name without the need for knowledge of the location of the machine(s) that will service it (see Figure 12.9). The RPC is sent as a request message to some remote system which is able to handle the request and return a reply message. Normally the calling process is suspended until a reply is received, as would be the case for a local procedure call. The paper on Xerox Courier RPC (Xerox Corporation, 1981) described the construction of remote systems and was followed by the classic paper by Birrell and Nelson (1984) which describes the RPC

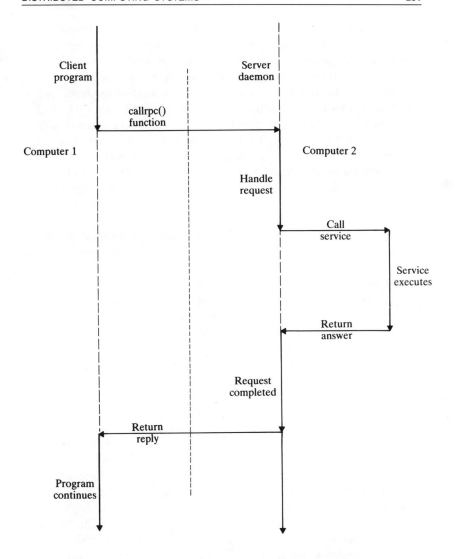

Figure 12.9   Network communication with the remote procedure cell

mechanism built using datagram communication over the Xerox
Internet.

In the C language, the procedure (function) is called with an optional
number of arguments. The arguments (parameters) are passed by what
is known as call by value, which means that the called function is passed
copies of the arguments which are used in the called function. The RPC
facility provides a mechanism whereby one process termed the client
process can have another process called the server execute a procedure
call, as if the caller process had executed the procedure call in its own
program environment. Since the client and the server are now two

separate processes, they no longer have to live on the same physical machine or even network but may communicate over an Internet.

The client requesting an RPC sends a request message to the server which receives the request and services the request by executing a procedure call local to that machine, and then the response from that call is encapsulated into another message which is sent as a reply to the requesting client. The server environment on the remote machine will not have access to the client programming environment in terms of data variables, etc., since it will be operating in a different address space. Thus the transfer of memory addresses (pointers) as arguments to RPCs is usually meaningless and hence programming in an RPC environment often requires complex data structure objects to be passed between client and server processes.

The passing of parameters in the form of messages must be carefully considered. In a homogeneous processor environment the messages may be sent and received in raw data format since the processors are of the same type and hence the messages will be interpreted in a consistent manner. However, it is more common for the environment to consist of mixed architectures of processors. Since data representation is different for various processor types then it is necessary to consider the encapsulation of the data into some processor independent form so that it may be correctly interpreted at both clients and servers.

Considerable research has been done on RPC mechanisms including some interesting developments within the UK Alvey project ADMIRAL (Wilbur and Bacarisse, 1987). The ADMIRAL RPC was designed to run under 4.2 BSD UNIX and the C programming language using datagrams for message passing. The system uses an interface language called 'Glue' and a binder to register the declaration of the interfaces. The interfaces require access to a set of RPC library functions which will handle the messages to be sent between processes. The interface name and its interface address in the form of an Internet address are recorded by the binder. The binder itself runs on a host on the Internet at a 'well-known address' which will be built into the RPC software. The client becomes bound to a server when it first initiates a call to a remote procedure and makes a request on the binder. The call requests the binder to return the address of the requested interface and henceforward there is no need to make requests upon the binder.

The interfaces between clients and servers are declared in terms of procedures, import and export statements. The Glue compiler is then used to compile the programs which include these interfaces and it generates C code as its output language which may now be compiled in the usual way. The performance statistics of the ADMIRAL RPC were found to be good. The UNIX kernel obscures the possible communication performance of the RPC mechanisms which might be obtained using other operating systems, or a modified UNIX kernel, which does not require the user space–kernel space data buffer copying which occurs in standard UNIX kernels and represents a significant overhead.

12.3.3 File servers and network services

The need for specialized network-based services has grown with the emergence of distributed environments. The need to share information for economic or logistic reasons has resulted in gateways linking separate networks into an internetwork. These internetworks are commonly used only for electronic mail and file transfer between parties, even though more exotic services could be supported. The most valuable server found in any distributed system is the file server which allows data and programs stored in files to be shared by the users of the distributed system.

All computers use files for permanent storage of information, usually on some magnetic media. The file systems themselves are designed such that the programs that access them use a set of functions which allow access to the files without knowledge of the full physical characteristics of the medium itself, providing a high degree of abstraction. For example, in the UNIX operating system the application program does not require detailed knowledge of the file system structure. UNIX provides a sequentially ordered byte storage for file data which may be randomly accessed, and the application program may build any abstractions required on top of this. Furthermore, the directory is another abstraction which allows the user a mapping between storage location and name of the file; in UNIX as in other systems we find the mapping applies to the familiar hierarchical tree structure of the file system and also applies various levels of access control at the directory level.

Network-based file servers can be justified since long-term storage of information is an integral part of modern computing. Without file sharing, the user is forced to copy files from machine to machine explicitly. Distributed program updates to all users, and concurrent access of data bases by several servers could be a nightmare. The cost–performance trade-offs for data storage have favoured large disks although this factor is less important now; however, the requirement for incremental expansion of storage capacity to meet demand requires distributed systems. Furthermore, in a large internetwork it would be painful if the exact location of files had to be known rather than a location implied by a naming abstraction.

File naming and the mapping from file names to information concerning location and identity are usually handled by separate services called *directory services*. The directory service is usually implemented as a client service of the file service since the mappings that it uses are stored by the file service itself. It may also be responsible for controlling the access information for the file concerning who has permission to do what actions on the particular file. The file service may be offered by one or more dedicated servers. The Xerox distributed file system (XDFS) (Mitchell, 1985) used a set of cooperating servers while the Sun NFS (see Section 12.8) and Locus (Popeck and Walker, 1985) have the file service software on every accessing computer. The services are usually offered as remote procedure calls.

The main difference between a centralized and distributed file system arises because there is no available state information for a computation process in the distributed system. Also there is a variable communication delay which exists in the network. In the centralized system, if a job should terminate, any resources are reclaimable by the operating system, whereas in the distributed system there is no way of knowing that a computation process has terminated. All that is determinable is that some communication process no longer obtains a response. It then has to be determined whether the remote machine or the network is malfunctioning. It is necessary to consider what happens if that process resumes at some later time. Thus it is necessary that the file server should be written in a robust manner that handles the interruption in service of the client, and equally the client should handle loss in service of the file server.

12.3.4 Concurrency and atomicity

Since we are building distributed systems where processes are running on independent computing systems, it is necessary to consider the concurrency of operations. If a file service is used by several clients who desire to read and write data to the same file, then without some form of control mechanism the result upon the file would be unpredictable because of the interleaved read and write operations that would be occurring.

Atomic transactions may be forced upon the client processes such that the operation of any client transactions upon the file service will be performed without interference from other competing processes. An atomic transaction on a file service may consist of several file operations which from the client's point of view can be deemed to be a single secure operation.

Consider the case of an airline seat reservation system being accessed by many travel agents. If one agent wishes to book seats on a particular flight then it is necessary to confirm that there are sufficient seats available for that booking to be accepted. While this information is being discussed with the customer another agent may make a similar reservation, perhaps removing all free seats. Thus the information obtained by the first agent is now incorrect and may not be used in the request to book seats. Atomic transactions might be enforced by allowing concurrent read transactions, but not concurrent writes or reads during write transactions. Thus a request for seat reservations will require associated data to be locked while the available seats information is read, modified to its new value and rewritten, this process representing a complete transaction.

It is also necessary to consider what should occur in the event of a client or server process terminating abruptly. It is essential that the atomic transaction must be recoverable such that the state will return to that which existed if the transaction had never commenced. It is likely that in a real environment the transaction may be composed of many elements accessing many files and changing several values. Thus the elements comprising the transaction must all be performed or else none at all. One commonly used mechanism is that of a *two-phase commit* transaction approach. During the first phase copies of all data items to be changed are made, and following a decision to commit, the second phase is entered where all copied data to be modified by the transaction is written back into the original files. In the event of an abort the second phase never occurs and the copies are deleted. The full details of this mechanism and others are beyond the scope of this chapter and may be pursued by the reader, who is referred to Gray (1978), Lampson (1981), Reed (1983) and Mullender and Tanenbaum (1985).

12.4 Window servers

The evolution of window servers has come about through the desire to share the high resolution (in excess of 800×800 pixels) screen of a workstation between processes that are running locally and those that may be remote to the workstation. Applications may hence be run remotely with the workstation acting as a graphical display system, i.e. window server.

The window environment of workstations is usually implemented through a process termed the *window manager* which is responsible for the updating of window outputs and handling the input sources such as the mouse and keyboard. The window manager may support the dragging of windows or resizing, and the consequential hiding and revealing of windows that overlap. The output associated with each window behaves as a separate display system and provides the user of the workstation with the opportunity of using the system at its simplest level to monitor the progress of several concurrent processes operating independently, each displayed in its own window. The application termed the client communicates with the window server process through the use of messages commonly implemented as RPCs.

Two window systems are competing, namely X windows (Scheifler and Gettys, 1986; Stern, 1987) and NeWS (Sun Microsystems, 1987), and several others including variants of X also exist. In the X system a process runs on a workstation to act as a server providing window management functions. A window is any defined area of the workstation screen and each window behaves like a separate screen. The application programs, which may run on this or any other workstation, access the server to demand window operations such as image update, etc. These applications are termed clients of the X window server and X provides an interface to the C language that can perform operations on the windows and their contents. The server and client communicate using the X protocol.

NeWS is accessed from the application program which sends sequences of program and data in a high level language called PostScript (Adobe systems Incorporated, 1986). The communication overhead between client and server must be kept to a minimum because of the large volumes of data that will occur. The requests on the server in these systems are made at regular intervals of around twenty-five times a second, making use of the human observer's deficiency in eye response to change and giving the impression of continual update. The protocols for communication are kept lightweight to maximize performance because it is usually assumed that these systems will be operating on a LAN, providing the necessary transfer characteristics. Considerable research including that at Essex is underway to support live image windows, driven by window servers, providing interesting developments for multimedia applications.

12.5 File servers

12.5.1 File server evolution

The earliest file servers were components of the host machines in the ARPANET (McQuillan and Walden, 1977). When accessed by the network they provided bulk transfer of files using the file transfer protocol (FTP) (ARPA, 1970, 1978). The first true file servers looked like the earlier ARPA hosts in as much as they provided FTP-like access. They had some intrinsic limitations since they did not allow concurrent sharing of files. The client machine was required to be able to hold the entire file since it copied the entire file to the local machine regardless of whether only a single data record had to be accessed. The client had to have secondary storage of its own.

The previous limitations led to the development of file servers which allow access to data at the byte or record level. Since clients could access files concurrently it was necessary to introduce some form of locking to mediate concurrent accesses. Various forms of atomic transaction, as discussed previously, were developed to permit this. The atomic transaction would allow either all or none of the changes to occur when a client made a request upon the file server. (Its use at the application layer of the OSI model is discussed in Chapter 5.)

12.5.2 File server design

A file service may be a distributed system in its own right. Services can be constituted from multiple servers each with its own file system. The client must know or be able to determine which server holds a specific file and must talk to that server to access the file. Alternatively, the servers may cooperate to provide client functions. The file system can be expanded by adding more storage media. However, more important is the possibility of addition of new processors to add more server processing power as the number of clients or traffic increases.

If there were only a small number of file servers then it might be possible for the client to probe each in turn until it locates the one with the file that is required. This strategy is poor and would be impossible in an internetwork of thousands of servers. One possible solution is to have the servers assist. A client requests a file from a server; if the server does not hold the file then it responds by suggesting where it may be found or informing the client that it does not exist if that is known. This means that only the servers or a directory service within them need understand the naming convention for files.

Alternatively, a broadcast strategy could be employed (Boggs, 1982) to request the location of the file. On a local network only one server may respond, resulting in one message exchange. If the file is not present then it is assumed that one server will respond that it does not exist. In an internetwork, a more sophisticated solution is the *remote broadcast* where the file is requested on the local network and one of the servers responds either with the file or with a message indicating which network should be tried for the file. Here, as before, it is assumed that the file will exist or that one of the servers will be able to respond with the information that the file does not exist and hence the client will not be passed from server to server or left waiting forever.

Errors in the network will exist and must be handled and should be tolerable to both client and server. Some file servers require that the clients send keep alive packets to indicate that they are still alive. If these keep alive packets are not received then it is assumed that the client has terminated and the session is aborted. Thus the server may reclaim resources associated with the server and that session's state.

A three-message protocol is used in the Xerox distributed file system (XDFS) (Mitchell, 1985). The client makes a request, the server responds and the client must acknowledge receipt of the server's response. This allows the server to throw away the data for a response, once it is assured that it has been received, or allows the server to repeat its response if the client duplicates its request.

Flow control must be exercised by the server to avoid exhausting the server's shared resources, i.e. specifically memory and processor. Servers in a local network environment can take advantage of its properties of flow control. In the Cambridge ring distributed system, the low level minipacket acknowledgement bit is used to inform a client that has begun to send a message that the server cannot accept the message at this time. The client can then retry some time later.

In an internetwork environment the server may simply ignore a request assuming that it will retry at some later time or abort a request already in progress, again assuming a retry will occur after some timeout period.

12.5.3 File structure

Files may hold two types of information, *data* and *attributes*. The data portion usually has a simple structure (e.g. a sequence of ordered bytes) on top of which may be constructed something more complicated, for example to implement perhaps a data base providing an indexed sequential access method, where each record is associated with a key by which it can be accessed. By providing a simple file structure the client is better able to make decisions but has the disadvantage that they must make their own structures built on the low level file access primitives. It is possible to build the data structure into the services offered by the server but at the cost of lower throughput due to the extra load on the file server processing system.

Attributes are associated with the file to record information concerning the data comprising the file. The attributes might include the length of the file, the time at which it was created, time last modified, the time it was last read and protection information. Some of the attributes may require different treatment with respect to concurrent access than is required for the data itself. Consider the case of the 'last time read' attribute; this has to be written even though the file itself is read only. Thus the correct way to handle such problems is to treat each attribute and the file data as a separate component for the purposes of access control, atomic transactions, etc.

Modern operating systems use names that relate to the file contents and directory facilities for naming files. However, some systems allow users to construct their own structures on top of the basic access mechanisms. The Cambridge file system (CFS) requires every file to be recorded in an *index*. Indices are objects just like files, which can be reached starting at a root index created by the file service. Indices may contain the *FileIDs* of files and indices. The structure makes it relatively easy to build directories by using the zeroth slot of the index to hold the *FileID* of a file containing the print names of files in that index and, associated with each name, the slot number in the index where the *FileIDs* can be found. The indices are used by CFS to form a graph to enable it to garbage collect files that are not reachable from any index and hence require deletion from the system.

The XDFS provides no directory service itself, but its implementors provide a client, the directory service, which other clients can use to associate print names and *FileIDs*. This service resides in the XDFS servers for convenience, but has no special privileges and is viewed as just another client by the file service itself.

If the storage hardware and the file system could be considered ultrareliable, then only the high level consistency problem of the data base would remain, which in itself is a difficult task. Atomic transactions are an attempt to provide a stable environment on which to build semantic consistency. Transactions must be atomic in two ways: firstly, they must be atomic with respect to update, all changes to files involved in the transaction occurring or none at all; secondly, they must be atomic with respect to concurrent access, appearing to execute either before or after conflicting transactions.

If any failure occurs during a transaction then the system will revert to the state prior to the transaction and will not leave the data base partially altered. CFS provides atomic update of individual files only, whereas XDFS provides atomic updates of a set of files when multiple servers are involved.

A transaction is invoked with an operation such as *Open Transaction[credentials]* where the credentials may be a name and password or capability, depending on the protection mechanisms used. The service returns a *transaction identifier (TransID)*, and subsequent requests contain both the *TransID* and the *FileID* of the file to be accessed. When a client issues a *Close Transaction* request to the server, it may ask for the transaction to be committed or aborted. If it requests that the transaction be committed, the server may respond that the transaction has been aborted due to outstanding concurrent access conflicts; however, this is expected to be a rare event following which it is expected that the client will repeat its process until a transaction has been committed.

There are many reasons why a temporary loss of service might occur. Clients should be able to deal with this loss of service. If the service supports atomic transactions then it is possible to recover by abandoning the current transaction and restarting a transaction when the server may be recontacted. A client can use a simple semaphore to tell if the transaction was committed. For example, an empty file could be created when the transaction is committed and henceforth the client can access that file; if it exists the transaction must have committed, if it does not exist the transaction is aborted. This is in practice somewhat inconvenient as each client must implement its own mechanism to deal with service interruption. XDFS therefore provides a mechanism by which clients enquire about the state of a transaction. The client can be informed whether the transaction is active, the transaction has aborted, the transaction has committed successfully or the server has no knowledge of the transaction.

Obviously it is necessary to provide some form of locking mechanism. However, the problem of deadlock can prove to be very serious.

Traditionally file systems place interlocks on a file at the time it is opened. A file may be opened for reading by any number of clients and may be opened for writing by at most one client, provided it is not already opened for reading or writing. In a data base system the locking may be applied to an individual record rather than a whole file. This reduction of granularity decreases the likelihood that separate clients will lock each other out.

Sometimes the system also applies a timeout to prevent infinite lockout. A lock is not broken when the timeout period expires but when some other transaction tries to gain access; this obviously results in a smaller number of broken transactions than would occur otherwise. A transaction that has had its locks broken should not then be able to commit since the data it is accessing may now have changed and the transaction must therefore be restarted.

XDFS allows transactions to complete even if some broken read locks have occurred. When a lock is broken, the relevant transaction is informed. If the transaction responds by voluntarily relinquishing the locks, it will not be prevented from successfully committing later because of these broken locks. After relinquishing the locks, the client can simply reread the data to regain the locks and recompute taking account of any changes that have occurred, and then try again to commit its transaction.

In XDFS, zero or more transactions may read a byte of a file along with at most one transaction intending to write the same byte. However, when that transaction commits, the 'intending to write' lock must become a standard write lock with no conflicting read locks. If there are read locks they are broken and the write lock is thence guaranteed. This works because write locks are only granted during a transaction commit when they cannot be broken for any reason.

12.6 The Cambridge distributed computing system

The Cambridge distributed computing system (CDCS) (Needham and Herbert, 1982), was built to provide a computing environment that had the traditional feel of a time-shared mainframe but which was constructed through the use of distributed components. The project was to use two technologies which were at that time recent advancements: local area networks (LANs) and relatively inexpensive processing engines of moderate processing power. The details of the Cambridge ring local area network used and its protocols are discussed in Chapters 7 and 8.

The advantages offered by personal computing over traditional time-sharing systems were seen to be that the load on a personal machine is under the control of its user and is not affected by the actions of other users. The user may choose the operating system and tailor the program environment in complete privacy since there is no external monitoring by the 'management' or even other users.

However, for a campus site there are advantages to be gained from access to a central service, particularly in respect of hardware. At the time of the CDCS, there was an economy of scale particularly associated with peripheral devices such as dish storage units. A common software environment allows the teaching and administration services to be managed more easily.

12.6.1 The CDCS architecture

A collection of small computers were made available for this project where some were available within offices as personal computers and others had 'public' access for more traditional roles such as teaching machines. Some computers were allocated dedicated tasks such as for administration functions of a typical centralized system. These latter machines are termed servers and are assigned statically to individual functions or services, even if that service is trivial.

The advantage of this philosophy is in the ability to increase the available computing power through the provision of more computers providing a smooth upgrade path. If any of the services offered on a server becomes overloaded then it may be replicated on one or more machines as necessary.

12.6.2 Processor banks and resource management

The CDCS environment was constructed such that a large amount of its computing power was made available through a collection of larger processors which were known as the processor bank. A user could request access to one of these processors from the management system handling the bank. If one with the suitable requirements was available then it would be exclusively allocated to the user while required. It would be possible to have more powerful processors accessible to users of the system to deal with computations beyond the 'power' of the personal or processor bank server computers (see Figure 12.10). A user could access the Cambridge system from a VDU connected to a terminal concentrator (or personal computer). To acquire a processing server the client opened a connection with the session manager which would provide a command level interface to the processor bank management system. The management system allocated a processor and returned a message containing capability information which allowed a user to access the nominated processor.

The resource manager was responsible for many management and control functions including policies for priority and privilege. It carried a table of resources available to particular users and the period of time

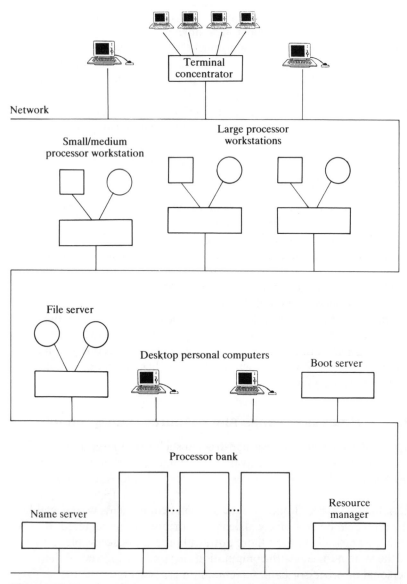

Figure 12.10 Cambridge distributed computing system

for which they may hold those resources. It was responsible for maintaining the allocation table for each of the machines in the processor bank. An algorithm found the processor which would just meet the requirements of the request, hence ensuring that a request did not tie up a processor which would be better suited to a more complex task in an alternate request. Timeouts would be given to the allocated processor so that a user could not keep it busy denying it to the resource manager indefinitely. The resource manager could withdraw a

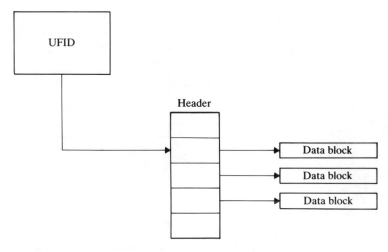

Figure 12.11   Cambridge file system UFID structure

processor from the processor bank for maintenance, etc., or under control of a privileged user a processor might be reallocated to the processor bank.

12.6.3 Small servers

A large number of small servers were connected to the network to provide services available within the standard time-share environment such as printer and plotter operation, authentication servers, etc. When the small servers were booted they accessed the nameserver to find the address of a loading service from which the server software was loaded and executed.

12.6.4 Cambridge file servers

The Cambridge file server (CFS) (Birrell and Needham, 1980) was designed to be simple, providing a low-level file server to enable shared access to what was then an expensive resource, namely disk storage. The CFS provides both files and indexes, where the CFS index was designed as a structure to hold the unique file identifiers (UFIDs) of files. Each UFID contains the address of the header block for the file that it refers to, and the header then contains the addresses of the data block (see Figure 12.11). The CFS provides fast access to small files and efficient use of disk space by using two sizes of data blocks, 512 and 2048 bytes. The larger blocks are used to store the data of files and the small blocks for file headers and the first part of data in file and indexes.

   The file service was designed such that a number of different file systems could be implemented on top of it. The CFS index is shown in Figure 12.12. When a client process creates a file its UFID is returned

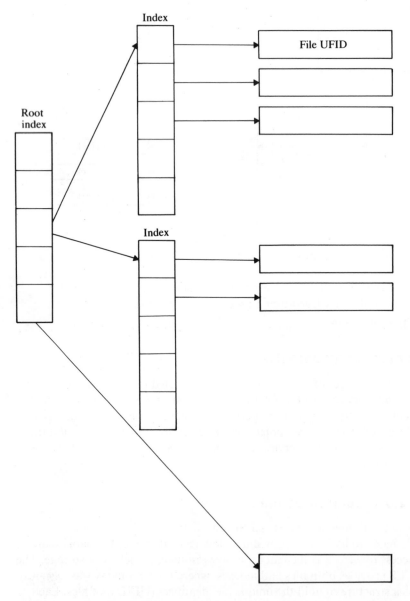

Figure 12.12 The structure below the root index in the Cambridge file system

to the client who may delete the file simply by removing the UFID from indexes. The client may store the UFID of the file it has created in any index known to that client. The root index UFID has access to all other indexes and is not supplied to the clients, each of whom may have its own subroot index to access files known only to that client. It is this technique that permits different clients to have different filing systems implemented upon the flat filing space of the CFS. The CFS allows a file or index to be recorded in more than one index allowing the sharing of files. If the file is to remain in existence all the time at least one reference from an index to the file exists. The CFS system can then run a garbage collection process periodically to reclaim any unreferenced data blocks which were formerly files or indexes.

Files when created may be marked as ordinary or transaction files. Transactions may refer only to a single transaction file and a simplistic atomic transaction mechanism is provided. The client requests a transaction open on a file or index by providing its UFID as an argument. A temporary UFID is returned as a transaction identifier. This transaction identifier is used in operations upon the file or index, and concurrent operations on the file by different processes are prevented since only one temporary UFID may be allocated per transaction file UFID. Deadlock may be avoided since there is a timeout mechanism on temporary UFIDs.

12.7 Interprocess communication in Berkeley UNIX

12.7.1 Introduction

The Berkeley UNIX system 4.2 BSD (Berkeley, 1981) (followed by 4.3) and its derivatives such as ULTRIX have become an industry standard for workstations. System V UNIX machines also usually include the Berkeley enhancements such as is the case for the Apple Macintosh UNIX operating system AUX. The prime reason for this is the all-important addition of interprocess communication (IPC) based on *socket* pairs.

Prior to 4.2 BSD the releases of UNIX had been weak in the area of interprocess communication; the only standard mechanism that allowed two processes to communicate were *pipes*. This method required that the two communicating processes must be related through a common ancestor, obtained by executing a *pipe* system call followed by a *fork* system call. The *pipe* system call provides a process with a pair of buffered unnamed asynchronous communication channels, one for input and one for output. The *fork* system call causes the process to be duplicated, creating an identical process known as the *child*. The parent process is returned a unique value, the process number of the child process, while the child is returned the value zero from the system call. Thus each process may continue and determine whether it is a child or parent process and may then, through conditional code, execute a separate processing task. The pipe channels set up prior to the fork call will exist both in the child and the parent processes, thus providing a method of communication between them in the classic producer–consumer manner.

12.7.2 Socket-based communication

The 4.2 BSD system provides extended mechanisms for interprocess communication over local and wide area networks as well as within the single computer environment. It uses the socket as a basic building block for communication. A socket is a reference point for communication to which a name is given or, using the correct terminology, may be *bound*. Each socket in use must be associated with one or more processes. In a single computer environment the name given to a socket is taken from the file naming space of that system, i.e. a path name for a file. However, within a network environment, a network name associated with network addressing must be used. One process sends messages through its socket to another process which uses it to receive messages. The socket has a type and a protocol which it will use. Any process may create a socket to communicate with any other process. The socket enables any messages from another process to be mapped directly to the receiving process. The protocol may be datagram (using the unreliable datagram protocol, UDP) or stream (using the transmission control protocol/Internet protocol, TCP/IP) which provides a reliable and sequenced message passing protocol (see Chapter 8). Each process is responsible for creating its own sockets and two processes may communicate using the two as a socket pair. The sender uses a socket to send to the socket of the other process, for which the sender must have an identifier so that access can be obtained.

Sockets exist in what are termed *communication domains*. To enable sockets with common properties to communicate, three domains are supported. In the UNIX communication domain, as used when communicating between processes in a single UNIX host, sockets are named with UNIX file path names. Any unique file path name may be used. The Internet domain is used by processes requiring to communicate using the DARPA standard communication protocols, and the XNS sequenced packet protocol domain is used by processes that communicate using the Xerox networking standard communication protocols.

Several types of sockets are supported. Most notable are *stream sockets* which provide for bidirectional, reliable, sequenced, and unduplicated data flow between processes and *datagram sockets* which support a bidirectional flow of data which is not promised to be sequenced, reliable, or unduplicated.

To create a socket the socket system call is used:

$s = socket$ (*domain, type, protocol*);

The domain is specified as one of:
 OAF__UNIX, OAF__INET, OAF__NS.
the socket type one of:
 *SOCK__STREAM, SOCK__DGRAM, SOCK__RAW, or
 SOCK__SEQPACKET.*

The call requests that the system create a socket in the specified domain, type and protocol. A protocol value of 0 will cause the system to select the protocol suitable for that domain. The value returned from the system call is a descriptor, an integer value, similar to that returned when a UNIX file is opened, and this is used as an argument in system calls which deal with that socket.

A socket is created nameless and a name must be bound to it if processes are to have a way to reference it. In the UNIX domain, an association is made using a path name of a non-existing file in the file space of the host computer. In the Internet and XNS domains, an association is composed of unique addresses and port numbers. The *bind* system call from the process specifies half of an association, < local address, local port > (or < local pathname >), while the connect and *accept* system calls described below are used to complete a socket's association.

Connections are usually asymmetric as one process assumes a client and the other a server role. The server binds a socket to a well-known address associated with the service and then passively listens on its socket. Thus an unrelated client process may rendezvous with the server; the client requests services from the server by initiating a connection to the server's socket with the *connect* call. For the server to receive a client's connection it must perform two steps after binding its socket. It must listen for incoming connection requests with the *listen* system call. Once listening, a server may *accept* a connection, and the system call will not return until a connection is available.

Once the connection is established, data flow may commence. For transmission and reception of data there are several system calls including the standard UNIX *read* and *write* system calls as used for file access. Connectionless communications are supported using datagram sockets, data is sent using the *sendto* system call which specifies the address of the intended receiver and reception of messages on an unconnected datagram socket is obtained with the *recvfrom* system call. It is the responsibility of the application program to perform all error handling when unreliable data transfer is performed.

A socket may be discarded by using the *close* system call with the socket descriptor as an argument, or if unread data is associated with the socket it may be shut down, discarding the data.

12.8 Sun's network file system

Sun as a workstation manufacturer decided to tackle the traditional mainframe market by providing high power workstations that could be connected using their network file system for shared access of disk files. Sun's network file system (NFS) (Sun Microsystems, 1985) is based on the Berkeley UNIX system and provides a distributed file service based

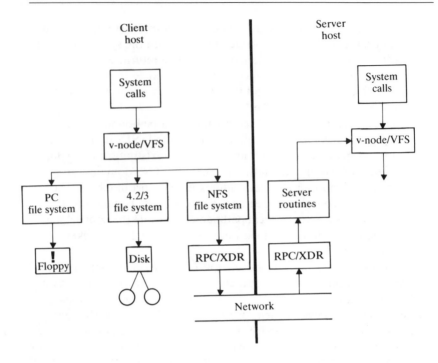

Figure 12.13 NFS architecture

on networked UNIX systems allowing users to treat remote files as if they were local by mounting directories across a network. The Sun NFS is an open standard; it has been ported to many vendor's UNIX systems and has been supported (as a client service) on MS-DOS and other operating systems.

Sun NFS permits client systems to access shared files on a remote system. Sharing is accomplished by *mounting* a remote file system, such that reading or writing of files may occur. Client machines request resources provided by other machines, called servers. Servers make particular file systems available by exporting them as part of their own file space and the client machines may then mount these as local file systems as if they were on the local machine. On each server there is a list of directories which may be exported in a file with a path name */etc/exports*. The exported directory may be the root of an entire file system or any part of a subfile system. Once the mounting procedure has occurred the users of the system may be unaware of the distributed nature of the files they are accessing. The mounting operations are often performed at boot time for the client and may be used to provide file systems for diskless nodes. NFS is not a distributed operating system but an interface to allow a variety of machines and operating systems to play the role of client or server.

The network file system (NFS) is composed of a modified kernel, a set

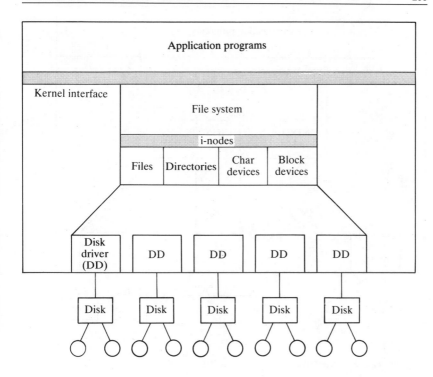

Figure 12.14 Traditional UNIX architecture

of library routines and a collection of utility commands. NFS flexibility
allows many configuration scenarios providing a variety of cost and
performance trade-offs as found in our various configurations at Essex
University. The open standard approach of Sun has resulted in the Sun
architecture being widely used and supported and might now be termed
a standard by acceptance.

The remote procedure call (RPC) mechanism used by Sun's NFS is
implemented as a library of procedures, plus a specification for portable
data transmission, known as the external data representation (XDR).
Both RPC and XDR are portable, providing a kind of standard I/O
library for interprocess communication. Thus programmers now have a
standardized access to sockets without having to be concerned about
the low level details of system calls themselves (see Figure 12.13).

12.8.1 NFS implementation

NFS is implemented as a mechanism using RPCs between UNIX
kernels and allowing any host to act as either a client, server or both.
The client application makes a request for a file operation in exactly the
same manner as it would for a truly local file. If the client kernel

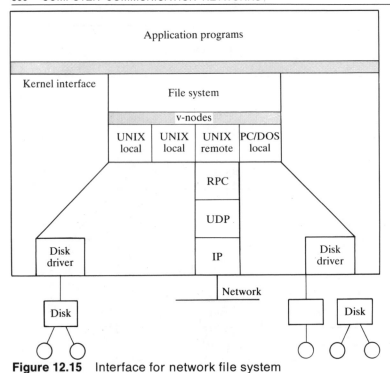

Figure 12.15 Interface for network file system

determines it is in a remote file system then it makes a request upon the remote host which is exporting that file's system.

The conventional UNIX file system uses *i-nodes* which are unique identifiers to access files within the file system (see Figure 12.14). The i-nodes exist on each UNIX file system and there is one for each file that stores information concerning attributes and permissions as well as references to the data blocks that comprise the file or directory. Every i-node is associated with a file system and has a unique number and files system identifier assigned to it. However, in the distributed file system there may be several files with the same unique numbers as the requested file. In the UNIX kernel the i-node structure of directories is fully understood since the directory structure is built into the kernel. To solve the problem of duplication of identifiers, Sun designed the virtual file system (VFS) (see Figure 12.15) based on the *v-node* which is a generalized implementation of i-nodes that are unique across file systems. The v-node consists of a triple comprising host_id, disk_id, and i-node. The v-node is not visible to the application since it is only used within the operating system itself.

The host acting as a server at any time is not responsible for retaining any state information on behalf of its clients. A client crash has no effect on the server, and a server crash results in nothing more than loss of service for a client. Kernels of servers are responsible for retaining a table of open files for each accessing program as for standard UNIX systems; however, the i-node is now replaced by the v-node number. Thus, the server kernel performs file operations for its clients by being passed the v-node number.

12.8.2 Sun's external data representation (XDR)

The data transferred using an RPC may be represented differently on different computers since not all machines store their data in the same bit order or use the same number of bytes to represent a quantity such as an integer. Thus for data to be transferred and correctly interpreted at the receiver it is necessary for the data to be packaged in an agreed format. The external data representation (XDR) standard is important for communication between mixed architecture machines and all of Sun's remote procedure calls transmit using the standard. The XDR was designed to work across different languages, operating systems and machine architectures. The XDR library solves data portability problems, allowing arbitrary C program constructs to be written in a consistent manner.

The Sun XDR library provides encapsulation routines for various data types including complex types such as structures. The programmer may use the basic routines provided in the libraries to encapsulate new objects that are more complex in nature.

12.8.3 Sun's administration service

Sun provides Network Information Services, formerly called *Yellow Pages* (YP), which is a network service to ease the job of administering networked machines. The YP is a centralized read-only data base containing sets of pairs of data, termed *maps*, in the form key and value. The YP is implemented on a master slave strategy, where updates of the data base are always performed on the master station first and then updates are propagated to the slaves. The master is master of a named domain and may have several slaves in that domain which may assume the role of master in the event of the master's demise. It is designed on the assumption that the number of updates is relatively small and the administered domain is largely static. A YP binder is used to locate the master server for the given domain. The YP protocol is used and includes operations to propagate maps and to query the timestamp data concerning maps. For a client in a particular domain on the network file system, the application's access to data served by the YP is independent of the relative locations of the client and the server. The YP data base on the server may provide password, group, network and host information to client machines but may be extended to provide other services as desired.

12.9 The Newcastle connection

The Newcastle connection (Brownbridge *et al.*, 1982) was a predecessor to the Sun approach of producing a distributed file system. Because of the similarity to Sun NFS a brief description is provided. The approach taken was simpler but required considerable modification to the UNIX operating system in order that applications could make use of the standard UNIX file primitives. The approach taken was to extend the standard UNIX file system tree hierarchy upwards to include a host name in the file system path. Thus a pathname to a file would become:

hostname/users/directory/file

Although the approach taken provides access to files of remote systems by mounting them into the local file system tree, it does not meet one of the state requirements of a distributed system, namely, it does not provide location transparency.

References

Adobe Systems Incorporated (1986) *PostScript Language Reference Manual*, Addison-Wesley, Reading, Mass.

ARPA (1970) 'Computer network developments to achieve resource sharing', in *Proc. AFIPS 1970 Spring Conf.*, L. G. Roberts and B. D. Wessler (eds), 543–549.

ARPA (1978) *ARPANET Protocol Handbook*, L. Feinler and J. Postel (eds), NIC 7104, Network Information Centre, SRI International, Menlo Park, California, January 1978.

Berkeley (1981) 'A 4.2BSD interprocess communications primer', supplied with UNIX Software Distribution, University of California at Berkeley.

Birrell, A. D. and R. M. Needham (1980) 'A universal file server', *IEEE Trans. Software Eng.*, **SE-6** (5), 450–453.

Birrell, A. D. and B. J. Nelson (1984) 'Implementing remote procedure calls', *ACM Trans. Computer Systems*, **2**, 39–59.

Boggs, D. R. (1982) 'Broadcast in an Internetwork', PhD Dissertation, Dept of Computer Science, Stanford University, California.

Brownbridge, D. R., L. F. Marshall and B. Randell (1982) 'The Newcastle connection or UNIXes of the world unite!', *Software—Practice and Experience*, **12**, 1147–1163.

Coulouris, G. F. and J. Dollimore (1988) *Distributed Systems, Concepts and Design*, Addison-Wesley, Wokingham.

Gray, J. (1978) 'Notes on database operating systems', in *Operating Systems: An Advanced Course*, R. Bayer, R.M. Graham and G. Seegmuller (eds), Lecture Notes in Computer Science, Vol. 60, Springer-Verlag, Heidelberg.

Lampson, B. W. (1981) 'Atomic transactions', in *Distributed Systems—Architecture and Implementation*, B.W. Lampson, M. Paul and H. J. Siegart (eds), Springer-Verlag, Heidelberg, 254–259.

Jones, S. T. and M. J. Clark (1990) 'Authentication in a UNIX network environment using smart cards', in *UK IEE IT Conference*, Southampton, March 1990, 32–37.

McQuillan, J. M. and D. C. Walden (1977) 'The Arpa Network Design Decisions', *Computer Networks*, **1**, August 1977, 243–289.

Mitchell, J. G. (1985) 'File servers', in *Local Area Networks: An Advanced*

Course, Lecture Notes in Computer Science 184, Springer-Verlag, Heidelberg, 221–259.

Mullender, S. (1989) *Distributed Systems*, ACM Press, New York.

Mullender, S. J. and A. S. Tanenbaum (1985) 'A distributed file server based on optimistic concurrency control', *ACM Operating Systems Review*, **19** (5), 51–62.

Needham, R. H. and A. J. Herbert (1982) *The Cambridge Distributed Computing System*, Computer Science Series, Addison-Wesley, Wokingham.

Popeck, G. and B. Walker (1985) *The LOCUS Distributed System Architecture*, MIT Press, Cambridge, Mass.

Reed, D. P. (1983) 'Implementing atomic transactions on decentralized data', *ACM Trans. on Computer Systems*, **1** (1), 3–23.

Scheifler, R. W. and J. Gettys (1986) 'The X window system', *ACM Trans. Graphics*, **5** (2), 76–109.

Sequent Computer Systems (1986) 'Balance technical summary', MAN-0 110–00, Sequent Computer Systems, Beaverton, Oregon.

Stern, H. L. (1987) 'Comparison of windowing systems', *Byte*, **12** (13), November, 256–272.

Sun Microsystems (1985) *Sun Network File System*, Sun Microsystems Inc., Mountain View, California. Complete details are to be found in the Sun NFS Protocol Specification Document.

Sun Microsystems (1987) *Sun Network-extensible Window System (NeWS)*, Technical Overview and Reference Manual, Sun Microsystems Inc., Mountain View, California.

Wilbur, S. and B. Bacarisse (1987) 'Building distributed systems with remote procedure call', *IEEE Software Eng.*, **2** (5), 148–59.

Xerox Corporation (1981) 'Courier: the remote procedure call protocol', in *Xerox Systems Integration Standards*, Xerox Corporation, Stamford, Connecticut.

13 Network management

ALWYN LANGSFORD

13.1 Introduction

Open systems interconnection (OSI) standards have been ten years in development, yet the subject of network management has become a significant part of this activity only relatively recently. It is not that the need has only just been realized. That was foreseen at the outset. Rather, the delay has been caused by the difficulty in agreeing:

1 What the scope of network management is.

2 Why it is needed.

3 Where its requirements can be supported through open management standards.

By seeking answers to these questions we shall obtain an overview of where OSI management standards are today and identify where further research and development is needed.

To be a viable facility, network management must be as appropriate for interconnected international networks (both public and private) to which many businesses are attached as it is for local area networks. In this context a business may be a national or multinational enterprise operating over one or several industrial/commercial sites, it can be government, an academic or similar research community. At each site we may postulate the existence of one or more local area networks, connected to each other and to networks at remote sites as dictated by local technical needs and management policies. This scenario helps to bring out the key features of network management:

1 There are many independent management domains which have to be able to interact meaningfully yet safely.

2 There are many areas which, while not truly independent have a measure of local autonomy.

3 There will be different communication technologies involved in the networking.

4 The networks can be large and they are a strategic component of many industries.

5   The networks will evolve over time.

Although much of this chapter concerns OSI management, (i.e. management of the communications environment), currently the subject of standards' activity, communications is but one of the services to be managed when information processing is distributed. Other information processing resources and activities need to be managed too, e.g. data storage, peripheral devices and processors. 'Network management' should properly include all these within its scope.

13.2 Management and networking

Business is dynamic. Consequently there is matching growth and decay of the networks that support the business. New requirements may lead to different types of equipment being connected to the network, or different networking technologies, with different behavioural characteristics (higher bandwidth, greater security, higher reliability), may be introduced to meet changing workloads. As old equipment becomes obsolete and harder (or less cost-effective) to maintain, it will be phased out to be replaced by newer technology. A network makes such evolution possible. Network management assists the evolutionary process and ensures that it is controllable.

Management concerns the *planning, organizing, monitoring, accounting* and *control* of resources and activities. Business enterprises establish policies and provide human structures to carry out (i.e. administer) the planning and organizational activities to effect a particular management policy. Although computers will probably be used to plan and organize, this use of information technology is usually regarded as lying outside the scope of network management. Rather, network management is more concerned with the day-to-day, operational aspects of maintaining a viable network environment; that is it is concerned with the accounting, monitoring and control functions of management.

A particular aspect of management policy in many systems concerns the availability and effective use of resources. The policy may also influence the security of information and its processing in the distributed processing environment. Where the policy demands availability, it is the task of network administrators to achieve this, for example by providing duplicate resources. Where the policy also demands economy, they will find cost effective ways of providing backup. Where security is important they will ensure that the network is operated in such a manner as not to compromise user and management data.

Whereas these tasks can in principle be carried out by people, the responsiveness demanded for monitoring and the complexity of management information is such that many of the tasks are handled by the distributed information processing system itself. The network management team interacts with the processing environment to effect management policies and to provide feedback to policy makers on network behaviour. In short, network management can be implemented as an interactive, distributed processing application where the application task is that of managing the distributed processing environment.

Not all parts of a large network need the same degree of management control, not all will be subject to precisely the same policies. The network environment is therefore composed of many domains of management. We define a *domain* as:

> A set of managed objects (i.e. resources and associated activities) that are grouped together, for the purpose of management organization, to support a particular aspect of management.

At the outermost level, domains are determined by the boundaries of different organizations. Within organizations there can be a hierarchy of subdomains, each responsible for a given subset of resources and activities. Domains may also be established on the grounds of delegated responsibility, security, operational reporting, resource naming, etc.

13.3 Modelling concepts in network management

A model for network management must cover three main aspects:

- The sphere of management responsibility.
- The delegation of management responsibility.
- The objects of management concern.

The principles of management authority and responsibility require that managers exercise control within their own management domain (and subdomains for which they have hierarchical responsibility) but that they cannot exercise control outside them. Yet, to provide effective management, managers will need information about activities in other domains which may influence their own domains. They will seek to access that information and even request the delegated authority to effect controlling actions on behalf of particular distributed applications. Management responsibility may be delegated within the management domain (ANSA, 1987; Sloman, 1987; Bacon *et al.*, 1988). A management entity lying within a domain and carrying out particular management activities is termed an *agent*.

These requirements lead to the following dialogue types:

- Manager to manager: to obtain information or (limited) authority for control.

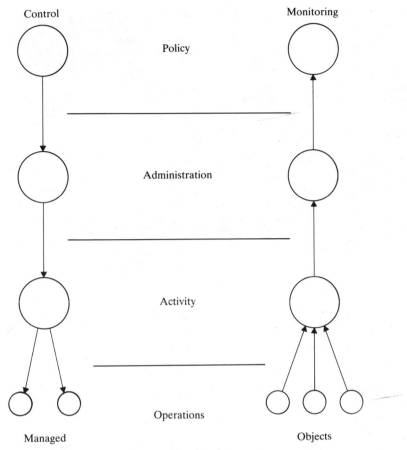

Figure 13.1 The ANSA four-layer management hierarchy

- Manager to manager: to request control be exercised on a manager's behalf.
- Manager to management agent: to invoke a management operation.
- Management agent to manager: to report the occurrence of a (novel) situation.

A collaborative project in the United Kingdom to develop an advanced network system architecture (ANSA, 1987) has adopted a four-level hierarchy of management in which administrators respond to the directives of policy makers and effect management activities. At the base of this hierarchy are the primitive management operations (Figure 13.1). Control is exercised down the hierarchy and management information flows in the reverse direction. This simple model can be further elaborated to allow for information filtering and for feedback at intermediate levels in the hierarchy (Figure 13.2).

When a domain is distributed, the interaction between managers and

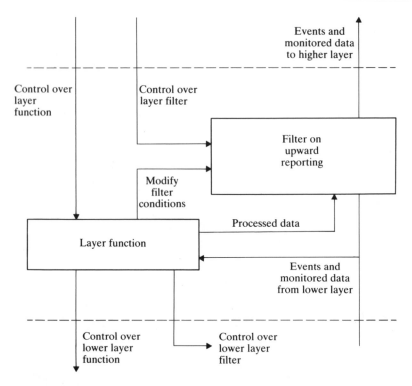

Figure 13.2 Information filtering and feed-back in the management model

agents becomes explicit. At the operations level there are no peer information flows but at higher levels there can be both peer communication and manager-to-agent communication.

At each level, control is exercised over managed objects and data values are obtained from them. Objects may be passive having values assigned to them or data read from them. Some objects may be active, i.e. changed by agencies external to management activity. Such objects may 'spontaneously' report changes in the values of their attributes to their managers.

13.4 Functional areas of management

It is generally accepted that network management covers five areas:

- Fault handling.
- Performance monitoring.
- Accounting.
- Security management.
- Configuration management.

Analysis of these areas identifies the management requirements, specific management functions and the types of managed objects involved. The following examples briefly illustrate this functional decomposition.

Fault handling is concerned with responding to situations when a system no longer meets its specification. For example, comparing a count of the successful number of transactions with a count of failed transactions will show if the error rate for a system has risen to unacceptable levels. A fault-reporting algorithm may contain a threshold to identify the level at which a fault report will be generated.

Performance monitoring records the actual behaviour of systems. For example, operations could monitor the values of the same two counters identified above to see what grade of service was being provided and if the system was meeting its management objectives. Alternatively, the rate at which transactions occur could be monitored to determine throughput.

Accounting management establishes the use that is made of the networked resources, essentially for the purpose of making a charge (though this could be a notional charge). Again, counters will be read that total the 'cost' of using a resource. Accounting management can also be used to exercise control over the use of resources by sanctioning only a given amount of resource to be used before warning that an accounting threshold has been reached (cf. the 'pay phone'). A further use of accounting management concerns the maintenance and communication of account records (cf. itemized phone bills).

Because management operations influence system and network wide behaviour, it is essential that they be properly controlled. Like many other distributed applications, network management will have to demonstrate that authority exists before certain operations can be performed. The need for access control and the authentication of management actions are not of themselves aspects of network management. Rather, *security management* is the set of activities that maintain the security of the environment by controlling the way access permissions are communicated and exercised. Security management also includes activities that monitor the success with which security policies are enforced by, for example, maintaining security audit trails.

Configuration management is needed because the network and the activities within it change. As the network evolves new resources are added and old ones are discarded. The detection and rectification of faults also affects the availability of networking components. Configuration management monitors and controls the following:

- The state of resources and activities.
- The relationships between resources.
- The introduction and removal of resources.

Status is characterized both by administrative and operational attributes. Administratively a resource can be *available*, *unavailable* or *under test*. Operationally it may be *active* or *inactive*. Relationships characterize resource alternatives and the interconnectivity of resources. In the networking environment, configuration management controls and reports upon the path or route taken to communicate between resources. These aspects of configuration management and the need to create or delete references to both types and instances of managed objects may involve passing considerable amounts of information between systems, for example to update control tables or specify revised management algorithms (e.g. as code updates).

These illustrative examples show that individual managed objects do not necessarily belong uniquely to a particular management functional area. Managed objects and their attributes, as typified by counters, gauges, reports, etc., are found in all aspects of management, and although some policy objectives may be satisfied by simple operations of the type described, many will require these simple operations to be compounded into more complex management procedures. For example, a report on the presence of a fault may lead to a check on performance. If this fell below the norm, it could lead to a change of configuration. The faulty component could be placed 'out of service' and made the subject of diagnostic tests to track down the nature of the fault in more detail.

13.5 Management standards

13.5.1 The scope of management standards

Users investing in distributed information technology want an open market for equipment in order to avoid being tied into a single vendor situation. They therefore welcome standardization which also brings the benefit of defined functionality and provides a yardstick against which to judge claims of conformance. These benefits of standardization apply equally to management standards though, because of the diversity of management requirement, the standards are both harder to define and to apply.

Standards apply to interfaces and functions, not to the way systems are implemented. Having identified management functions, what are the interfaces? Network management is a distributed processing activity which requires a symbiosis between people and their computing systems. We identify three interfaces of concern to network management:

1 User–system.

2 System–system.

3 Application program–system.

The first two require some form of command and response language (Beech, 1985). Although the semantic requirements of such languages are similar, the syntax of the first interface is oriented towards human expression. The second interface can be optimized for efficient communication between computer systems. The third interface also concerns language but this time it is the binding between conventional programming languages and the objects of concern to management. These allow systems programmers to write programs to carry out monitoring and controlling operations. All three interfaces provide mechanisms to express the activities which a system is required to perform but, whereas the first and third are handled locally, the second interface involves the communication of the management information between systems. Thus, system–system interfaces require us to consider two aspects of standardization: (1) the end system activity and (2) the open communications between systems. The latter are expressed through open systems interconnection standards.

This distinction between the standards needed to specify the management objects and their associated operations, on the one hand, and the communication standards, on the other, is being developed in a standard document which gives an *overview* of the set of OSI management services (ISO 10040). This standard is essentially concerned with the services to manage the OSI environment, that is the communications environment. However, research suggests that the principles it embodies are applicable to the wider aspects of management which are the subject of this chapter. An analysis of the requirements of the open communications environment shows that the OSI basic reference model (ISO 7498/1) has to be augmented to handle management. This extension to the reference model is specified in the OSI management framework (ISO 7498/4).

13.5.2 The OSI management framework

The purpose of this standard is twofold:

1 It augments the protocol architecture of the OSI basic reference model by describing the principles of communication in the OSI environment when applied to communication management activities.

2 It is prescriptive of some of the management standards that will be needed.

The basic reference model encapsulates the concept of protocol exchange between peer entities at seven layers. The service provided by each layer adds value to the communication service provided by lower layers and so provides an enhanced service to a higher layer. The highest layer specifies the application specific protocol. Network management is a distributed processing application so it is not surprising that the system–system communications discussed above should be conveyed by application layer protocol exchanges. This aspect of OSI management is defined as *systems management.* However,

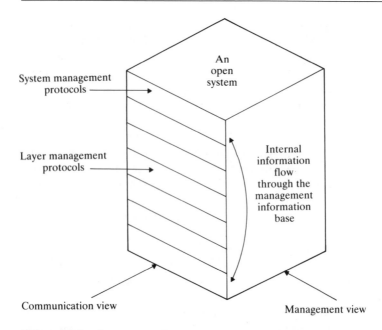

Figure 13.3 Systems management, layer management and the management information base

some management information is of necessity communicated also in lower layer protocols. This may occur during failure conditions, during system bootstrap or reconfiguration. To handle these management communications, the OSI management framework defines the concept of *layer management*. It specifies the circumstances when layer management is appropriate and the constraints upon its use. For example, a network layer operating the X.25 protocol can provide accounting information relevant to a particular instance of communication as a normal part of the layer protocol. Alternatively, in a local area network that offers a broadcast facility, it is appropriate to use data link layer protocols to identify which of the attached systems is acting as manager to control the operation of the network at that level. Systems management is however specified as the norm.

Whereas normal protocol exchanges ultimately serve interapplication communications, layer management exchanges are only between the systems connected at that level of protocol. However, the information being transmitted may be destined for a further system or may be transmitted as a result of a request from a distant system. Consider the example of system A requesting system B to place a remote modem (itself a system and an object of concern to management) into the loop-back test mode. The communication between A and B is a systems management exchange, that between B and the modem occurs through layer management protocol. How the information is handled within system B is an implementation matter. However, the OSI management framework provides an important modelling concept, the *management information base* to cover this type of communication. The management information base is a conceptual repository of information which is accessible for management purposes (see Figure 13.3).

13.5.3 Management information services

OSI management standards encompass the following:

- Those parts of layer protocol standards that help to maintain layer activity on behalf of all users of the layer service.
- Specifications of managed objects that are the subject of systems and layer management communications.
- A naming structure to identify managed objects.
- Abstract definitions of management information together with the specification of operations on the information.
- Procedures that may be invoked to carry out management activities.
- Operations to perform management tasks for a specific management functional area (such as fault handling).
- The definition of a common application layer service for communicating management operations between open systems.
- A protocol specification for that communication service.
- Mappings between the definition of managed objects and their operations onto their representation in protocol exchanges.

Of these standards, those for the common (application layer) management information service and protocol are the most advanced (ISO 9595, 9596). These standards set out the rules for open communication between a manager and a management activity operating as a management agent. The standards are independent of specific management procedures and are applicable to all management functional areas. They distinguish between communications that notify managers of event reports in the agent system and operations that the manager requests of the agent. The general form of the latter is:

< operation type > < managed object identifier > < access right token > < value >

where the operation type is currently defined as one of the following:

- *Get* the value of the identified object.
- *Set* the value of the identified object.
- Perform a given *action* on the identified object.
- *Create* a managed object of a specified type.
- *Delete* the identified object.

Other operations, now being appended to the above list, are those to *cancel* a get operation and to *add* and *remove* members from a set of managed objects. The *cancel get* operation has been introduced to deal specifically with the prospect that a large amount of data may be returned to a manager who has requested a *get* operation of an agent. When network traffic is particularly heavy or the manager does not wish to be overburdened, this operation allows the original *get* to be terminated. The operation may be cancelled at any time. The protocol will ensure that the manager will know, unambiguously, whether the *get* finished normally or prematurely. Because a *get* operation leaves the agent system unchanged, there is no uncertainty concerning the state of the agent system after an early cancellation. There are also proposals to be able to cancel *set* and *action* operations, but these are meeting with resistance because they would leave the state of the agent undefined. Although the common management information protocol is seen as the principal mechanism for the transfer of management information, there is no reason in principle why file transfer protocols should not be invoked for the transfer of very large volumes of information.

All operations and notifications may require the agent system to respond to the communication to acknowledge that the operation has been performed or the notification has been dealt with. However, these standards are not concerned with the consequences of the semantics of the end system operation; these are defined by the specific management information services for each functional area and by the definitions of managed objects.

Some of the standards relating to the fault and configuration management functional areas are well advanced and draft proposals for international standards have recently been registered. One of the difficulties of standardization in this highly integrated area of OSI management standards is establishing an effective structure for the standards documents and meaningful cross-references between standards. It would be impractical to publish the whole of OSI management as a single standard since management is a potentially open-ended activity and new aspects of the standards could be added continually. Even if a reasonable subset were to be identified, publishing all those together represents a major editorial task which would delay stable parts of the standard while the less developed parts were refined.

Two sets of documents help to maintain the coherence of this set of standards. One, the management information services overview, has already been mentioned. The other set recognizes that many objects and operations are common to several different management functions (see the examples given in Section 13.4). A group of documents defines the structure of management information (ISO 10165), the common information types, their attributes and the operations upon them. The documents provide a prescriptive method for specifying new managed object types. For each type the following information is needed:

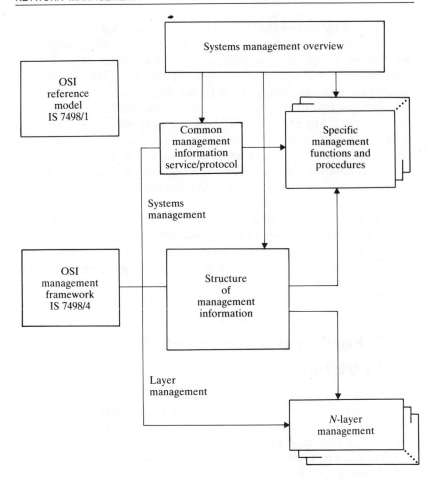

Figure 13.4 Interrelationship between OSI management standards

- The individual *items* of information that can be invoked by management.
- The *permitted* operations on those items.
- The *inherent* properties of the object.
- The *implicit* relations with other information elements.
- The *specific* properties for each instance of an object type.

The documents define such objects and attributes as: status, counter, gauge (meter), threshold, event and log.

The relationship between OSI management standards is shown diagrammatically in Figure 13.4. One aspect of OSI management not included in this figure is the way conformance to management standards is defined. This area still needs further exploration since not only must implementations of the standard conform to the rules of the protocol exchanges but they must also conform to the definitions of the semantics of operations that take place in the end systems where management operations are performed. This presents an interesting problem since tests for conformance cannot expect common implementations of particular management operations. Therefore, indirect methods are needed to test for conformance to a distributed management algorithm and demonstrate that the desired operations have been performed correctly. The difficulties of testing for conformance are further compounded by the nature of the management operations themselves. For example, when testing for conformance to standards related to fault reporting, it is necessary to cause a system to simulate faults to carry out the specified procedure. However, responses may be generated spuriously when the system under test is itself faulty. Also, how may one test for performance without compromising the performance one is seeking to test? Even more significantly, how does one test security management functions without compromising system security? Finally, will it be possible to check reconfiguration procedures in such a way that recovery is still possible even if the implementation of the standard under test fails to conform to requirements?

13.6 Further developments in open management

Those developing OSI management standards are well aware of the above issues and that they are only scratching the surface of the possible set of management activities that may need to be expressed as standards. So far, most operations and most objects are very simple as are their mappings onto management protocol. Also, they represent only those objects of concern to the open communications environment. It is the author's contention that, in the long run, all aspects of management have to be handled in a coherent manner whether they relate to communications objects, processing resources and activities or to data storage. It will not be possible for managers to monitor and control their distributed processing environments effectively if these areas of their responsibility are handled separately.

Research into distributed systems is now sufficiently mature as to show the appropriate architectures for distributed systems. This is now being complemented by the development of standards for open distributed processing (ODP). More recent work (Bacon *et al.*, 1988) has extended these architectural approaches to encompass management by implementing distributed systems management activities in a wide area networked test bed (Holden *et al.*, 1988; Van Renesse *et al.*, 1988). As a result, there is growing confidence that the current approach to OSI management is capable of handling all aspects of management. This requires that the current standards are developed with an eye to their eventual extension. Failure to do so runs the risk of constraining the development of open distributed management standards or, worse, of developing mutually incompatible sets of standards requiring different implementation approaches. The consequences for manufacturers would be increased development cost and this would be reflected to users in the costs of products.

What then must be done to extend OSI management? Firstly, the set of objects of concern to all of management and the operations upon them must be defined. Secondly, an extended set of management procedures is needed, though these are still likely to fall within the five functional areas of management that have been established. Then the objects, their operations and the procedures have to be mapped onto the services needed to communicate the management requirement between systems. This will call for extensions to the set of common management information services and protocol. One type of extension will allow for typical control structures to be communicated, for example to express conditional constructs or to request that a set of operations be repeated. Since these constructs can be nested, the syntax of information transfer will take on some of the aspects of a block structured language. A second type of extension will concern the conditions under which groups of activities will be performed. It will define relative priorities for remote operations and whether a group of operations shall be processed as an atomic action (i.e. all operations in the group proceed to completion or none are done) or whether a more relaxed processing regime is permitted. Many of these extensions will add to the functionality of the local operating system enabling its functionality to be extended over the distributed processing environment. This will assist in realizing the concept of a distributed operating system when interworking between existing systems with conventional operating systems.

One further benefit which is implicit in this approach to the development of open distributed management is that it handles the open endedness of management requirements and the need to allow managers to manage their systems according to their own policies; that is there is no requirement to specify a standard way in which management is performed. Drawing upon the analogy of a programming language we see that, by adopting a language structured approach to management communications—with a well-defined syntax and semantics—we can generate any management procedure *in a standard way* and have it communicated and interpreted *according to a standard* so that it is understood, without having to standardize the procedure itself (except in the case of particularly common procedures). With such a language, procedures for remote operation can be specified for immediate action or transferred as named procedures to be invoked by a subsequent *action* primitive.

13.7 Current practice

Although progress towards management standards has been slow and there are, as yet, only proposed functional standards for OSI management, manufacturers have recognized the need for some management tools to handle their proprietary networking products. These provide facilities for gathering performance related data, carrying out tests and performing reconfiguration of network resources. The products are largely related to managing the communications infrastructure though some local area networking products for personal computer and workstation systems do permit their users to manipulate remote processing and storage resources.

Space does not permit a full analysis of existing products (and such a list would soon be out-dated as new products emerge). However, the current ranges of products tend not to integrate their management functions into a single package. Rather, management is performed by a loosely coordinated group of software packages. Nor is communications management coordinated with distributed processing or distributed data management. Many of the current products are concerned solely with the local area communications network on which they reside. They are frequently installed within an IBM personal computer (or equivalent) and observe network traffic as it passes; that is on a CSMA/CD network the interface would be instructed to operate in a 'promiscuous' mode, listening to all packets on the network cable. A similar operation is possible on token ring networks and though the packet is not processed by the management computer, it is able to monitor all traffic. Such management devices are well able to produce network traffic statistics, report packet collisions and detect ring failures. They are essentially monitoring devices. Some control can be exercised where such systems are used as filtering bridges between network segments. Also, where networks provide mechanisms for terminal multiplexing, some manufacturers offer proprietary products that restrict the services that may be accessed from a given terminal or by a given user at a terminal.

In some manufacturer's offerings the different aspects of management may have to be performed from different network management devices. For example, some of the products that support the interconnection of local area networks (e.g. Ethernets and token rings) through fibre optic cable only allow the cable environment to be managed from a special management terminal. Management of the attached local networks must be done from other terminals while management of attachments (e.g. terminal concentrators) must be done from yet another control system.

These are implementation matters, for even if there is a coherent set of management standards, manufacturers will not be obliged to place all their management functionality into one box. However, offering a coherent set of standards increases the likelihood of their doing so and, having a compatible set of management standards, allows the users to purchase networking products knowing them to be accessible, for management purposes, in a common way.

13.8 Summary

Within this short chapter, it has only been possible to touch upon the main features of network management. More detail can be found in the documents that have been referenced. The aim has been to convey three main points about management:

- That management is a complex, highly interrelated set of concepts which is being supported by a highly interrelated set of standards.
- That more research is needed to show the best ways to realize network management.
- That although much remains to be done to realize those standards and the products that will conform to those standards, rapid progress is being made.

We commented in the introduction that network management has been slow to arrive. However, it is well to recall that research into the basic mechanisms of computer networks went on for many years, supported by several national and international research networks (e.g. ARPANET (Kahn, 1972) and the European Informatics Network (Barber, 1976)) before the layered structure of OSI was recognized and an international consensus was obtained. Only recently has there been the corresponding intellectual underpinning of network management by international research teams from which the corresponding consensus is rapidly forming.

With such a rapid rate of development, any detailed description of the specifics of management, as it is today, must rapidly become outdated. With this as rationale, the author has concentrated upon the structure of network management: for though details may change with the realization of management standards and their supporting products, the principles of network management will remain constant.

3.9 References

ANSA Reference Manual, (1987) ANSA, Poseidon House, Castle Park, Cambridge, June 1987.

Bacon, J. C. Horn, A. Langsford, S. J. Mullender and W. Zimmer (1988) 'MANDIS: An architectural basis for management', *Proc. European Teleinformatics Conf. (EUTECO'88) on Research into Networks and Distributed Applications*, Vienna, North-Holland, Amsterdam, 975.

Barber, D. L. A. (1976) 'The European Informatics Network—achievements and prospects', in *Proc. ICCC76 Conf.*, Toronto, 1976.

Beech, D. (ed.) (1985) *Concepts in User Interfaces: A Reference Model for*

Command and Response Languages, Lecture Notes in Computer Science, Springer—Verlag, Heidelberg.

Holden, D. B., O. Anhus, T. Fallmyr, J. Hall, R. Van Renesse, J. M. Van Staveren and G. Skogseth (1988) 'An approach to monitoring in distributed systems', *Proc. European Teleinformatics Conf. (EUTECO'88) on Research into Networks and Distributed Applications*, Vienna, North-Holland, Amsterdam, 811.

ISO 7498/1 (1984) Information Processing, 'Open systems interconnection—basic reference model'.

ISO 7498/4 (1989) Information Processing, 'OSI management framework'.

ISO 9595 (1990) Information Processing, 'Common management information service'.

ISO 9596 (1990) Information Processing, 'Common management information protocol'.

ISO 10040: ISO/IEC JTC1/SC21 N4865 (1990) 'OSI management overview', June.

ISO 10165: (three parts) (June 1990):

ISO/IEC JTC1/SC21 N5252 'Structure of management information—Part 1: Management information model'.

ISO/IEC JTC1/SC21 N4867 'Structure of management information—Part 2: Definition of management information'.

ISO/IEC JTC1/SC21 N4852 'Structure of management information—Part 4: Guidelines for the definitions of managed objects'.

Kahn, R. E. (1972) 'Resource-sharing computer communications networks', *Proc. IEEE*, **20**, 1397.

ODP: Document Register and Bibliography (1989) ISO/IEC JTC1/SC21 N4021 (December), gives detailed references to this developing standard.

Sloman, M. (ed.) (1987) 'Distributed systems management', available through the author from the UK Special Interest Group in Distributed Systems Management.

Van Renesse, R., J. M. Van Staveren, J. Hall, M. Turnbull, B. Janssen, J. Jansen, S. J. Mullender, D. B. Holden, A. Bastable, T. Fallmyr, D. Johansen, S. Mullender and W. Zimmer (1988) 'MANDIS/Amoeba: A widely dispersed object oriented operating system', *Proc. European Teleinformatics Conf. (EUTECO'88) on Research into Networks and Distributed Applications*, Vienna, North-Holland, Amsterdam, 823.

14 Voice/data integration—some performance issues

JOHN ADAMS

14.1 Introduction

14.1.1 Overview of this chapter

Using efficiency and delay criteria, different LAN access protocols, i.e. the CSMA/CD random access bus, token passing rings and slotted rings, are compared. The need for hybrid LAN designs which use a mixture of dedicated access for voice and polled access for data sources is also considered.

In terms of efficiency and delay it is shown that a slotted ring incorporating destination deletion (i.e. the slot is emptied at the destination) is more effective than token passing rings or CSMA/CD. Its efficiency compares favourably with hybrid LAN designs, but the absence of voice packetization delays on a hybrid LAN always gives the hybrid a superior voice delay performance.

Examining the influence of the emerging standards for broadband ISDN (B-ISDN), it is clear that there is a *window* for hybrid LANs which will close when the narrowband network disappears and voice is routed entirely within the B-ISDN. The delay advantage of hybrids then disappears and it is likely that only pure asynchronous transfer mode (ATM) LANs, perhaps based on the slotted ring principle, will be used as the basis of multiservice integration.

14.1.2 Factors influencing LAN access protocols

A local area network (LAN) may be defined as a network linking users within a limited area which supports some type of communications processing and transparent information transfer. LANs have been described in detail in Chapter 7, and an introduction to their performance for data sources is given in Chapter 8. A general characteristic of most LANs is that a mixture of services with different bit rates can share the available bandwidth through the operation of an appropriate media access protocol. The choice of which access protocol to use has been an active study area for some years, taking into account such factors as:

1 The network topology.

2 The range of service bit rates.

3 The delay tolerance of each service.

4 The error tolerance of each service.

5 The total load to be carried.

6 The degree of imbalance in the load distribution.

7 The requirements of overload control.

8 The efficiency of bandwidth utilization.

9 The number of attached nodes.

The number of factors involved reflects the large range of uses that have been identified for the LAN. At one end of the range are LANs that are isolated from other networks and that are required to handle purely delay-tolerant data connections with modest bit rates. Such LANs may utilize transmission rates that are of the order of 1 Mbit/s. Increasing the transmission rate by an order of magnitude implies that a larger number of terminals can be attached to the network and a wider range of service bit rates can be accommodated. Because there are more terminals, these LANs are also more likely to have a bridge or gateway connection to another network, giving rise to a potential long-term imbalance in the load distribution. Increasing the transmission rate to around 100 Mbit/s takes us into the realm of high-speed LANs, for which the principal interests are in having high point-to-point bandwidth availability and an integrated services capability, including the ability to handle a large number of voice connections.

The motivation for studying LAN access protocol performance is to determine the area of applicability of a particular protocol taking into account each of the factors mentioned above. The access protocols that will be discussed here can be grouped into three categories, namely:

1 *Polled access protocols*, implemented either using a circulating token or using circulating empty slots.

2 *Random access protocols*, especially CSMA/CD.

3 *Dedicated access protocols*, where each connection is assigned a fixed amount of bandwidth and this capacity is unavailable to other connections for the duration of the call.

Case 1: no collision—short acquisition time

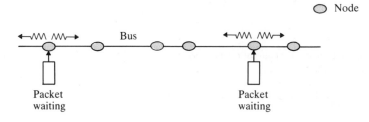

Few active nodes = few collisions = smaller acquisition time

Case 2: collision—acquisition time incomplete

More active nodes = more collisions = greater acquisition time

Figure 14.1   Random access acquisition time

This last type of access is normally employed in a *hybrid LAN*, i.e. a type of LAN that uses different access protocols for different services. Usually it combines dedicated access for delay-sensitive services and polled access for other services, especially data services.

For different choices of LAN speed, number of attached nodes and level of imbalance, the above classes of access protocol will be compared with respect to delay and efficiency criteria.

14.2 Efficiency considerations

Once bandwidth has become free an interval of time will elapse before the protocol has established which node can transmit next. This delay reduces the amount of useful load that can be carried and leads to a measure of the efficiency of the protocol, which an be expressed as

$$E = \frac{\text{transmission time}}{\text{transmission time} + \text{acquisition time}} \qquad (14.1)$$

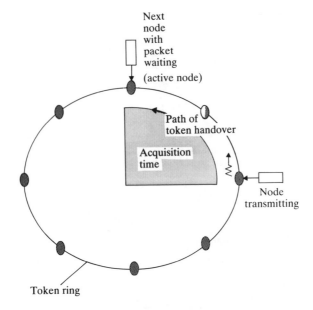

More active nodes = shorter path for token handover
= shorter acquisition time

Figure 14.2 Polled access acquisition time

In this expression, the *acquisition time* refers to the expected time required to transfer an entitlement to transmit from one active node to the next. The 'transmission time' refers to the expected time required to complete each transmission.

For random access protocols, the acquisition time lengthens as the number of nodes waiting to transmit increases (see Figure 14.1), and this produces a corresponding deterioration in efficiency. This loss of efficiency becomes significant if the acquisition time approaches a value similar to the transmission time. With CSMA/CD it can be shown that when many nodes have packets to send the acquisition time is approximately $5.4L$, where L is the end-to-end propagation delay (Metcalfe and Boggs, 1976). Thus, for a given packet length of P bits, the acquisition time and transmission time will be equal when

$$\frac{P}{C} = 5.4L \qquad (14.2)$$

where C is the channel transmission rate in bits per second. This formula shows that efficiency problems become significant on high-speed LANs and on LANs that have a long end-to-end propagation delay.

In contrast, for polled access protocols the acquisition time reduces as the number of nodes waiting to transmit increases (see Figure 14.2).

This is true of both slotted and token systems. For example, consider a token ring where N nodes each transmit a packet of length P bits (including the token). If the ring is of length R bits the token completes one cycle of the ring in a time S given by

$$S = \frac{NP + R}{C} \tag{14.3}$$

However, the fraction of this time spent transmitting the N packets is just NP/C. Hence the efficiency E is:

$$E = \frac{NP}{NP + R}$$

that is

$$E = \frac{P}{P + R/N} \tag{14.4}$$

The second term in the denominator represents the acquisition time and exhibits the characteristic that it reduces with increasing node activity. However, its dependence on R shows that efficiency is again a significant problem on high-speed rings and on long rings.

For both random access and token access protocols, equation (14.2) or (14.4) implies that an increase in P (i.e. the transmission length) improves the efficiency. This fact is usually exploited in the design of high-speed channel access protocols to overcome the efficiency problems created by the acquisition time. This means that a node on a high-speed token ring or random access bus is allowed to transmit several packets before releasing the channel to other nodes. However, this introduces a certain amount of deliberate *hogging* into the protocol, i.e. a node is allowed to continue transmitting packets that have just arrived, even though it is delaying other nodes from gaining access where packets have been waiting longer. The success of this technique for overcoming inefficiency is therefore limited by delay considerations. This aspect will be examined in more detail in the next section.

For slotted rings, the transmission length depends on the size of each empty slot. Slots are usually small to avoid excessive packetization delays (especially for voice) and so maintaining efficiency on high-speed rings requires further examination. Consider first the Cambridge slotted ring protocol (Wilkes and Wheeler, 1979), where if a node fills a slot it must wait for it to return before being allowed to seize another empty slot. Although this simplifies the retransmission of rejected *minipackets* (the term often used for describing the contents of each slot) it also reduces slot utilization because of the enforced reduction in node activity (i.e. the effective size of N in equation (14.4)). This effect becomes more significant on long rings or high speed rings.

To minimize the acquisition time on high-speed slotted rings it is necessary to remove some of the restrictions on slot access. For

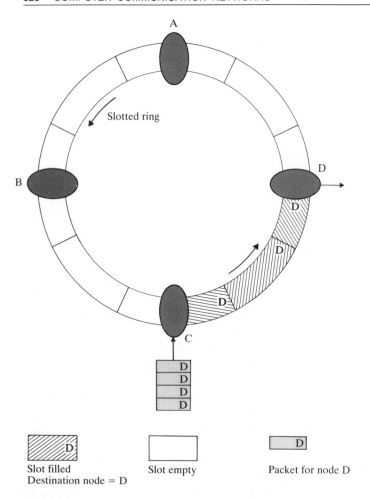

Figure 14.3 Destination deletion on a slotted ring

example, it is not considered necessary or desirable to retransmit voice packets, firstly, because of the extra delay introduced given that voice has a poor delay tolerance, secondly, voice coded at 64 kbit/s exhibits a relatively high degree of error tolerance. For voice minipackets, at least the access protocol should be such that a node is not restricted from seizing a second empty slot if it already has a voice minipacket circulating round the ring.

While the removal of such restrictions has important efficiency benefits, the major improvement comes from a different concept, namely *destination deletion*. This is where the slot is emptied at the destination node allowing it to carry more than one minipacket in any one ring cycle (see Figure 14.3). This means that, unlike token and random access LANs, the combined transmission rate from all nodes can exceed the channel rate. An example of this is the Orwell protocol

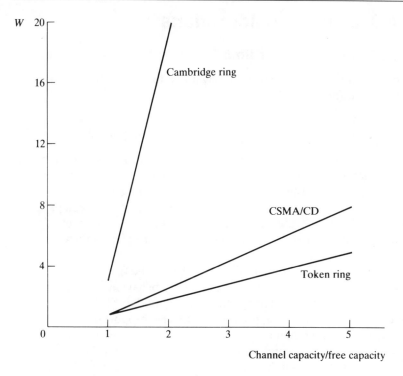

Figure 14.4   Transfer delay characteristics (for a channel rate of 1 Mbit/s and packet length of 1000 bits)

(Adams and Falconer, 1984) described in Section 14.3.3. Simulation results have shown that the ring is able to support a combined transmission rate in excess of 1.4 times the channel rate.

A hybrid ring-shaped LAN with high efficiency would combine time slot access for voice with destination deletion slot access for data. The time slots would be available for reuse by each destination node on the return direction of all voice connections. However, the reason for choosing a hybrid arrangement would not be because of any extra improvement in efficiency since the benefits would be marginal. The reasons are to do with delay, as discussed in the next section. A particular efficiency problem for hybrids is the control of the capacity allocated to the voice side. It may be difficult to reassign to data connections any unused capacity allocated for voice and this reduces the net efficiency.

14.3 Delay considerations

14.3.1 Mean transfer time

To compare the delays introduced by different access protocols a suitable performance measure is

$$W = \frac{\text{mean transfer time}}{\text{mean packet transmission time}} \tag{14.5}$$

where the *mean transfer time* is defined as the time interval from the generation of the packet at the source node until its reception at the destination node. The *mean packet transmission time* refers to the time needed to transmit a packet whose length is equal to the mean packet length, but not including any allowance for the transmission of packet overheads introduced by the access protocol. The ideal is therefore $W = 1$.

With this measure, systems like CSMA/CD, the Cambridge ring and token rings may be compared for their delay behaviour as a function of increasing load. Using mathematical modelling, Bux (1981) shows that in each case W varies in approximately inverse proportion to the free capacity (the latter quantity is, more precisely, the difference between the channel rate and combined node transmission rate). The constant of proportionality is such that, at a channel rate of 1 Mbit/s, CSMA/CD and the token ring have very similar values of W over the entire load range. However, the Cambridge slotted ring produces significantly larger transfer delays (see Figure 14.4) for two reasons:

1 *Large packet overheads* Control and routing information must be carried in every minipacket that increases the length of the transmission, given that a complete packet consists of several minipackets.

2 *Restricted slot access* At any node each minipacket is delayed until the slot containing the node's previous minipacket has returned.

The question is whether delays are greater on any type of slotted ring, compared with token rings or CSMA/CD. This is particularly important for the higher-speed LANs which are assumed to be carrying voice. To answer this question it is important to first understand what changes must be made to LAN design to accommodate voice.

14.3.2 Voice packets

It was stated earlier that small packets are necessary for voice connections to keep packetization delays low. The main reason for this is because there is a quality degradation due to echo which is perceived as worse when the delay in the echo path increases. On long distance international calls this degradation is so severe that it must be removed by an echo control device at the international switching centre. Such devices are effective for any customer making an international call provided the customer's equipment meets specified delay limits, otherwise the customer's equipment must include its own echo control. Typically the private branch exchange (PBX) or LAN is limited to about 5 ms mean one-way delay between the telephone and the connection to the public network (if the equipment does not include echo control).

Allowing for other delays within the customer's premises (e.g. propagation), this suggests that 64 kbit/s voice should be carried in packets with about 24 octet samples as a maximum (excluding the header), to minimize the cost of echo control. Given that the choice of the slot length for any new standard for a voice/data LAN is likely to be based on broadband ISDN (48 octets of information plus a 5 octet header), the slots will only be half filled with 24 octets of voice. Other types of LAN, e.g. a token ring, can avoid the inefficiency of half filling because of their variable packet length structure. Hence the slotted ring loses some of its efficiency gains (i.e. the gains from destination deletion) if the dominant part of the load is voice. This was mentioned as one of the two factors giving rise to a larger value of W on slotted systems, but its main effect is to reduce the gains from destination deletion back to about the same efficiency as the token ring. What about the second factor influencing W, i.e. delays due to access control?

The access control on a voice LAN must meet the tight limits for voice delay mentioned above. Of course there is a trade-off between the packetization delay, network size (i.e. propagation delay) and access delay within an overall limit of, say, 5 ms mean one-way. However, because of the desire to limit the inefficiency caused by partial filling of a 48-octet information field and to allow for large-sized networks, the access delay must be very short. A mean access time of about 10 μs or less is desirable since the effect of such a small delay is to reduce not only the switching delay but also the delay variability end-to-end. Delay variability becomes added into the delay budget at the far end where a buffer is needed to smooth out the rate of packet arrivals and play back a constant 64 kbit/s. A short mean access time of about 10 μs implies that this buffer is equivalent to only a few hundred microseconds of extra delay.

Low access delays should hold even when there is a significant level of imbalance in the load distribution. It should also hold when the LAN is subjected, temporarily, to heavy bursts of load from other services. The latter point implies that it is not enough to rely on the underutilization of the shared channel to meet voice delay requirements. Instead there is a need to make changes in the access protocol to accommodate the case of general mixed services.

The most appropriate form of delay control is to impose a threshold, T, such that the sum of the transmission times of all nodes is less than

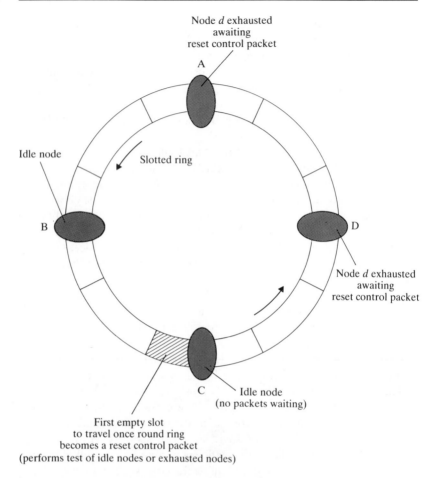

Figure 14.5 Orwell slotted ring—access control

T. This allows each node to have a different share of the bandwidth depending on its needs while ensuring that all nodes have some opportunity to transmit in any interval of length T. If this procedure is to include all nodes, rather than all currently active nodes, then an allowance of a certain minimum amount of bandwidth for every node must be deducted from the total. The remaining bandwidth is then allocated according to the above flexible principle. A node can only increase its allocation if this does not cause T to be exceeded. This allows other nodes to retain their current bandwidth allocations.

14.3.3 Orwell slotted ring protocol incorporating delay control

The Orwell protocol controls delay by allowing each node to transmit up to d minipackets, where d varies from node to node. After all nodes have had the opportunity to transmit (see Figure 14.5), the d counters are reset by a control minipacket which travels once round the ring. A feature of the protocol is the fact that the counters are reset at least once in each interval T by controlling the magnitude of the values d. Thus load can be added and d subsequently increased on a node provided this does not cause T to be exceeded.

A fuller description of the protocol follows. Nodes recognize slots as either full or empty according to the setting of a control field in the slot header. Up to d empty slots may be seized by any node. In each case the control field is converted to 'full' and a destination address is inserted in the slot header. The destination address is derived either from look-up tables in each node (especially for connection-oriented services) or via signalling to a central look-up table (e.g. for connectionless services).

The slot control field is converted to 'empty' as it passes a node matching the address in the destination part of the header. The empty slot is immediately available for other nodes to seize (destination deletion), but also performs a second important function. This is the concept of a 'trial' or 'test' for packets waiting for access.

If the empty slot goes fully round the ring, the value in its destination field is once again recognized by the node that originally emptied it. The fact that it remained empty for the entire ring cycle is an indication of either of the following:

- There were no packets waiting at other nodes.
- Where packets were waiting, the node had used up its d value.

Hence if the destination node also has no packets waiting (or has used up its d value) it converts the empty slot into a 'reset' slot by changing the control field in the header. This slot now circulates round all nodes and back to the destination node again. As it passes each node the d allocation is restored. On arrival back at the destination node it is finally converted to an empty slot. Thus the node responsible for initiating a 'reset' minipacket is fully distributed as a function performed by all nodes.

A node may increase its d allocation provided that the current reset rate (observed independently by each node by noting the frequency of passing 'reset' slots) is such that it exceeds the minimum frequency with some safety margin. The Orwell protocol thus provides feedback to each node on the current level of loading (derived simply from the reset rate) so that call acceptance can also be distributed to each node.

14.3.4 Comparison of delay control in token and slotted rings

On token passing rings delay control is applied by ensuring that the token comes back at least once in each interval T. The length of time until the token returns to a node is known as the token rotation time

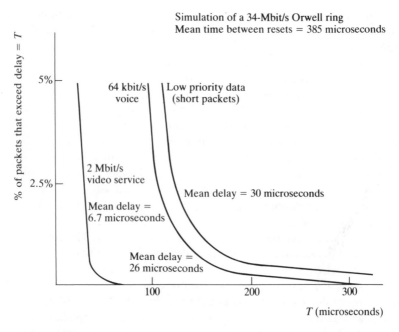

Figure 14.6 Orwell delay characteristics

(TRT) and is controlled by setting the maximum length of time that the token can be held by each node. The token holding time on any node is directly comparable to the Orwell d value since it can have a different value to other nodes according to demand, and it is set long enough to allow all of the delay-sensitive packets to be transmitted (using the peak bit rate of such services to determine what the setting should be).

An important performance difference between the slotted and token ring versions described above is the access delay time. This was the second factor to be considered in the delay expression W quoted earlier. For any given value of T, the limiting case for the token ring is just one reappearance of the token every T milliseconds. Each node's transmissions are then spaced T milliseconds apart. For the slotted ring, the limiting case of one reset of the d counters every T milliseconds does not imply that transmissions are spaced T milliseconds apart. Instead, transmissions from any one node occur throughout the interval T, on the basis of one slot at a time whenever an empty slot can be found after the next minipacket has arrived. The result is that, when small packets arrive (i.e. the packet length is equal to the slot length), the access waiting time is significantly reduced below the value $T/2$ (this being approximately the mean access waiting time on the token ring when packets arrive randomly throughout the interval T, given that each node must wait T milliseconds between successive arrivals of the token).

From simulation results of a 34-Mbit/s Orwell ring (see Figure 14.6), when the mean time between successive resets of the d counters is

385 μs, the mean access waiting time for voice packets is only 26 μs. Less than 0.5 per cent of all voice packets were delayed by more than 150 μs. For a token ring with a mean token rotation time of 385 μs, the mean packet delay time would be approximately 190 μs, and some packets will experience delays approaching 385 μs.

Some explanation is necessary for the differing delay results for 'video' packets and 'data' packets in Figure 14.6. In addition to the features of Orwell already described, this type of slotted ring operates a priority mechanism such that, when an empty slot arrives, a packet with the highest priority is selected by the node for transmission. This is why the video packets (highest priority in this simulation) have the smallest delay and data packets (lowest priority) have the largest delay.

Actually the packet selection mechanism is more complex since each service queue (e.g. the video queue) also has a d allocation in addition to a priority. This prevents a bursty service with intermediate priority from stealing all the bandwidth from the lowest priority queue (this is the concept of 'minimum guaranteed bandwidth' per service queue). Thus a packet is selected for transmission from the highest priority queue with some remaining d allocation.

These considerations have shown that for high-speed multiservice LANs carrying a considerable amount of voice, slotted rings out-perform token rings or CSMA/CD in terms of access delays. There remains the question of the slotted ring versus hybrid debate for high-speed multiservice LANs. The advantages of the hybrid are reduced delays, because there is no packetization delay and no variation in delay to be smoothed out. Given that slotted rings do not carry the voice load as efficiently as the hybrid, why should hybrids not emerge as the preferred solution to high-speed voice/data LANs? This question is taken up in the next section.

14.4 Broadband ISDN issues

A slotted ring is one example of an asynchronous transfer mode (ATM) switching fabric, i.e. it performs routing decisions on the basis of the connection number read from each packet header, where the packets are all of fixed length. The packets are multiplexed together statistically rather than being assigned fixed positions in a frame structure. Much work has lately gone in to the study of ATM as the solution for integrating a wide range of services, both broadband and narrowband, within a single new digital public network, i.e. the so-called broadband ISDN (B-ISDN).

B-ISDN is being studied as a means for providing:

- Switched connections for high-speed or bursty services.
- Variable-capacity cross-connection (e.g. between two LANs). This concept is explained in more detail below.
- A flexible network for the introduction of new services. Not only

broadband services but also low bit rate bursty services are foreseen (e.g. some video services).

- High-speed relaying of signalling messages using destination-addressed ATM packets. This concept supports highly distributed processing environments as may emerge, for example, from the greater use of mobile terminals involving the storage and retrieval of a customer's current location.
- A single network for all services (including existing services).

Some further explanation is necessary on what is meant by the concept of variable-capacity cross-connection. If only part of the connection number field in the ATM packet header is used for routing the packet through switches, it is easy to set up multiple connections following the same routing (e.g. from one LAN to another).

Thus, if another connection is to be established between two LANs, it is only necessary to choose a connection number with the correct routing subfield (this could equally be a separate field to the connection number in the ATM packet header). B-ISDN switches can be preprogrammed to switch some ATM packets only on the basis of the value of this subfield.

Multiple connections can therefore be established and correctly routed without providing individual connection routing information to the network. The advantage is greater simplicity in terms of the network administration coping with changes in the number of connections a customer requires, and faster more flexible response to his or her needs. This could be done without ATM by establishing, say, a 2-Mbit/s cross-connection between the two LANs using conventional switches and then allowing individual connections to be established or cleared down within this pipe. However, the customer has to pay for 2 Mbit/s even if he or she is only using a fraction of this capacity at any one instant. Also there may be occasions when 2 Mbit/s is not enough capacity. This leads to the concept of variable-capacity cross-connections.

Because ATM switches are not fixed on any one bit rate they instantly take account of variations in the customer's needs. Therefore there is no need to set up fixed-capacity pipes between two LANs and a customer can be charged more closely on the basis of actual utilization. Using ATM, a variable-capacity cross-connection can be set up simply by choosing the correct routing label and possibly a minimum level of guaranteed bandwidth. (This concept was also discussed in relation to the Orwell protocol which can be adapted to provide a minimum guaranteed bandwidth for cross-connections using d allocations in the same way that it provides minimum guaranteed bandwidth per service queue—see Section 14.3.4.)

In summary, the major advantage of ATM is flexibility, the ability to support both existing bit rates and new services with different fixed or

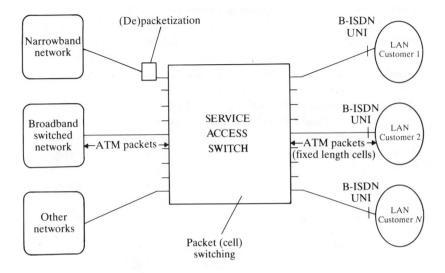

UNI User network interface

Figure 14.7 Introduction of broadband ISDN

variable bit rates chosen over a very wide range. A key element is its
support of variable bit rate services, since the first demand for
broadband services in many countries is likely to be high-speed data.
Later on there may be new variable bit rate video codecs suitable for
video telephony.

Following CCITT recommendation I.121, ATM has been chosen as
the target solution for implementing a B-ISDN. This implies that the
interface between the LAN and the broadband network will consist of
an ATM *cell* (i.e. packet) organized payload for all services.

One way of carrying voice within an ATM network is to assemble
composite packets, i.e. each octet of the cell information field is used by
a different 64 kbit/s connection. These can be routed within B-ISDN via
a service access switch (SAS) to the narrowband network (see Figure
14.7) where the connections are individually switched to different
destinations. The advantage of this is the reduced packetization delay
(i.e. only 125 μs to obtain 1-octet samples from several separate
64-kbit/s connections). Although it was stated in Section 3.2 that
half-filled packets would satisfy the delay budget of up to 5 milliseconds
one-way within the customer's premises network, this does not make
much allowance for propagation delays. Thus the use of composite
packets is an option that is appropriate for very large customer's
premises networks.

A hybrid LAN which is compatible with the B-ISDN interface would
combine both cell switching and octet switching, where the latter is

done within specially recognized cell-sized composite packets. Conditions are clearly most favourable for such a hybrid LAN while the octet-switching telephone network and cell switching B-ISDN coexist. It is thought likely that the deployment of B-ISDN may take one or more decades from about 1995 onwards. In the first decade beyond the year 2000, voice packets may be routed partly within B-ISDN, perhaps starting with connections involving digital mobile voice terminals, since such terminals already introduce the need for echo control. However, it may be many decades before the telephone network completely disappears. Does this imply that the cell-switching/octet-switching hybrid LAN is the best solution for the foreseeable future?

Pure ATM LANs can compete strongly earlier than this timescale suggests because of their simpler and more flexible bandwidth management for all services. Thus, while pure ATM LANs do not allow several nodes to share a composite packet by writing/reading particular octets in the information field (which a hybrid LAN can do), some composite packet arrangements are nevertheless possible. An individual node on a pure ATM LAN can multiplex several voice connections into one packet prior to placing it on the ring. The efficiency of such an arrangement is certainly lower than on the hybrid LAN, which allows other nodes to switch octets in and out of shared cells and hence fill any holes created when a call clears down. However, this point is offset by the more flexible 'boundary' on pure ATM LANs between voice and other services competing for available bandwidth. Furthermore the packetization delay of a composite packet created in an individual node of a pure ATM LAN is very small and the hybrid offers no advantage here. Many customers require only a small-sized LAN where the need for composite packets is not justified, i.e. voice can be carried in half-filled cells without exceeding the delay budget. Often such a customer demands a wide range of services and bit rates (e.g. a small advertising company who may require some fast high-quality still-image transfer). The pure ATM LAN is well suited to this market.

In conclusion it is expected that the pure ATM LAN will emerge soon after the introduction of B-ISDN, as the marketplace demands a LAN that flexibly allocates bandwidth and flexibly supports different delay control options for a range of services including appropriate packetization methods.

14.5 Summary

We have seen that CSMA/CD and token passing rings have very similar performance characteristics at low channel rates (i.e. for channel rates of approximately 1 Mbit/s). CSMA/CD is ideally suited to the situation where the load consists of long packets between only a few nodes where the acquisition time is insignificant. However, polled access protocols in general are better suited to situations where there are a large number of attached nodes. At the higher transmission rates, the need to take into account a sustained high bandwidth demand from a node also favours polled access. Both slotted and token polled access methods can be applied. However, the slotted ring is better suited to situations where the dominant load consists of short packets, especially voice, where it offers a significant reduction in access delays. For these reasons, the slotted ring is a good choice for the development of multiservice LANs. While the slotted ring cannot compete with the hybrid LAN in terms of efficient use of composite packets, this factor may be of less significance compared with the flexible bandwidth allocation it provides to different services.

References

Adams, J. L. and Falconer, R. M. (1984) 'Orwell: a protocol for carrying integrated services on a digital communications ring', *Electronics Lett.*, **20** (23), November, 970–971.

Bux, W. (1981) 'Local area subnetworks: a performance comparison', *IEEE Trans. Commun.*, **COM-29**, (10), October, 1465–1473.

Metcalfe, R. M. and D. R. Boggs, (1976) 'Ethernet: distributed packet switching for local computer networks', *Commun. ACM*, **19** (7), July, 395–404.

Wilkes, M. W. and Wheeler, D. J. (1979) 'The Cambridge digital communications ring', in *Proc. Local Area Communications Network Symp.*, Boston, Mass., May 1979, 47–60.

15 Advances in networking

GILL WATERS AND JOHN LIMB

15.1 Introduction

In earlier chapters, we have discussed a variety of approaches to networking and protocols. We have also seen how the OSI model provides a framework not only for understanding and combining protocol techniques but also for defining standards. The fact that we have arrived at a current position where the description and implementation of standards enables many different systems to exchange information effectively is a tribute to the research work of a large number of people, and we have referred to their work earlier.

There still remain many interesting problems to be solved. The impetus for further research can be divided into three main categories:

1   Technology is changing: VLSI, all-digital communications and optical transmission offer new challenges in the form of higher speeds (multi-gigabit/s in the case of optical fibres), lower bit error rates and cheaper access interfaces. Easier access to radio and satellite systems enables new forms of communication with both fixed and mobile users.

2   Many new services are emerging. When computers were first introduced in the 1950s only a few specialized applications were possible; now their use is widespread and varied. Similarly, as new networking techniques become available and packet networks emerge that can carry real time services such as speech (as well as computer data), we can envisage new services such as multimedia information retrieval and distance learning tools.

3   There is the need to implement networks in the most efficient way. This can be helped by performance studies, network design systems and the use of formal tools both for protocol specification and for the design and implementation of communication software.

In this chapter, we shall be looking at some of the work that is being carried out now as a result of all the above influences.

15.2 Network architectures

15.2.1 Introduction

Many networks have grown in an ad hoc way and some are much larger than was originally envisaged. There is therefore continuing work into efficient routing strategies (e.g. Khanna and Zinky, 1989) and into the best topological approaches (e.g. Orda and Rom, 1990). Other architectural issues arise out of the type of transmission medium being used. In this section we look at two important transmission media, optical fibres and the radio spectrum, which are becoming increasingly important, and then at some projects that combine different communications media. The possibility of including links with extremely low error rates indicates that existing protocols, where error recovery is provided at a very low level, may not be appropriate; indeed, as we shall see, for many types of traffic it is better for the network either to drop corrupted packets or to indicate corruption but not to undertake recovery.

15.2.2 Optical fibres

When those involved in traditional computer networking applications encounter optical communication, they have to reevaluate their use of the term 'high speed'. To a traditional network user a WAN with links running at tens of kilobits per second is high speed; to an optical fibre expert 1 gigabit/s is an acceptable figure. Multimode fibres offer speeds in excess of 100 Mbit/s and these are now being employed in high-speed LANs such as FDDI (see Section 15.4). Monomode fibres will offer much higher speeds, presenting a problem for current techniques in electronic processing and switching, even for some of the simpler access techniques. There is therefore much work to be done on the interaction between high-speed optical fibre techniques and the flexible techniques offered by packet networks. Fibre optic techniques are discussed in special issues of the *IEEE Journal of Lightwave Technology* (*IEEE JLT*, 1985) and the *IEEE Journal on Selected Areas in Communications* (*IEEE JSAC*, 1985, 1986, 1988). A recent paper by Herskowitz and Warms (1988) also describes suitable topologies, access protocols and standards.

15.2.3 The use of radio frequencies

Established networks use satellite links for point-to-point communication (for example in the ARPANET network); more recently the use of satellites has been explored for connection of multiple sites where it has advantages for sites widely spread or separated by hostile terrain such as mountains or oceans. We shall be looking at networks with very small aperture terminals in Section 15.6. Earlier examples of satellite networks are the DARPA wideband research project (Falk *et al.*, 1983), the UK Universe project (Burren and Cooper, 1989) and the European Satine project (Hine, 1988). One of the major problems is to ensure sufficiently fast and flexible access to the satellite channel where there is about a 250 ms propagation delay from earth station to earth station via a satellite in geostationary orbit. Access protocols are generally based on time division multiple access schemes, but these can be made more flexible and efficient by allocating more capacity to stations that are generating a large amount of traffic and by viewing the traffic from sites as an overall trend, thus reacting to slower changes where a packet by packet adaptation would obviously be inappropriate. Satellites can be used to transfer very large files quickly (Owings, 1983).

Cellular radio systems are now becoming widely used for telephony, and data transfer over packet radio also has advantages for people who are mobile and yet wish to convey or enquire about accurate information from a base location (Ewen-Smith, 1985). Garcia-Luna-Aceves (1986) describes techniques for routing table updates in multihop packet radio networks, where a mobile may rapidly switch between different transceiving stations. Radio offers the prospect of mobility not just in terms of travel but also within buildings, and opens up interesting challenges in providing directory services that can locate people who are mobile (see, for example, Cox, 1987).

15.2.4 Interconnection of dissimilar networks

With the availability of improved transmission media it is important that networks can be constructed flexibly using several of the media and that the problems of interconnection are overcome. Two recent UK projects have demonstrated that this is possible, but much work remains to be done to ensure a unified approach. Both projects involved collaboration between industry and academic institutions. The Universe project expanded the concept of local area networking into a wider area by connecting sites, each of which had one or more LANs (mainly Cambridge rings), by means of a shared satellite channel. The channel offered 2Mbit/s which was half-rate encoded to provide 1 Mbit/s at a bit error rate of about 1 in 10^9, thus enabling the network to use lightweight protocols (connection setup without low level flow control) and to carry voice and images as well as data (Adams *et al.*, 1982; Cooper, 1984; Adams and Ades, 1985). The Unison project (Clark, 1986), with a slightly different set of partners, has continued and expanded this work but the links between sites are now 2Mbit/s ISDN primary rate channels. The Cambridge fast ring (CFR) acts as an exchange between the LANs at each site and the ISDN connection (Tennenhouse *et al.*, 1987). A special issue of *IEEE Journal on Selected Areas in Communications* was devoted to heterogeneous computer network interconnection in January 1990 (*IEEE JSAC*, 1990).

15.3 OSI—continued work and new applications

A good review of the state of the OSI standards and their relationship to the market place and the PTTs is given by Pouzin (1986). He argues that the transport layer was a useful tool in that it enables higher layers to be unaware of the details of the network thus allowing them to use new networks such as the ISDN as they become available. The services of the session and presentation layers, on the other hand, might have been better provided as application service elements (in the same way as concurrency control) as an open-ended set of functions, and would have fitted more naturally into typical implemented software interfaces. Carpenter (1989) advocates concentrating effort in specific areas of OSI in order to ensure some success in achieving its objectives. Studies of various aspects continue, e.g. transport protocols (Sabnani and Netravali, 1989).

Now that all layers of the OSI model have been defined, we can expect much more work on particular services and applications. One example is the provision of facilities for group communication using messaging systems (Palme and Speth, 1986). It is interesting to note that different application services offer alternatives for the same task. For example, file transfer may be achieved by the FTAM or by use of a Teletex protocol or by use of X.400 messaging. Many teleservices such as videotex, teletex, facsimile and messaging can make good use of the OSI model, but often these are required in unsophisticated terminals and personal computers, and their interaction with the complexities of the full seven-layer model needs further attention.

15.3.1 Office document architectures

Much computing power is used in processing and formatting documents electronically. However, document processing systems have been developed in an ad hoc manner and the format of the information of one system will generally be totally incompatible with that of other systems. The advantages of interchanging documents electronically within an organization are already with us, but standards are required if we are to exchange documents electronically across wide area networks. In order to meet this demand an office document architecture (ODA) is the subject of the ISO 8613 proposals. This architecture addresses not only text documents but multimedia documents containing text, images, graphics and voice, so offering additional facilities such as voice annotation. Documents exchanged using the ODA must be formatted in an office document interchange format (ODIF) which uses abstract syntax notation one (see Chapter 5). ODIF has two components, both of which are logical hierarchical structures. The first is the logical structure which includes chapters, paragraphs, headings, footnotes, etc.

The second is the layout structure which determines the position of the information on the page. A profile is also included which specifies how the document is to be treated as a whole.

Obviously the integration of the ODA with the X.400 messaging facilities will facilitate its use by a wide community. However, the complexity of processing required to implement the standard interfaces imposes overheads on what are very often quite simple systems—so it is essential to break down the problems in a modular manner so that full use can be made of the advantages of electronic document interchange (Ma, 1987, suggests such a scheme). IBM has its own approach to these problems called document content architecture (DCA). Electronic publishing is the subject of a special issue of *The Computer Journal* (1989).

15.3.2 Electronic funds transfer at point of sale

The ability to use plastic money is convenient, and for the banks and other financial institutions it can be very useful to capture information electronically at the time when a transaction is completed. Consequently, there are now many systems for electronic funds transfer at point of sale (EFT/POS). The traffic is characterized by very large numbers of terminals, each of which is used infrequently to initiate a transaction conveyed in a few packets. Also, because the transactions involve money, security and reliability are important. A number of countries have invested in pilot projects to bring together those wishing to make use of the system. The general scheme is to provide concentrators for up to, say, 60 terminals which then interface to a packet switching network. At the other side of the network users are authenticated before the bank's records are updated. Examples are described for Sweden in Jansson *et al.* (1986) and Australia in Lockwood *et al.* (1986).

15.4 Integrated service networks

One research area to which much effort is being devoted is the design of networks capable of handling many different services (e.g. see *IEEE Communications Magazine*, December 1986). There are a number of reasons why a single network for all communication needs is preferable to the current situation, where a number of information networks such as the existing telephony network, packet switched networks and the telex network are running side by side. A single network could be easier and cheaper to install and update. Digital transmission and the high speeds offered by optical fibres also enable new services to be carried such as interactive video, and ideally the single communications network will be capable of carrying new services as they are devised. A fully integrated network would also allow access to the network for services that consist of a mix of traffic types, such as information retrieval from multimedia data bases. Examples of such services are given in Irven *et al.* (1988).

An integrated network must therefore be efficient, flexible and adaptable. Packet networks have already demonstrated these three qualities for low-speed data applications. Packet interleaving offers an economical use of a communications network. Packet networks offer flexibility: information is transmitted transparently and it is possible for systems of different manufacture, speed or size to communicate with each other. Many different bit rate requirements can be met including variable bit rates. As we shall see, packet networks can also be adapted for higher speeds and for real time services such as voice traffic. A packet approach to integration allows the network to use bandwidth made available by silence suppression in speech conversations and allows service quality to be adapted to network availability. An early paper by Hughes and Atkins (1979) proposed a switching technique for virtual circuits which could offer 'bandwidth on demand' to new telecommunication services. Recent work has looked at the interaction of variable bit rate video coding with integrated service networks (Ghanbari *et al.*, 1988). This flexibility for multiservice networks has already been emphasized by a number of authors including Gallagher (1986).

15.4.1 Packet mode access to the integrated services digital network (ISDN)

Recognizing the large community that uses packet switching, packet mode access is being introduced to the ISDN in a staged manner. In the first stages, CCITT recommendation X.31 describes how the ISDN can offer support for existing X.25 terminals both for switching calls to existing public packet switched networks and within ISDN itself. The next stage offers two additional packet mode services—frame relay and frame switching (Cooper, 1988). In both cases a virtual call is set up. In frame relaying, information is transferred in frames of variable length between the two endpoints of the call; frames are checked and may be discarded by the network but no acknowledgements are sent. This mode of operation is suitable for high-speed applications such as LAN interconnection. Frame switching adds error recovery and flow control to this scheme.

The next stage is the broadband ISDN (B-ISDN) and for this the CCITT is investigating the use of the asynchronous transfer mode (ATM), already discussed in Chapter 14. ATM uses fixed length packets called cells which contain headers carrying virtual circuit identification. The source rather than the network determines the period between successive cells of a virtual circuit, thus enabling a variety of bit rates and variable bit rate services to be carried (Fisher *et al.*, 1988). Some of the techniques proposed owe much to previous work on integrated service packet networks which will be discussed in the next few sections, but we first look at another large body of research on integrated broadband switch fabrics.

ATM switch fabrics

A comprehensive categorization of the wide variety of approaches to broadband integrated switching architectures is given in Daddis and Torng (1989). This covers the way in which links are shared, the distribution of routing intelligence, the use of buffering and a variety of switching techniques. The techniques described have developed both from telecommunication circuit switching and from experience with computer networks.

A survey of four fast packet switching methods is given by Jacob (1990). The performance of the four methods is compared with regard to buffering, speed and hardware complexity. Buffering can be undertaken either at the input to the switch or at the output lines from the switch, or at intermediate switch elements. Output buffering has been shown to be more efficient than input buffering (Karol *et al.*, 1987) in terms of its capacity to handle the offered load. The required speed can be obtained from either a single high-speed link or from several lower-speed internal links used in parallel. The hardware complexity of a switching method depends not only on the simplicity of operation but also on the size of switching matrices and the buffering capacity required.

To give a feel for fast packet switch design we now describe one of the four systems reviewed by Jacob. The knockout switch (Yeh *et al.*, 1987) connects N input links, each carrying fixed-length packets in time slots, to N similarly operating output links (see Figure 15.1). No filtering of packets is done at the input side, but each packet is passed directly to a concentrator in the appropriate output switch. The concentrator then uses a shifter to queue packets in sequence on m first in first out (FIFO) queues (where m is the number of time slots on the outgoing link). The packets are then taken cyclically one from each of the FIFO queues and transmitted to the link. Packets that cannot be copied to the FIFOs by the concentrator are dropped.

The knockout switch is reasonably easy to implement because it is fully interconnected and output switch elements are identical. It uses output buffering and can offer a total capacity of 50 Gbit/s with 1000 inputs, each of 50 Mbit/s. Although the technique is simple, a large number of output switching elements are needed and in order to guard against unacceptable packet loss each output element typically needs about 50 kbytes of buffer memory.

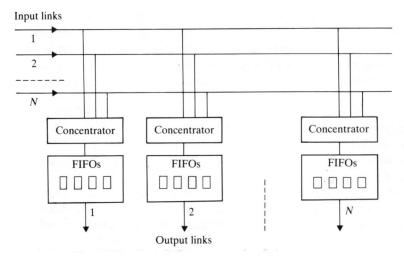

Figure 15.1 The knockout switch

15.4.2 Requirements of data, voice and video transmission

Before looking at medium access techniques for integrated service packet networks it is useful to look at the traffic they are intended to carry. A comparison of the transmission rate, nature, toleration of errors and toleration of delay for a number of services is given in Table 15.1, indicating some of the typically orthogonal requirements of information that have conventionally used circuit switched techniques and of data.

Table 15.1. Typical service requirements

Type	Rate	Nature	Errors tolerated	Delays tolerated
Data (e.g. file transfer remote log-in electronic mail)	0–2 Mbit/s	Bursty	None	Yes
Speech	64 or 32 kbit/s	Periodic	Some	No
Facsimile	0–64 kbit/s	Bursty	Some	Yes
Fixed bit rate video	140 Mbit/s	Periodic	Some	No
Variable bit rate video	0.5–30 Mbit/s	Bursty	Some	No
Video conferencing	6.4 kbit/s–2 Mbit/s	Bursty	Some	No

Data communication services have a very bursty traffic pattern—the information tends to come in short sequences with long silent periods, making it feasible for large numbers of simultaneously active virtual circuits to be carried by existing packet switched networks. Error detection and recovery are important in order to maintain data integrity.

Real time speech traffic, which is still the principal user of bandwidth on national networks, has completely different requirements. Coded in the simplest way it has a regular bit rate requirement and a small amount of loss can be tolerated, but it is sensitive to delay, particularly variable delay. Fixed bit rate coded video has similar characteristics but much higher bandwidth requirements. Packet video is likely to require a proportion of guaranteed bandwidth with extra capacity requirements depending on changing picture information. Both packet speech and video are discussed in a special issue of *IEEE Journal on Selected Areas in Communication* (1989).

Many of the newly envisaged services have requirements that fall between these two extremes. Examples are variable bit rate video and the retrieval of remotely stored voice messages. Some individual services will generate several types of traffic at the same time, for example separately stored subtitles for television images, interpersonal communication using a mix of traffic types or the transmission of textual documents with voice annotation. Information retrieval systems will want access to video sequences, still pictures, spoken messages and graphs.

15.4.3 Techniques for integrated packet networks

Access protocols for integrated service packet networks must take into account the high speeds offered by optical fibres and be efficient and manageable. Gruber and Le (1983) studied the performance requirements for combined voice and data networks and concluded that: packet size should be small to reduce packetization delays, a small amount of packet loss may be tolerable (up to about 1 per cent), regular bandwidth should be available where required and error detection should be an option (correction being the responsibility of higher layers of protocols where reliability is essential) (see also Gruber and Strawczynski, 1985).

Protocols for conventional packet networks are not suitable for an integrated network. Wide area network protocols are complicated and relatively slow. Local area network access schemes are more suited to high speeds, but have disadvantages for real time traffic. These include poorer performance for short packets than long packets due to the time required to process each packet (CSMA/CD and token ring), high overheads (especially on slotted rings) and variable delays compared with fixed TDM schemes (although variability is generally much less than on WANs).

Medium access protocols for integrated service packet networks generally treat synchronous and bursty traffic differently. For the greatest flexibility there will be a movable boundary between synchronous and asynchronous information, unused slots in the synchronous portion being made available for bursty traffic if needed. The allocation of capacity is best achieved in a distributed manner whenever possible, with centralized control of overall capacity requirements (see, for example, Limb and Flamm, 1983).

Packet overheads depend on the number of bits required for the access protocol and the number of bits in the address field. Fixed reservation for synchronous traffic requires no address field, but capacity cannot be reused. Carrying a full network address of, say, 12 decimal digits in all packets is wasteful. Virtual circuits reduce the size of the address field while retaining flexibility. For very high speed networks (e.g. $\geq$ 1 Gbit/s) absolute variability in the time spent waiting to access the network is low. Here virtual circuits may not be needed, but priority may be used as in IEEE 802.4 and FDDI. The *IEEE Journal on Selected Areas in Communications* devoted an issue to packet switched voice and data communication in 1983.

15.4.4 Examples of integrated service packet networks

Among the many examples of integrated local area networks are MAGNET (Lazar *et al.*, 1985), Expressnet (Tobagi *et al.*, 1983), Fasnet (Limb and Flamm, 1983), FDDI-2 (Ross, 1986), Orwell (Adams and Falconer, 1984), DQDB (Newman *et al.*, 1988) and the Cambridge fast ring (CFR) (Hopper *et al.*, 1986). All of these offer some amount of bandwidth to both bursty and real time traffic with varying amounts of flexibility for reuse. Orwell has been described in Chapter 14 and DQDB will be discussed in the next section on metropolitan area networks. Here we examine the suitability of FDDI and look briefly at two other approaches—Fasnet and the CFR.

FDDI

FDDI was described in Chapter 7. The suitability of FDDI for carrying both voice and data traffic has been studied by Watson and Frontini (1988), using both analytical and simulation techniques. Voice traffic was based on 32 kbit/s adaptive differential pulse code modulation (ADPCM). To ensure no packet loss, the maximum token rotation time must be less than the time between successive packets of a voice call. Under these conditions, using simple analysis, a 10 km ring with 100 stations could support 376 voice calls where each call produces a packet every 10 ms. However, Watson and Frontini's simulations which

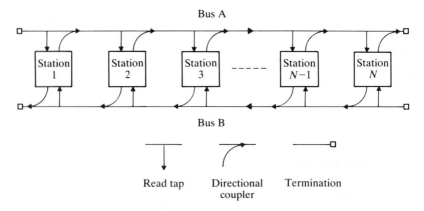

Bus A

Bus B

Read tap   Directional   Termination
            coupler

Figure 15.2   Physical configuration of Fasnet

also include asynchronous traffic show that for the same size network, the same number of stations and the same packet production rate it is possible to almost double the number of voice calls whilst suffering a packet loss of only 0.5 per cent. Because voice traffic takes priority, insufficient bandwidth was available for the simulated asynchronous traffic, but further simulations showed that a 10 km FDDI ring with 100 stations could support 600 voice calls and 30 Mbit/s of asynchronous data. Their results show that FDDI is suitable not only for backbone networks but also for carrying a mix of voice and data.

A later version called FDDI-2 which has a more complex protocol and carries circuit switched traffic (carrying, for example, 8 bits of a 64 kbit/s voice channel every 125 μs) is receiving renewed interest.

Fasnet

(Limb and Flores, 1982; Limb and Flamm, 1983)

The physical configuration of a Fasnet link is shown in Figure 15.2. It consists of stations that all have an attachment to two unidirectional transmission lines. The links may be a twisted pair, coaxial cable or optical fibres. For a coaxial system, a read tap precedes a passive directional coupler used for writing. A station writes to the line by adding energy to the existing signal on the line. If a station wishes to send information to another station on its right, it uses the upper line; if the information is destined to a station on its left, it uses the lower line. The protocol ensures that only one station can transmit at any one time on a line. Larger networks may be formed from a number of Fasnet links.

Fasnet uses the frame structure illustrated in Figure 15.3. The source and destination address are within the link layer protocol header (contained within the PACKET field). Access to the link is controlled by the START, END and BUSY fields. We first look at the case where each field is one bit long. Access takes place by implicit token passing.

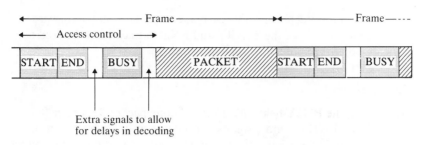

Figure 15.3   Fasnet frame structure

If a station wants to transmit to a station on its right, it waits until a frame with an empty BUSY field arrives and writes into it. This simple protocol would favour stations at one end of the link, so the START and END bits are used to form a cycle of frames; the head station marks the start of a cycle with the START bit set to 1. Having seen START = 1, each station is permitted to write into a given maximum number of slots. When an empty frame reaches the other end of the link, a new cycle can begin; this is signalled by setting the END = 1 bit in a frame travelling along the reverse link. When the head station detects END = 1 it sets START = 1 on the forward link.

A longer START field offers scope for the support of different traffic types. A simple example for a 10 Mbit/s link is illustrated in Figure 15.4. Here START = V indicates the start of a voice cycle and START = D indicates the start of a data cycle. The time between successive START = V's is 10 ms and each voice call is coded at 64 kbit/s, thus filling 640 bits of a packet every 10 ms. Assuming a packet size of 700 bits (including overheads) 142 packets can be accommodated in 10 ms, so that if, as in the figure, 50 voice calls are active, 92 packets will be available for data before a new voice cycle starts.

In fact, the protocol can be used in a more sophisticated way and can, for example, take advantage of silence detection by allocating slots only to active voice stations. When no more voice slots are needed a data cycle can begin. The head station uses information in the END field on

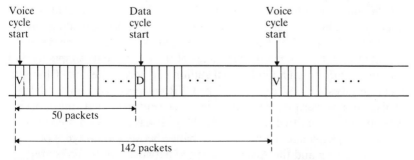

Figure 15.4   Fasnet—example of cycles for voice and data

the reverse link to determine the use made of packets on the forward link. Further values of the START and END flags can also be used to indicate other types of traffic such as video.

Error recovery

An error in the BUSY field will affect only one packet. Errors in START and END are more significant. However, it can be shown that an additional START has only a temporary effect on the system, and an additional END will simply result in an additional START.

Because Fasnet has two links it is possible to detect faults which occur with a station's attachment to one of the links. For example, if a fault has occurred on one link, the intact link could be used by the end station to demand a response on the link where the fault has occurred. The fault would lie with the station that did not reply. Fasnet can also be configured in a ring where reconfiguration is possible; this technique is used in the IEEE 802.6 MAN architecture (see Section 15.5).

Performance

More detailed accounts of the performance are given in the references. Capacity is not wasted in collisions, since these cannot occur. Simulation of Fasnet using silence detection for voice circuits has shown that approximately 100 voice stations with an activity level of 40 per cent may comfortably share 50 virtual circuits with an acceptable value (Gruber and Le, 1983) of less than 2 per cent of the talk spurts clipped by more than 50 ms. (Clipping is the amount of useful voice traffic lost by a silent voice station when it becomes active before it can access the network again.)

The Cambridge fast ring

(Hopper *et al.*, 1986)

The Cambridge fast ring (CFR) is a faster version of the Cambridge slotted ring (up to 100 Mbit/s). It has a longer packet size (256 information bits) and thus lower overheads. The access protocol is unchanged, allowing fair access to all stations, but not providing any guaranteed bandwidth. The efficiency is high as very little time is spent waiting for slots. The main advantages of the CFR are its simplicity and the speed with which an individual station can access the ring. Chip sets and interfaces are now available. The CFR has been used for the Island project as a modern PABX, with one station ensuring that ring capacity was not overcommitted (Ades *et al.*, 1986). It has also been used in the Unison project where it acts as an exchange between both types of Cambridge ring and the Alvey high-speed network, whose links are ISDN channels (Tennenhouse *et al.*, 1987).

15.4.5 Summary of integrated packet networks

The above techniques are just a few of the many integrated service packet networks devised and demonstrated. As with conventional data networks, there are plenty to choose from. There will probably always be a choice for locally installed and maintained networks. Where packet networking techniques are also being considered for much wider use in national or international digital networks, it is crucial that standards be agreed and that they be based on sound principles A realistic choice can be made only as a result of practical experience and performance comparisons.

15.5 Metropolitan area networks

15.5.1 Development of MAN standards

Local and wide area networks are both well established now and standards are available for them. However, high speed communication is seen as a need not only in the local area but also in the metropolitan area, where many branches of the same firm or different large organizations wish to communicate (*IEEE Communications Magazine*, 1988; Mollenauer, 1988). Having seen a need for metropolitan area networks (MANs) the IEEE project 802 recognized in 1981 that standards were required—a need that was only later seen by operating organizations. MAN standardization is therefore the remit of the IEEE 802.6 committee.

The committee's view of a MAN is like a very large LAN in that it has a shared medium but whose access protocols are less sensitive to network size. It may be of any appropriate speed (the IEEE 802 LAN maximum speed was 20 Mbit/s) and the architecture is optimized for a distance of between 5 and 50 km. Traffic is expected to include a large amount of digital voice and some video with the consequent requirements discussed in Section 15.4.

MANs may be private or public. The former connects sites of the same organization where the need for security is not very high; the latter is operated centrally by a public operator and therefore requires agreed standards and safeguards against accidental or deliberate misuse. Public MANs would not be acceptable if they passed through company premises and a two-segment architecture is envisaged as shown in Figure 15.5, where access to the public MAN is provided at a bridge from the company's LAN or MAN.

The standardization process has gone through a number of stages, first considering time division access. A slotted ring was proposed by Burroughs, Plessey and National Semiconductor in 1984 and was developed, but lost ground in 1986. FDDI-2 was brought in as a new proposal, but at the same time an interesting dual bus arrangement was

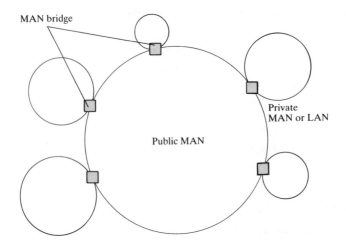

Figure 15.5 Architecture for a metropolitan area network

proposed from Australia called the queued packet and synchronous exchange (QPSX), and this is now the focus of attention.

15.5.2 The current IEEE 802.6 solution—DQDB

(Newman *et al.*, 1988)

The QPSX is now known as the distributed queueing dual bus (DQDB) MAN. The configuration (shown in Figure 15.6) employs the same dual bus as Fasnet. Each station is connected to each of two buses (which will normally be optical fibre) by a read and write tap. The stations are arranged in the form of an incomplete ring where the last station on each bus does not repeat the signals. This arrangement introduces

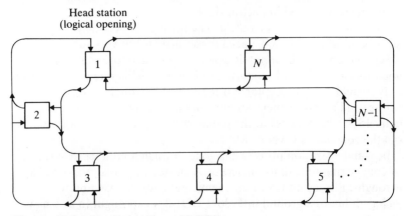

Figure 15.6 Configuration of DQDB

Head station
(opening is now closed)

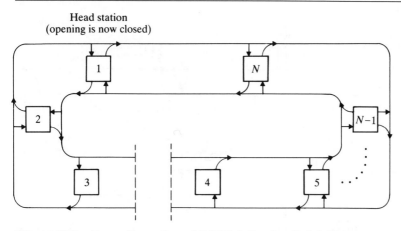

Figure 15.7   Reconfiguration of DQDB following link failure

reliability; in the case of link failure the buses can be terminated in the
nodes at either side of the failed link and completed through the
original end station, as shown in Figure 15.7.

Fixed size slots of 64 octets (each carrying 48 octets of data) are
created by the master station on each bus and information is ORed onto
the slots on the appropriate bus according to the medium access
protocol. Each slot may carry either isochronous or asynchronous
traffic. A cycle of an integral number of slots is created every 125 μs.
Isochronous slots may carry octets from different nodes; the quantity,
slot position and slot offset(s) are determined centrally for each
isochronous circuit to guarantee the required bandwidth. Asynchronous
traffic follows a protocol called *distributed queueing*.

Distributed queueing is achieved by use of two control bits in each
slot call REQ and BUSY which indicate request and full slots
respectively (see Figure 15.8) A counter *count1* is incremented for each
request seen on the uplink and decremented as an empty slot passes on
the downlink (but not below zero). At any time this counter represents
the number of stations ahead of the station on the downlink that have
made a request to transmit. When a station wishes to transmit on the
downlink it sends a REQ on the uplink and then copies the value of
count1 into a second counter, *count2*, and continues to adjust *count1* as
before. Then *count2* is decremented for each passing empty slot on the
downlink. When *count2* becomes zero, station A is allowed to use the
next empty slot as it has waited for all requests ahead of it in the queue
to be granted. (A similar queue is kept for transmission on the uplink.)

This form of access is very efficient and responsive at both high and
low loads because the time spent waiting is due primarily to use of the
bus by other stations. Access delay is not significantly affected by either
network length or bit rate. Asynchronous traffic of different priorities
can be accommodated by providing a request bit for each priority level

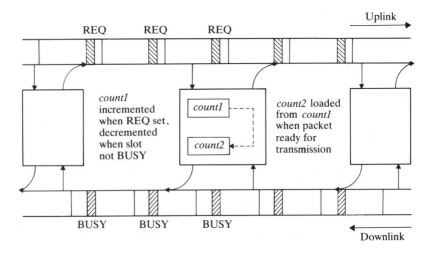

Figure 15.8   Distributed queueing on DQDB

and counting requests separately for each priority level. The slotted structure and access mechanism of the DQDB MAN has many similarities with the requirements of ATM for broadband ISDN. Analysis of the DQDB asynchronous protocol is rather difficult. Simulations show that under certain conditions the protocol exhibits unfair behaviour with stations near the head end of the bus seizing a large percentage of the capacity (Wong, 1989). Schemes to correct this have been proposed to the 802.6 committee.

15.6 Very small aperture terminal (VSAT) networks

Recent advances in satellite and earth station technology, deregulation in the telecommunications industry and demand for services that can quickly be installed have all led to an increasing number of developments in the use of very small aperture terminals (VSATs) (*IEEE Communications Magazine*, 1988/89). VSATs are characterized by the their small size (antennas are typically 1.2 to 1.8 m diameter for Ku band frequencies), low power requirements, low cost and ease of installation.

VSATs have a wide variety of applications such as broadcast services—video or news and weather data—point-to-point communication, or use in connection with ISDN either as part of an ISDN network or through a gateway. Interactive systems of two or more VSATs can employ packet techniques to provide information network capabilities (on which we now concentrate). Such networks are particularly useful for swift connection of geographically dispersed sites,

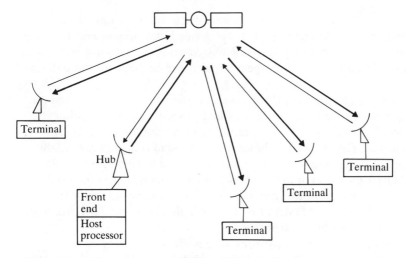

Figure 15.9 Typical VSAT network configuration

for networks crossing international boundaries and for file distribution. A wide variety of installed VSAT networks for financial, leisure and manufacturing industries are described in Sundara Murthy *et al.* (1989).

15.6.1 Network architectures

The majority of VSAT networks are configured in a star network with one site acting as a *hub* (see Figure 15.9). The hub usually has a more powerful earth station with a larger antenna. Typical bit rates are 56 kbit/s from VSAT to hub and 256 kbit/s from hub to VSAT (after forward error correction). The hub is often connected to a host computer enabling the VSAT systems to access it for typical transaction processing applications such as credit card verification. A logical mesh network may also be formed. Information from one VSAT to another is first sent to the hub and then redirected; higher level protocols such as SNA or X.25 can then be mounted. VSAT networks are particularly suitable for broadcasting or multicasting over a wide area (see Section 15.7).

15.6.2 Channel access protocols

VSAT terminals must share the channel on a multiaccess basis, and channel access protocols must be selected with a view to throughput, delay characteristics, congestion avoidance, error recovery and operational procedures.

Raychaudhuri and Joseph (1988) discussed a range of possibilities and compared their performance. These were ALOHA, slotted ALOHA, unslotted selective reject ALOHA and demand assigned

TDMA. The first two techniques were discussed in Chapter 2. In selective reject ALOHA, packets are divided into subpackets; if packets collide they are unlikely to completely coincide so that some subpackets will not be affected, while repeat requests can be made for those that are destroyed.

In demand assigned TDMA, the time allocated to users is based on requests that are of short duration and generally use a contention algorithm; successful requests result in longer duration slots being allocated. This is a better choice where the load is high as potentially more of the channel capacity can be used. It is less efficient than the ALOHA protocols at low loads because of satellite delays imposed by making a request to the hub site and receiving a response from it. Demand assigned TDMA also imposes higher processing overheads and is more susceptible to channel errors.

The performance comparisons were made for a typical enquiry/response application. Selective reject ALOHA provided better throughput/delay characteristics than both ALOHA and slotted ALOHA. (Here the choice of subpacket size is a key issue.) Demand assigned TDMA has a minimum delay of about 0.75 s (three one-way geostationary satellite delays), but is much more stable as load increases. Up to 250 VSATs could be supported by demand assigned TDMA—fewer for the other protocols.

Other channel access protocols employ code division multiple access (CDMA), a technique that counters possible interference by spreading the frequency spectrum of the signal, and other varieties of demand assigned TDMA in which different classes of traffic are treated separately (see, for example, the CODE project of Nesbeth *et al.*, 1989).

15.6.3 Future trends

Future VSAT networks are likely to take advantage of other technological advances such as multiple spot beams and on-board processing and switching (Golding *et al.*, 1989). On-board processing can considerably reduce the delays inherent in reservation protocols, and thus opens up the possibility of integrated voice and data applications. VSATs are also likely to be used for LAN interconnection and for mobile services.

15.7 Multiparty protocols

Computer networks have traditionally been used for communication between two systems or individuals regardless of the type of network used. More recently there have also been developments in the provision of services which enable simultaneous communication between three or more parties. In this section we first discuss enabling protocols. There are two approaches—replicating the information or using broadcast facilities directly where they are inherent in the operation of a network. We then look at multiparty applications.

15.7.1 Information replication

Where inherent broadcasting facilities are not available, multiparty communication involves message replication. Participants can be arranged in a tree structure or a loop to reduce the overheads, or for simplicity relaying can be done by the network. Whether connectionless or connection-oriented techniques are used will depend on the nature of the network and of the application.

15.7.2 Broadcasting and multicasting

Broadcasting is a technique that enables a single transmitted message to reach all the stations attached to the same network. In *multicasting*, a single transmitted message reaches a selected group of destinations on a network.

Many networks, in particular LANs and networks based on satellite or radio technology, are inherently capable of broadcasting or multicasting and in principle it would be possible for application programs to broadcast packets at no extra transmission cost. This is not always allowed in practice. This is because there is an overhead at receiving stations in discarding unwanted messages, and provision must be made in terms of logical channel numbers for users of specific broadcast facilities and protocols for group formation and maintenance. The IEEE 802 MAC layer standards for CSMA/CD, token ring and token bus all provide for both group addresses (multicast) and broadcast addresses. The logical link control (LLC) layer offers an unacknowledged connectionless service which enables higher layers to make use of point-to-point, muticast or broadcast addressing.

Broadcast channel numbers were provided in project Universe, where protocols were devised to allow applications to make use of the broadcast nature of the satellite (Waters *et al.*, 1984). An example of group formation and maintenance can be found in the distributed V-system (Cheriton and Zwaenepoel, 1985). Reliable broadcast networks are discussed in Tseung (1989).

15.7.3 Applications

Broadcasting offers a number of advantages to applications: it reduces the overhead on the transmission channel and broadcasting is quick and simple (destinations are reached by the shortest possible route and delays are uniform). It is also the only way of reaching an unknown address or an unknown number of recipients. Multicasting has similar advantages and reduces the processing overhead for stations that do not wish to receive the information.

File distribution can be achieved economically by broadcasting. If there is no return channel, forward error correction can be used to improve reliability. When an acknowledgement path is available file distribution techniques are modified versions of point-to-point techniques such as stop and wait and selective repeat. See, for example, Youssef *et al.* (1984), Sabnani and Schwartz (1985) and Daka and Waters (1988).

Broadcasting is also useful for detecting *network status*, and if each system periodically broadcasts its status (as is done, for example, using the Internet suite of protocols over Ethernet) other systems can keep track of which services are available.

Multiparty protocols can be used for *group communication* to support conferencing and electronic meetings. For example, at Xerox PARC communicating workstations are being used to develop an environment to support collaborative problem solving in meetings (Stefik *et al.*, 1987).

15.8 Formal definition of protocols

(*IEEE Transactions on Software Engineering*, 1988; Rudin, 1985)

As we have seen throughout the book, protocols are complex; they have to cope with indeterminate sequences of events while providing meaningful information flow. We have also seen the importance of standards which encourage all implementations of a protocol to exhibit the same behaviour. Most standards documents combine semi-formal specifications such as finite state machines with natural language, but any use of natural language can easily lead to ambiguities. If a protocol can be defined formally, and without ambiguities and inconsistencies, we can then be sure that its behaviour can be reproduced or at least we can test conformance.

For these reasons several techniques for formal specification of protocols have been developed. Once a formal specification of a protocol is available it can in principle be used in a number of ways. Exactly how this can be done is a fruitful area of research; techniques are emerging but there is plenty of scope for improvement. A protocol can be shown to meet certain conditions, e.g. no deadlock or no unreachable states (validation), or shown to provide some specific functions (verification), in particular that the service offered to the next higher layer is in fact provided by the protocol and its use of lower layer services. A formal specification can also help in the production of a simulation model to estimate the performance of a protocol—both its logical behaviour and its ability to meet requirements. A desirable goal is automated implementation; although parts of some protocols have been implemented automatically it is likely that some hand coding will always be necessary. Formal specification can also help in the generation of test sequences and conformance testers.

Formal techniques have arisen from a number of other techniques such as the communicating finite state machines concept that broadens the finite state machine to cover several concurrently communicating processes. The formalism is provided by a process algebra which defines the communication between the different processes. Two algebras—the calculus of communicating systems (CCS) (Milner, 1980) and communicating sequential processors (Hoare, 1985)—form the basis for most formal description techniques used for protocols. Two important formal techniques are the language of temporal ordering specification (LOTOS) which became an ISO standard in 1988 (ISO 8807) and specification and description language (SDL) described in CCITT recommendation Z.100. Unfortunately we do not have space to describe them here; informal overviews are given for LOTOS by Van Eijk *et al.* (1989) and for SDL by Rockström and Saracco (1982).

15.9 Other research areas

There are many other interesting research topics in computer communication networks. The following paragraphs indicate some of the key areas and give suggestions for further reading.

15.9.1 Performance studies

Everyone is interested in knowing the limits to which one can push a network and where problems might arise in large networks or in LANs with many active stations. Also it is very easy to propose a new protocol, but it will need justification before others agree that the protocol is a good one. A major aspect of performance assessment is therefore to use analytical or simulation tools to determine aspects such as network utilization or delay versus offered load. Consequently there are many published papers on performance issues. Schwartz (1987) gives a comprehensive introduction to performance modelling for both packet switched and circuit switched networks. A wide coverage of LAN performance is given in Hammond and O'Reilly (1986). A recent issue of the *IEEE Journal on Selected Areas in Communication* is devoted to performance evaluation of multiple access networks (IEEE JSAC, 1987). Of course theoretical performance is only one aspect, even the best protocol architecture on paper may have problems in the field, for example implementation may be complex; so it is also interesting to find studies of existing networks such as that given on the measured capacity of an Ethernet by Boggs *et al.* (1988).

15.9.2 Planning methods and tools

For large networks, the siting of nodes and links can be assisted by good planning tools. Grout and Sanders (1988) discuss existing algorithms for communication network optimization and look at likely future areas of interest. The necessary software for implementation of communication systems can become complex; this can be helped by a unified approach to software architecture such as the Fuzzball (Mills, 1988). An example of software architecture for an X.400 electronic mail system is given in Hammer (1988).

15.9.3 Distributed systems

Distributed systems are possible because of widely available and cheap computer communications, and have been covered in Chapter 12 by Mark Clark. This is an area set to expand as applications become more varied. An annual international conference on distributed computing systems is sponsored by the Computer Society of the IEEE. Recent papers representative of current research include global time estimation in distributed systems (Duda *et al.*, 1987), distributed data bases (Gretton-Watson, 1988), reliable data storage (Bernstein, 1985), software solutions (Lorin, 1990) and parallel distributed processing using a master/slave model (Sullivan and Anderson, 1989).

15.9.4 Network management

There is still much research to be done in the area of network management and the reader is referred to Chapter 13, where Alwyn Langsford discussed the topic in detail.

15.9.5 Human factors, social and policy aspects

The majority of the book has been concerned with the technical aspects of achieving intercomputer communication, but we should never forget the social framework within which these tools are used. There is no point in creating tools if they are unfriendly or unreliable or if the job could have been done more efficiently or more pleasantly using conventional procedures. The economic aspect should not be overlooked and often has a major influence on the choice of options over and above technical considerations (Thompson, 1986). The impact of technologies on people are discussed in MacLean and Wilson (1986) and Taylor and Katmbwe (1988).

15.10 Summary

We have covered a wide selection of topics in this chapter from high-speed integrated networks to satellite networks and formal techniques. In the course of the book we have progressed from the background of basic packet networking techniques to look in detail at wide and local area networks, at the way in which networks can be interconnected and managed, at security issues and at the applications ranging from simple access to a remote host to complex distributed systems with shared filestores and processing elements. Widespread installation of systems that use the techniques covered in this book have an influence on the lives of many people and new forms of interworking at the application level will ensure that this continues to be so.

A huge standardization effort has enabled us to reach a stage where computer communication networks have become cheap, usable and interworkable using today's technologies. Emerging technologies and techniques will require a similar amount of rigour and practical experience before new sets of standards can be made available. The future of computer communication networks is likely to be just as exciting as the past.

References

Adams, C. and S. Ades (1985) 'Voice experiments in the Universe project', *Proc. IEEE Conf. on Communications*, New York, 927–935.

Adams, J. L. and R. M. Falconer (1984) 'Orwell: a protocol for carrying integrated services on a digital communications ring', *Electronics Lett.*, **20** (23), November, 970–971.

Adams, C. J., G. C. Adams, A. G. Waters, I. Leslie and P. Kirk (1982) 'Protocol architecture of the Universe network', in *Pathways to the Information Society, Proc. of ICCC*, London, North-Holland, 379–383.

Ades, S., R. Want and R. Calnan (1986) 'Protocols for real time voice communication on a packet local network', *Proc. IEEE Int. Conf. on Communications*, Toronto, Canada, vol. 1, 525–530.

Bernstein A. J. (1985) 'A loosely coupled distributed system for reliably storing data', *IEEE Trans. Software Eng., SE–11* (5), May, 446–454.

Boggs, D. R., J. C. Mogul and C. A. Kent (1988) 'Measured capacity of an Ethernet: myths and reality', *Proc. ACM Sigcomm*, Stanford, California, 222–234.

Burren, J. W. and C. S. Cooper (1989) *Project Universe: An Experiment in High Speed Computer Networking*, Clarendon Press, Oxford.

Carpenter, B. E. (1989) 'Is OSI too late?', *Computer Networks and ISDN Systems*, **17**, (4/5), October, 284–286.

Cheriton, D. R. and W. Zwaenepoel (1985) 'Distributed process groups in the V-kernel', *ACM Trans. Computer Systems*, **3** (2), May, 77–107.

Clark, P. (1986) 'The Unison project', *Computer Commun.*, **9** (3), June, 126–127.

Computer Journal, The (1989) Special issue on 'Electronic Publishing', **32** (6), December.

Cooper, C. S. (1984) 'Managed file distribution in the Universe project', *Proc. ACM Sigcomm*, Montreal, pp. 10–17.

Cooper, N. J. (1988) 'Packet mode services for ISDN', *British Telecom Technol. J.*, **6** (1), January, 14–22.

Cox, D. C., H. W. Arnold and P. T. Porter (1987) 'Universal digital portable communications—a system perspective', *IEEE J. Selected Areas of Commun.*, **SAC-5** (5), June, 764–773.

Daddis, G. E. and H. C. Torng (1989) 'A taxonomy of broadband integrated switching architectures', *IEEE Commun. Mag.*, **27** (5), May, 32–42.

Daka, J. S. J. and A. G. Waters (1988) 'A high performance broadcast file transfer protocol', *Proc ACM Sigcomm*, Stanford, California, 274–281.

Duda, A., G. Harrus, Y. Haddad and G. Bernard (1987) 'Estimating global time in distributed systems', *Proc. 7th Int. Conf. on Distributed Computing Systems*, Berlin, September 1987, 299–306.

Ewen-Smith, B. M. (1985) 'Data protocols for cellular and private mobile radio', in *Proc. Cellular and Mobile Communications*, Online Publications, London, 193–204.

Falk, G., S. Groff, W. Milleken, M. Nodine, S. Blumenthal and W. Edmond (1983) 'Integration of voice and data on satellite networks', *IEEE J. Selected Areas in Commun.*, **SAC-1** (6), December, 1076–1083.

Fisher, D., P. Maynard and S. Alexander, 'Broadband ISDN', *Computer Commun.*, **11** (4), August, 185–190.

Gallagher, I. D., (1986) 'Multi-service networks', *British Telecom Technol. J.*, **4** (1), January, 43–49.

Garcia-Luna-Aceves, J. J. (1986) 'Analysis of routing table update activity in multi-hop packet radio networks', in *New Communication Services: A Challenge to Computer Technology, Proc. ICCC*, Munich, 648–653.

Ghanbari, M., C. J. Hughes, D. E. Pearson and J. Xiong (1988) 'The interaction between video encoding and variable bit-rate networks', *IERE Conf. on Digital Processing of Signals and Communications*, Loughborough University of Technology, September 1988, 125–131.

Golding, L. S., A. J. Viterbi, R. W. Jeshin, J. N. Pelton, B. G. Evans, J. Rinde, P. P. Nuspl, P. Bartholome and K. M. Sundara Murthy (1989) 'VSATs: expert views on future trends', *IEEE Commun. Mag.*, **26** (7), May, 58–64.

Gretton-Watson, P. (1988) 'Distributed database development', *Computer Commun.*, **11** (5), October, 275–280.

Grout, V. M. and P. W. Sanders (1988) 'Communication network optimization', *Computer Commun.*, **11** (5), October, 281–287.

Gruber, J. G. and N. H. Le, (1983) 'Performance requirements for integrated voice/data networks', *IEEE J. Selected Areas of Commun.*, **SAC-1** (6), December, 981–1005.

Gruber, J. G. and L. Strawczynski (1985) 'Subjective effects of variable delay and speech clipping in dynamically managed voice systems', *IEEE Trans. Commun.*, **COM-33** (8), August, 801–808.

Hammer, D. K. (1988) 'Software architecture of an X.400 electronic mail system', *Computer Commun.*, **11** (3), June, 130–135.

Hammond, J. L. and J. P. O'Reilly (1986) *Performance Analysis of Local Computer Networks*, Addison-Wesley, Reading, Mass.

Herskowitz, G. J. and J. G. Warms (1988) 'Fibre optic LAN topology, access protocols and standards', *Computer Commun.*, **11** (5), October, 227–233.

Hine, M. (1988) 'Satine-2: an experiment in high speed wide area networking', in *Research into Networks and Distributed Applications, Proc. Int. Symp.*, Graz, Austria, organized by European Space Agency, Elsevier Science, Brussels, 577–605.

Hoare, C. A. R. (1985) *Communicating Sequential Processes*, Prentice-Hall, Englewood Cliffs, New Jersey.

➤ Hopper, A., S. Temple and R. Williamson (1986) *Local Area Network Design*, Addison-Wesley, Wokingham Chapter 8.

Hughes, C. J. and J. W. Atkins (1979) 'Virtual circuit switching for multi-service operation', in *Int. Switching Symp.* (ISS), Paris, 344–350.

IEEE Communications Magazine, (1986) 'Voice/data integration', 24 (12), December.

IEEE Communications Magazine, (1988) 'Metropolitan area networks', 26 (4), April.

IEEE Communications Magazine, (1988/89) Special series on 'VSAT communication networks: technology and applications', *IEEE Communications Magazine*, commencing May 1988 and continued in July 1988, September 1988, February 1989 and May 1989.

IEEE Journal of Lightwave Technology (1985) 'Optical fiber local area networks', LT-4 (3), June.

IEEE JSAC (1983) *IEEE J. on Selected Areas in Commun.*, 'Packet switched voice and data communications', SAC-1, (5), December 1983.

IEEE JSAC (1985) *IEEE J. on Selected Areas in Commun.*, 'Fiber optics for local communications', SAC-3 (6), November 1985.

IEEE JSAC (1986) *IEEE J. on Selected Areas in Commun.*, 'High speed fiber optic communication systems', SAC-4 (9), December 1986

IEEE JSAC (1987) *IEEE J. on Selected Areas in Commun.*, 'Performance evaluation of multiple access networks', SAC-5 (6), July 1987

IEEE JSAC (1988) *IEEE J. on Selected Areas in Commun.*, 'Fiber optic local and metropolitan area networks', SAC-6 (6), July 1988

IEEE JSAC (1989) *IEEE J. on Selected Areas in Commun.*, 'Packet speech and video', 7 (5), June 1989

IEEE JSAC (1990) *IEEE J. on Selected Areas in Commun.*, 'Heterogeneous networks interconnection', 8 (1), January 1990

IEEE Transactions on Software Engineering (1988) Special issue on 'Tools for computer communication systems', 14 (3), March.

Irven, J. H., M. E. Nilson, T. H. Judd, J. F. Patterson and Y. Shibata (1988) 'Multi-media information services: a laboratory study', *IEEE Commun. Mag.*, 26 (6), 27–43.

ISO international standard 8807, (1988) LOTOS.

Jacob, A. R. (1990) 'A survey of fast packet switches', *ACM Computer Commun. Rev.*, 20 (1), January, 54–64.

Jansson, L., B. Akeson, P. Lindstroem, and C. Eng, (1986) 'A general basic system for post-transactions', in *New Communication Services: A Challenge to Computer Technology, Proc. ICCC*, Munich, 131–135.

Karol, M. J., M. G. Hluchyi and S. P. Morgan (1987) 'Input vs output queuing on a space-division packet switch', *IEEE Trans. Commun.*, COM-35, (12), December, 1347–1356.

Khanna, A. and J. Zinky (1989) 'Revised ARPANET routing metric', *Proc. ACM Sigcomm*, Austin, Texas, 45–56.

Lazar, A. A., A. Patir, T. Takahashi and M. el Zarki (1985) 'MAGNET: Columbia's integrated network testbed', *IEEE J. Selected Areas in Commun.*, 3 (6), November, 859–871.

Limb, J. O. and C. Flores (1982) 'Description of Fasnet, a unidirectional local area communications network', *Bell Systems Tech. J.*, 61, (7) Part 1, September, 1413–1440.

Limb, J. O. and L. E. Flamm (1983) 'A distributed local area network packet protocol for combined voice and data transmission', *IEEE J. on Selected Areas in Commun.*, SAC-1 (5), November, 926–934.

Lockwood, J. A., P. Frueh, and J. L. Snare (1986) 'Electronic funds transfer at the point of service (EFT/POS) in Australia', in *New Communication*

Services: A Challenge to Computer Technology, Proc. ICCC, Munich, 142–147.

Lorin, H. (1990) 'Application development, software engineering and distributed processing', *Computer Commun.*, **13** (1), January/February, 4–15.

Ma, Z. (1987) 'New tools to support ODA and ODIF', *Computer Commun.*, **10** (1), February, 15–29.

MacLean, D. J. and W. T. Wilson (1986) 'New computer communication services: a challenge to people technology', in *New Communication Services: A Challenge to Computer Technology, Proc. ICCC*, Munich, 200–204.

Mills, D. L. (1988) 'The Fuzzball', *Proc. ACM Sigcomm*, Stanford, California, August 1988, 115–122.

Milner, R. (1980) *A Calculus for Communicating Systems*, Lecture Notes in Computer Science, **92**, Springer-Verlag, Berlin.

Mollenauer, J. F. (1988) 'Standards for metropolitan area networks', *IEEE Commun. Mag.*, **26** (4), April, 15–19.

Nesbeth, T. L., F. P. Coakley, B. G. Evans and C. Smythe (1989) 'Selection of an optimum protocol for a VSAT wide area network', *Proc. 2nd IEE National Conf. on Telecommunications*, York, April, 1989, Conf. publication 300, 116–121.

Newman, R. M., Z. L. Budkis and J. L. Hulett, (1988) 'The QPSX MAN', *IEEE Commun. Mag.*, **26** (4), April, 20–28.

Orda, A. and R. Rom (1990) 'Multihoming in computer networks: a topology design approach', *Computer Networks and ISDN Systems*, **18** (2), February, 133–141.

Owings, J. L. (1983) 'High speed data transfer over satellites', in *Proc 27th IEEE COMPCON 83*, Fall, Arlington, Virginia, 66–70.

Palme, J. and R. Speth (1986) 'Group communication in message systems: the AMIGO project', in *New Communication Services: A Challenge to Computer Technology, Proc. ICCC*, Munich, 98–102.

Pouzin, L. 'OSI progress and issues', in *New Communication Services: A Challenge to Computer Technology, Proc ICCC*, Munich, 154–158.

Raychaudhuri, D. and K. Joseph (1988) 'Channel access protocols for Ku-band VSAT networks: a comparative evaluation', *IEEE Commun. Mag.*, **26** (5), May, 34–44.

Rockström, A. and R. Saracco (1982) 'SDL—CCITT specification and description language', *IEEE Trans. Commun.*, **COM-30** (6), June, 1310–1317.

Ross, F. E. (1986) 'FDDI—a tutorial', *IEEE Commun. Mag.*, **24** (5), May, 10–17.

Rudin, H. (1985) 'An informal overview of protocol specification', *IEEE Commun. Mag.*, **23** (3), March, 46–52.

Sabnani, K. K. and A. N. Netravali (1989) 'A high speed transport protocol for datagram/virtual circuit networks', *ACM Computer Commun. Rev.*, **19** (4), September, 146–157.

Sabnani, K. and M. Schwartz (1985) 'Multidestination protocols for satellite broadcast channels' *IEEE Trans. Commun.*, **COM-33** (3) March, 232–240.

Schwartz, M. (1987) *Telecommunication Networks—Protocols, Modeling and Analysis*, Addison-Wesley, Reading, Mass.

Stefik, M., G. Foster, D. Bobrow, K. Kahn, S. Lanning and L. Suchman (1987) 'Beyond the chalkboard: computer support for collaboration and problem solving in meetings', *Commun. ACM*, **30** (1), January, 32–47.

Sullivan, M. and D. Anderson (1989) 'Marionette: a system for parallel distributed programming using a master/slave model', in *Proc. 9th Int. Conf. on Distributed Computing Systems*, June 1989, 181–188.

Sundara Murthy, K. M., J. Alan, J. Barry, B. G. Evans, N. Miller, R. Mullinax,

P. Noble, B. O'Neal, J. J. Sanchez, N. Seshagiri, D. Shanley, J. Strateigos and J. W. Warner (1989) 'VSAT user network examples', *IEEE Commun. Mag.*, **27** (5), May, 50–57.

Taylor, J. R. and J. M. Katmbwe 'Are new technologies really reshaping our organizations?', *Computer Commun.*, **11** (5), October, 245–252.

Tennenhouse, D., I. Leslie, R. Needham, J. Burren, C. Adams and C. Cooper (1987) 'Exploiting wideband ISDN: the UNISON exchange', in *Proc. IEEE Infocom*, Washington DC, March/April, 1018–1026.

Thompson, G. B., (1986) 'Information economic aspects of new services', in *New Communication Services: A Challenge to Computer Technology, Proc ICCC*, Munich, 190–193.

Tseung, L. C. N. (1989) 'Guaranteed, reliable, secure broadcast networks', *IEEE Network*, **3** (6), November, 33–37.

Tobagi, F. A., F. Borgonovo and L. Fratta, (1983) 'Expressnet: a high performance integrated services local area network', *IEEE JSAC*, **SAC-1** (5), November, 898–913.

Van Eijk, P. H. J., C. A. Vissers and M. Diaz (eds), (1989) *The Formal Description Technique LOTOS—The Results of the Esprit SEDOS (Software Environment for the Design of Distributed Open Systems) Project'*, North-Holland, Amsterdam.

Waters, A. G., C. J. Adams, I. M. Leslie and R. M. Needham (1984) 'The use of broadcast techniques in the Universe project', in *Proc. ACM Sigcomm 84 Symp. on Communications Architectures and Protocols*, Montreal, 3–11.

Watson, G. and M. Frontini (1988) 'An investigation of packetized voice on the FDDI token ring', in *Int. Zurich Seminar on Digital Communications*, 171–178.

Wong, J. W. (1989) 'Throughput of DQDB networks under heavy load', in *Proc. EFOC/LAN 1989*, 146–151.

Yeh, Y. S., M. G. Hluchyi and A. S. Acampora (1987) 'The knockout switch: a simple, modular architecture for high performance packet switching', *IEEE J. Selected Areas of Commun.*, **SAC-5**, October, 1274–1283.

Youssef, H., C. Huitema and F. Kamoun (1984) 'Performance evaluation of NADIR bulk data transmission protocol for multipoint satellite link' in *Performance of Computer Communication Systems*, W. Bux and H. Rudin (eds), North-Holland, Amsterdam, 367–382.

Index